AF616329

HOT SPOT POLLUTANTS: PHARMACEUTICALS IN THE ENVIRONMENT

HOT SPOT POLLUTANTS: PHARMACEUTICALS IN THE ENVIRONMENT

Edited by

Daniel R. Dietrich
Environmental Toxicology
University of Konstanz, Germany

Simon F. Webb
Procter & Gamble
Bruxelles, Belgium

Thomas Petry
The Weinberg Group
Bruxelles, Belgium

ELSEVIER
ACADEMIC
PRESS

AMSTERDAM • BOSTON • HEIDELBERG • LONDON
NEW YORK • OXFORD • PARIS • SAN DIEGO
SAN FRANCISCO • SINGAPORE • SYDNEY • TOKYO

Elsevier Academic Press
30 Corporate Drive, Suite 400, Burlington, MA 01803, USA
525 B Street, Suite 1900, San Diego, California 92101-4495, USA
84 Theobald's Road, London WC1X 8RR, UK

This book is printed on acid-free paper. ♾

Chapters 1–18 in this volume were originally published in volume 131, issues 1–2, and volume 142, issue 3 of Elsevier's journal *Toxicology Letters*. As the field is moving fast, authors were invited to update their work as necessary, and six articles have been revised for publication in this book. One new article has also been included.

Library of Congress Cataloging-in-Publication Data
Application submitted

British Library Cataloging in Publication Data
A catalogue record for this book is available from the British Library

ISBN: 0-12-032953-0

For all information on all Elsevier Academic Press publications
visit our Web site at www.books.elsevier.com

Printed in the United States of America
04 05 06 07 08 9 8 7 6 5 4 3 2 1

Contents

Concentrations of the UV Filter Ethylhexyl Methoxycinnamate in the Aquatic Compartment: A Comparison of Modeled Concentrations for Swiss Surface Waters with Empirical Monitoring Data

Jürg Oliver Straub

Exposure Simulation for Pharmaceuticals in European Surface Waters with GREAT-ER

Diederik Schowanek and Simon Webb

Indirect Human Exposure to Pharmaceuticals via Drinking Water

Simon Webb, Thomas Ternes, Michel Gibert, and Klaus Olejniczak

PART III Effects

Morphological Sex Reversal Upon Short-Term Exposure to Endocrine Modulators in Juvenile Fathead Minnow (*Pimephales promelas*)

M. Zerulla, R. Länge, T. Steger-Hartmann, G. Panter, T. Hutchinson, and D. R. Dietrich

Determination of Vitellogenin Kinetics in Male Fathead Minnows (*Pimephales promelas*)

T. Schmid, J. Gonzalez-Valero, H. Rufli, and D. R. Dietrich

Integrated *In Vivo* and *In Vitro* Assessment of Reproductive and Developmental Effects of Endocrine Disrupters in Invertebrates

Thomas H. Hutchinson

How Can Toxic Effects of Pollution of the Aquatic Environment on the Immunocompetence of Fishes Be Detected? A Discussion on the Relevance of the Biomarkers Philosophy

B. Köllner, B. Wasserrab, U. Fischer, G. Kotterba, and M. van den Heuvel

Aquatic Ecotoxicology of Fluoxetine: A Review of Recent Research

Bryan W. Brooks, Sean M. Richards, James J. Weston, Philip K. Turner, Jacob K. Stanley, Thomas W. La Point, Richard Brain, Elizabeth A. Glidewell, A. Rene D. Massengale, Whitney Smith, C. LeRoy Blank, Keith R. Solomon, Marc Slattery, and Christy M. Foran

Aquatic Ecotoxicity of Pharmaceuticals Including the Assessment of Combination Effects

Michael Cleuvers

PART IV Principal Considerations

Environmental Risk Assessment of Pharmaceutical Drug Substances—Conceptual Considerations

Reinhard Länge and Daniel Dietrichy

Pharmacodynamic Activity of Drugs and Ecotoxicology—Can the Two Be Connected?

Jürg P. Seiler

Proposed Development of Sediment Quality Guidelines Under the European Water Framework Directive: A Critique

Mark Crane

Contributors

Numbers in parentheses indicate the pages on which the authors' contributions begin.

Marc Adam (11) Federal Institute for Risk Assessment, 14195 Berlin, Germany

Paul A. Blackwell (37, 255) Cranfield Centre for EcoChemistry, Cranfield University, Shardlow, Derby, DE72 2GN, United Kingdom

C. LeRoy Blank (165) Department of Chemistry and Biochemistry, University of Oklahoma, Norman, Oklahoma

Alistair B. A. Boxall (37, 253) Cranfield Centre for EcoChemistry, Cranfield University, Shardlow, Derby, DE72 2GN, United Kingdom

Richard Brain (165) Centre for Toxicology, University of Guelph, Guelph, Ontario, Canada

Bryan W. Brooks (165) Department of Environmental Studies, Baylor University, Waco, Texas

Romina Cavallo (37) Institute for Risk Assessment Science, Faculty of Veterinary Medicine, Utrecht University, 3584 CL Utrecht, The Netherlands

Michael Cleuvers (189) Department of General Biology, Aachen University of Technology, D-52056 Aachen

Mark Crane (235, 289) Crane Consultants, Faringdon, Oxfordshire, SN7 7AG, United Kingdom

Andy Croxford (255) Environment Agency, National Centre for Ecotoxicology and Hazardous Substances, Wallingford, Oxon, OX10 8BD, United Kingdom

Joop A. de Knecht (271) National Institute for Public Health and the Environment (RIVM), NL-3720BA Bilthoven, The Netherlands

D. R. Dietrich (3, 97, 115, 205) Environmental Toxicology, University of Konstanz, 78457 Konstanz, Germany

U. Fischer (143) Federal Research Centre for Virus Diseases of Animals, Institute of Infectology, Insel Riems, Germany

Lindsay A. Fogg (255) Cranfield Centre for EcoChemistry, Cranfield University, Shardlow, Derby, DE72 2GN, United Kingdom

Christy M. Foran (165) Department of Biology, West Virginia University, Morgantown, West Virginia

Michel Gibert (79) VEOLIA Environment, 75008 Paris, France

Elizabeth A. Glidewell (165) Department of Environmental Studies, Baylor University, Waco, Texas

J. Gonzalez-Valero (115) Syngenta Crop Protection AG, Ecological Sciences, 4002 Basel, Switzerland

Thomas Heberer (11) Institute of Food Chemistry, Technical University of Berlin, 13355 Berlin, Germany and Federal Institute for Risk Assessment, 14195 Berlin, Germany

Thomas H. Hutchinson (97,131) AstraZeneca Brixham Environmental Laboratory Freshwater Quarry, Brixham, Devon, TQ5 8BA, United Kingdom

B. Köllner (143) Federal Research Centre for Virus Diseases of Animals, Institute of Diagnostic Virology, Germany

Paul Kay (37, 255) Cranfield Centre for EcoChemistry, Cranfield University Shardlow, Derby, DE72 2GN, United Kingdom

G. Kotterba (143) Federal Research Centre for Virus Diseases of Animals, Institute of Infectology Insel Riems, Germany

Reinhard Länge (97, 205) Schering AG, Experimental Toxicology, Research Laboratories, Berlin, Germany

Thomas W. La Point (165) Institute of Applied Sciences, University of North Texas, Denton, Texas

Carol Long (289) Veterinary Medicines Directorate, New Haw, Addlestone, Surrey, KT15 3LS, United Kingdom

A. Rene D. Massengale (165) Department of Biology, Baylor University, Waco, Texas

Mark H. M. M. Montforts (271) National Institute for Public Health and the Environment (RIVM), NL-3720BA Bilthoven, The Netherlands

Klaus Olejniczak (79) BfArM—Federal Institute for Drugs and Medical Devices, D-53175 Bonn, Germany

G. Panter (97) CEFIC-EMSG, AstraZeneca, Global Safety, Health & Environment, Brixham, United Kingdom

Emma J. Pemberton (255) Environment Agency, National Centre for Ecotoxicology and Hazardous Substances, Wallingford, Oxon, OX10 8BD, United Kingdom

Thomas Petry (3) The Weinberg Group, Bruxelles, Belgium

Sean M. Richards (165) Department of Biological and Environmental Sciences, University of Tennessee at Chattanooga, Chattanooga, Tennessee

H. Rufli (115) Syngenta Crop Protection AG, Ecological Sciences, Basel, Switzerland

T. Schmid (115) Environmental Toxicology, University of Konstanz, 78457 Konstanz, Germany

Diederik Schowanek (63) Procter & Gamble Eurocor, B-1853 Strombeek-Bever, Belgium

Jürg P. Seiler (217) ToxiConSeil, CH-3475 Riedtwil, Switzerland

Marc Slattery (165) Environmental Toxicology Research Program, School of Pharmacy, University of Mississippi, University, Mississippi

Whitney Smith (165) Department of Chemistry and Biochemistry, University of Oklahoma, Norman, Oklahoma

Keith R. Solomon (165) Centre for Toxicology, University of Guelph, Guelph, Ontario, Canada

Jacob K. Stanley (165) Institute of Applied Sciences, University of North Texas, Denton, Texas

T. Steger-Hartmann (97) Schering AG, Experimental Toxicology, Berlin, Germany

Jürg Oliver Straub (51, 299) EurProBiol CBiol MIBiol Corporate Safety & Environmental Protection CSE, F. Hoffmann-La Roche Ltd, CH–4070 Basel, Switzerland

Thomas Ternes (79) Bundesanstalt Gewaesserkunde, D-56068 Koblenz, Germany

Johannes Tolls (37) Institute for Risk Assessment Science, Faculty of Veterinary Medicine, Utrecht University, 3584 CL Utrecht, The Netherlands

Philip K. Turner (165) Institute of Applied Sciences, University of North Texas, Denton, Texas

M. van den Heuvel (143) Forest Research, Rotorua, New Zealand

B. Wasserrab (143) Environmental Toxicology, University of Konstanz, Konstanz, Germany

Simon F. Webb (3, 63) Procter & Gamble Eurocor, B-1853 Strombeek-Bever3, Belgium

Simon Webb (79) CEFIC—European Chemistry Industry Council, Brussels B-1160, Belgium

James J. Weston (165) Environmental Toxicology Research Program, School of Pharmacy, University of Mississippi, University, Mississippi

M. Zerulla (97) Schering AG, Experimental Toxicology, Berlin, Germany

Preface

In the 1960s and 1970s, alarm bells began to ring around the world, particularly in several European states. Europe's inland waterways were suffering under the burden of contamination with industrial and domestic waste, mainly in the form of large phosphate deposits. The overt eutrophication led to anti-pollution initiatives (Urban Wastewater Treatment Directive; for integrated pollution and prevention control, see: http://www.themes.eeaa.eu.int/Specific_media/water/links) in several countries, culminating in the reduction of point-source discharges of organic matter, phosphates, and eventually nitrates. Currently, approximately 80% of European Union waste water passes through a sewage treatment plant (STP), before discharging into rivers and lakes (see: http://ecb.jrc.it/cgi-bin/reframer.pl?A=ECB&B=/tgdoc/). This strategy has been quite successful in improving the water quality of these water bodies. Although this is quite an achievement, there is little cause for sitting back and enjoying the laurels of the technological success. In reality, there is even more to worry about now. We are currently looking at another group of potential water pollutants, namely human pharmaceuticals, endocrine active substances in particular, that are considered to be responsible for a number of detrimental effects in the aquatic environment including reduced reproductive success of aquatic organisms. Consequently, this area has become a subject of intense and costly research for the last decade, and most likely will be for the next one as well, if not longer. Simultaneous to the basic research efforts, the European Union via the EMEA (European Agency for the Evaluation of Medicinal Products) has proposed a guideline for the environmental risk assessment of pharmaceuticals. Amongst the pharmaceuticals to be regulated are some that are intentionally tailored to the regulation of the hormonal status of patients and thus, by definition, are potential endocrine disrupters in the environment. This new draft guideline, following consideration of the numerous commentaries by industry, government organizations, non-government organizations, and academia, is currently under revision and will become a legally binding guideline by the end of 2004.

Both views, namely the basic research needs and current state of the art on real effects and affects assessment of pharmaceutical compounds in the environment as well as the EMEA draft guideline are being presented and discussed in detail in this book. It is interesting to note the vast discrepancy between the many areas considered uncharted territory by the "basic" researchers and the rather rough assumptions made by regulators when developing the draft guideline. Indeed, even in cases where known synthetic hormones or hormone synthesis inhibitors have been investigated, the results obtained from fish or other species cast quite a bit of doubt on whether these results can be directly extrapolated to the "aquatic environment" per se, or whether this perceived risk is more likely to be restricted to point sources of xenobiotic discharges such as the outflow of sewage treatment plants. If the latter is true, then the question must be raised of whether the guideline approach by the EMEA is the best way forward, or if a general rethinking of the whole situation, including the use and efficiencies of our currently operating sewage treatment plants, should be considered and other strategies sought. I will leave it to the reader of this book to come to his or her own conclusions in this rapidly changing, complex, and demanding area of research and regulation. Finally it remains for me to thank the authors and co-authors of the chapters of this book for their hard work and patience. Last but not least, I would like to express my deepest gratitude to Silke, Larissa, Tim, and Sam for their unrestricted support, understanding, and patience. This wouldn't have been possible without you.

Daniel R. Dietrich
Konstanz, Germany, May 10, 2004

PART I

Editorial

Daniel R. Dietrich*, Simon F. Webb[†], and Thomas Petry[‡]

*Environmental Toxicology
University of Konstanz, Germany

[†]Procter & Gamble
Bruxelles, Belgium

[‡]The Weinberg Group
Bruxelles, Belgium

Hot Spot Pollutants: Pharmaceuticals in the Environment

Pharmaceuticals are important and indispensable elements of modern life. They are employed in human and veterinary medicine, agriculture, and aquaculture. Until the 1990s, however, relatively little consideration was given to the likely fate, occurrence, or effect of pharmaceuticals on the environment following normal use. This apparent lack of scientific interest in pharmaceuticals as contaminants of the aquatic environment is somewhat puzzling. Common over-the-counter (OTC) drugs such as paracetamol or aspirin are sold in quantities comparable to high production volume (HPV) materials, close to or exceeding 1000 tons/year in European countries such as the UK and Germany (Ternes, 2001a; Webb, 2001). Total use of human prescription drugs in such countries is even greater (Webb, 2001). Drugs are also inherently biologically active and often exquisitely potent. They are often resistant to biodegradation, as metabolic stability is necessary to pharmacological action. Certain pharmaceuticals or their metabolites are also highly water soluble. When combined with a lack of biodegradation, removal during wastewater treatment will consequently be limited for such compounds. These compounds will then enter the aquatic environment, resulting in exposure of aquatic biota.

Contributions to our knowledge of pharmaceuticals in the environment (PIE) that predate this period include the observations by Aherne *et al.* (1985) on compounds such as ethinyl oestradiol, diazepam, theophylline, erythromycin, tetracycline, and methotrexate in various environmental matrices as a consequence of normal patient use. Richardson and Bowron (1985) likewise report on analytical studies. They also detail the development of simple modeling techniques aimed at predicting likely concentrations in surface waters following normal use by the patient. This pioneering work included a consideration of national usage patterns, human metabolism, fate during wastewater treatment, and surface water dilution of effluents.

The last decade has seen a marked growth in the literature relating to observations of PIE at concentrations that result from normal use by the patient. At least 60 compounds have now been reported from aquatic matrices (Heberer and Stan, 1997; Hirsch *et al.*, 1999; Stumpf *et al.*, 1996a,b; Ternes, 1998, 2001a,b). Such observations necessitate a consideration of any potential risk. This in turn requires knowledge of the effects of pharmaceuticals upon relevant aquatic biota. This requirement is now being addressed for many classes of compounds such as selective serotonin reuptake inhibitors (SSRIs) (Brooks *et al.*, 2003; Fong *et al.*, 1998), steroids (Länge *et al.*, 2001) and antihyperlipoproteinemics (Köpf, 1995). Concurrently, there have been various regulatory developments in the United States and Europe relating to requirements for risk assessment of new actives as part of their registration process, as elegantly presented by Straub in Chapter 19. The risk assessment of existing pharmaceuticals has also been attempted (Halling-Sørenson *et al.*, 1998; Stuer-Lauridsen *et al.*, 2000; Webb, 2001).

Of particular concern is speculation that the presence of pharmaceuticals in the environment may be leading to subtle but hitherto unrecognized or undetected effects leading to irreversible damage of the ecosystem (Daughton, 2001, 2003a,b; Daughton and Ternes, 1999). This requires empirical research aimed at thoroughly understanding the effects of these biologically active materials at the low exposure levels occurring in the environment (Pfluger and Dietrich, 2001). Equally important is the need to develop solid and scientifically sound approaches to assess the associated risks.

This book contributes in our efforts to extend our knowledge vis-à-vis the occurrence and fate of pharmaceuticals in the environment, their effects, and potential risks. It represents a concerted effort of academic, regulatory, and industry scientists to bring greater understanding to the PIE issue. Together, these 18 papers reflect the state-of-the-art as presented at the Statuskolloqium in Environmental Toxicology in Konstanz, Germany (November 2001), the Special Session at the Society of Environmental Toxicology and Chemistry (SETAC) Europe Annual Meeting in Vienna, Austria (May 2002), and the 11th European Congress on Biotechnology in Basel, Switzerland (August 2003).

In the section "Occurrence and Fate," the first three contributions from Heberer and Adams Boxall *et al.*, and Straub all reflect analytical studies on the presence and fate of pharmaceutical residues in the environment. These include compounds such as antibiotics and UV filters. Chapter 5 by Schowanek and Webb details exposure modeling of common pharmaceuticals using the GREAT-ER software, while Chapter 6 by Webb *et al.* deals with the probability of human exposure to pharmaceuticals via drinking water.

Within the "Effects" section, the four papers from Zerulla *et al.*, Schmid *et al.*, Brooks *et al.*, and Köllner *et al.*, detail work on the responses of fish species to endocrine modulators (including steroids), immune modulators, and other compounds. Hutchinson presents work on the *in vivo* and *in vitro* responses of invertebrate species. In addition and in contrast to the more singular effect assessment of the previous authors, Cleuvers presents data on the combinatorial effects of pharmaceuticals.

The first of the three contributions in the section "Principle Considerations" is by Länge and Dietrich, who deal with various conceptual aspects of environmental risk assessment as it relates to pharmaceuticals. In the second, Seiler speculates on whether the established knowledge relating to the pharmacodynamic activity of pharmaceuticals can be of use in ecotoxicologial risk evaluation. The third of this group is a critique of the proposed sediment quality guidelines under the European Water Framework Directive by Crane.

Until recently, pharmaceuticals were not subject to environmental risk assessment as part of the registration process. In the last section, "Risk Assessment," there are four papers by Boxall *et al.*, Montforts and Knecht, Long and Crane, and Straub dealing with developments regarding EU regulatory requirements for the environmental risk assessment of new veterinary and human pharmaceutically active compounds.

Overall, this book aims to critically discuss the knowledge on PIE, their potential impact on the environment, and consequently, the most proper and sensible steps for risk assessment. In combining the views from academic, industry, and regulatory scientists, a balanced presentation of the most pressing issues and gaps of knowledge is emphasized. This is especially important in view of the efforts in regulating environmental testing and risk assessment within the EU. We hope that this book will help the interested scientist gain easy entry to this hot spot of current research, foster discussion among scientists, stimulate additional efforts in addressing the knowledge gaps identified, and thus provide for a better scientific basis of dealing with pharmaceuticals in the environment.

Acknowledgments

We would like to thank Karin Rieder for organizing this book and the various reviewers of the enclosed publications for doing such an excellent job in the little time that was available.

References

Aherne, G. W., English, J., and Marks, V. (1985). The role of immunoassay in the analysis of microcontaminants in water samples. *Ecotoxicol. Environ. Saf.* **9**(1), 79–83.

Brooks, B. W., Foran, C. M., Richards, J., Weston, P. K., Turner, J. K., Solomon, K. R., Slattery, M., and La Point, T. W. (2003). Aquatic toxicology of fluoxetine. A review of recent research. *Tox. Lett.* **142**(3), 169–184.

Daughton, C. G., and Ternes, T. A. (1999). Pharmaceuticals and personal care products in the environment: Agents of subtle change? *Environ. Health Perspect.* **107**(6), 907–938.

Daughton, C. G. (2001). Illicit drugs in municipal sewage: Proposed new non-intrusive tool to heighten public awareness of societal use of illicit/abused drugs and their potential for ecological consequences. *In* "Pharmaceuticals and Personal Care Products in the Environment: Scientific and Regulatory Issues" (C. G. Daughton and T. Jones-Lepp, Eds.), Symposium Series 791, pp. 348–364. American Chemical Society, Washington, D.C.

Daughton, C. G. (2003a). Cradle-to-cradle stewardship of drugs for minimizing their environmental disposition while promoting human health. I. Rationale for and avenues toward a green pharmacy. *Environ. Health Perspect.* **111**(5), 757–774.

Daughton, C. G. (2003b). Cradle-to-cradle stewardship of drugs for minimizing their environmental disposition while promoting human health. II. Drug disposal, waste reduction, and future directions. *Environ. Health Perspect.* **111**(5), 775–785.

Fong, P. P., Huminski, P. T., and d'Urso, I. M. (1998). Induction of potentiation of parturition in fingernail clams (Sphaerium striatinum) by selective serotonin re-uptake inhibitors (SSRIs). *J. Exp. Zool.* **280**(3), 260–264.

Halling-Sørenson, B., Nors Nielsen, S., Lanzky, P. F., Ingerslev, F., Holten-Lutzhøft, H. C., and Jørgenson, S. E. (1998). Occurrence, fate and effects of pharmaceutical substances in the environment—A review. *Chemosphere* **36**(2), 357–393.

Heberer, T., and Stan, H.-J. (1997). Determination of clofibric acid and N-(phenylsulfonyl)-sarcosine in sewage river and drinking water. *Int. J. Environ. Anal. Chem.* **67**, 113–124.

Hirsch, R., Ternes, T., Haberer, K., and Kratz, K.-L. (1999). Occurrence of antibiotics in the aquatic environment. *Sci. Total Environ.* **225**, 109–118.

Köpf, W. (1995). Effects of endocrine substances in bioassays with aquatic organisms (abstract). Presented at the 50th Seminar of the Bavarian Association for Waters Supply, (German).

Länge, R., Hutchinson, T. H., Croudace, C. P., Siegmund, F., Schweinfurth, H., Hampe, P., Panter, G. H., and Sumpter, J. P. (2001). Effects of the synthetic oestrogen 17α-Ethinylestradiol over the life-cycle of the fathead minnow (*Pimephales promelas*). *Environ. Toxicol. Chem.* **20**(6), 1216–1227.

Pfluger, P., and Dietrich, D. R. (2001). Pharmaceuticals in the environment—an overview and principle considerations. *In* "Pharmaceuticals in the Environment" (K. Kümmerer, Ed.), pp. 11–17. Springer Verlag, Heidelberg.

Richardson, M. L., and Bowron, J. M. (1985). The fate of pharmaceutical chemicals in the environment. *J. Pharm. Pharmacol.* **37**, 1–12.

Stuer-Lauridsen, F., Birkved, M., Hansen, L. P., Holten-Lutzhøft, H. C., and Halling-Sørenson, B. (2000). Environmental risk assessment of human pharmaceuticals in Denmark after normal therapeutic use. *Chemosphere* **40**(7), 783–793.

Stumpf, M., Ternes, T. A., Haberer, K., and Baumann, W. (1996a). Nachweis von natürlichen und synthetischen ostrogenen in kläranlagen und fliessgewässern. *Vom Wasser* **87**, 251–261. (German).

Stumpf, M., Ternes, T. A., Haberer, K., Seel, P., and Baumann, W. (1996b). Nachweis von Arzneimittelrückständen in Kläranlagen und Fließgewässern. *Vom Wasser* **86**, 291–303. (German).

Ternes, T. A. (1998). Occurrence of drugs in German sewage treatment plants and rivers. *Water Res.* **32**(11), 3245–3260.

Ternes, T. (2001a). Pharmaceuticals and metabolites as contaminants of the aquatic environment. *In* "Pharmaceuticals and Personal Care Products in the Environment—Scientific and Regulatory Issues" (C. G. Daughton and T. L. Jones-Lepp, Eds.), Symposium Series 791, pp. 39–54. American Chemical Society, Washington D.C.

Ternes, T. A. (2001b). Analytical methods for the determination of pharmaceuticals in aqueous environmental samples. *Trends Analy. Chem.* **20**(8), 419–434.

Webb, S. F. (2001). A data-based perspective on the environmental risk assessment of human pharmaceuticals II-Aquatic risk characterization. *In* "Pharmaceuticals in the Environment—Sources, Fate, Effects and Risks (Kümmerer, Ed.), pp. 203–219. Springer Verlag, Heidelberg, Germany.

PART II

Occurrence and Fate

Thomas Heberer*,[†] and Marc Adam[†]

*Institute of Food Chemistry
Technical University of Berlin
13355 Berlin, Germany

[†]Federal Institute for Risk Assessment
14195 Berlin, Germany

Occurrence, Fate, and Removal of Pharmaceutical Residues in the Aquatic Environment: An Extended Review of Recent Research Data

I. Introduction and Background

In recent years, the occurrence and fate of pharmaceutically active compounds (PhACs) in the aquatic environment has been recognized as one of the emerging issues in environmental chemistry (Andreozzi *et al.*, 2003a; ARGE, 2003; Daughton and Jones-Lepp, 2001; Daughton and Ternes, 1999; Halling-Sørensen *et al.*, 1998; Heberer, 2002a; Kümmerer, 2001; Stan and Heberer, 1997; Verstraeten *et al.*, 2002). The disposal of unused medication via the toilet seems to be of minor importance, but many of the pharmaceuticals applied in human medical care are not completely eliminated in the human body. Often they are excreted only slightly transformed or even unchanged, mostly conjugated to polar molecules (e.g., as glucoronides). These conjugates can easily be cleaved during sewage treatment; the original PhACs will then be released into the aquatic environment, mostly by effluents from municipal sewage treatment plants (STPs). Several investigations have shown some evidence that substances of pharmaceutical origin are often not eliminated during wastewater treatment and also not biodegraded in the environment (Daughton and Ternes, 1999; Heberer, 2002b; Khan and Ongerth, 2002; Ternes, 1998; Zwiener *et al.*, 2000). Under recharge conditions, residues of PhACs may also leach into groundwater aquifers (Verstraeten *et al.*, 2002). Thus, PhACs have already been reported to occur in ground and drinking water samples from waterworks using bank filtration or artificial groundwater recharge downstream from municipal STPs (Brauch *et al.*, 2000; Heberer, 2002b; Heberer and Mechlinski, 2003; Heberer and Stan, 1997; Heberer *et al.*, 1997, 2001b, 2002b; Kühn and Müller, 2000; Sacher *et al.*, 2001; Ternes, 2001; Verstraeten *et al.*, 2002).

The presence of PhACs from human medical care in groundwater may, however, also be caused by other sources such as reuse of sewage by soil-aquifer treatment (SAT) (Drewes *et al.*, 2002, 2003), landfill leachates (Ahel and Jelicic, 2001; Eckel *et al.*, 1993; Holm *et al.*, 1995; Ternes, 2001) or manufacturing residues (Reddersen *et al.*, 2002). Nowadays, especially in industrialized countries, strong regulations and advanced manufacturing practices prevent such spills. In the past, regulations were not as strong and, in several cases, the release of production residues was either tolerated or even accepted. Such spills could result in Superfund sites, which may be responsible for today's findings of PhAC residues in the environment (Reddersen *et al.*, 2002). But the occurrence of pharmaceutical residues in the environment may also be caused by agriculture applying large amounts of PhACs as veterinary drugs and feed additives in livestock breeding. Figure 1 shows possible sources and pathways for the occurrence of PhAC residues in the environment. This chapter will give an overview of the current state of scientific knowledge on the occurrence, fate, and removal of PhACs in the aquatic environment. It is restricted to PhACs originating from human application, and compiles the most recent data and information from some scientific studies and projects, mostly carried out in Europe and the United States.

II. Occurrence of PhACs in Sewage, Surface, Ground, and Drinking Water

The occurrence of PhACs in the aquatic environment has been investigated in several studies in Austria, Australia, Brazil, Canada, Croatia, Czech Republic, England, France, Germany, Greece, Italy, Spain, Switzerland, Sweden, The Netherlands, and the United States. Through 2004, more than 100 PhACs from various prescription classes had been detected up to the μg/L-level in sewage, surface, and ground water. The total number of detected PhACs and their concentrations are decreasing from sewage effluents to

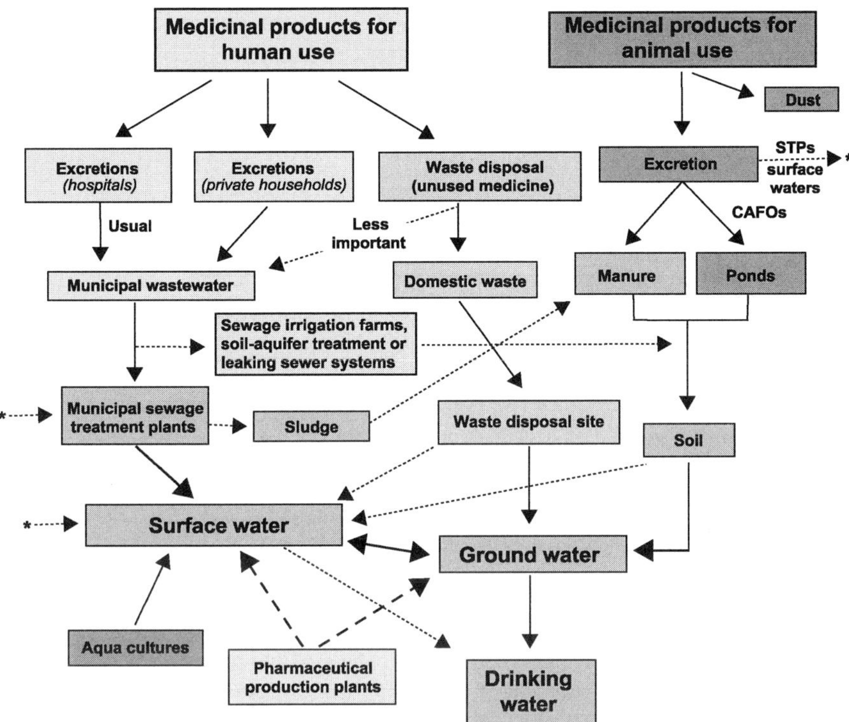

FIGURE 1 Scheme showing possible sources and pathways for the occurrence of pharmaceutical residues in the aquatic environment (modified according to Heberer, 2002a). The asterick indicates that the residues in excretions from animals may also be reaching the municipal STPs or surface waters. (See Color Insert.)

surface, ground, and drinking water, where now only a few compounds have been detected at low ng/L concentrations. This decrease can be explained by dilution and natural attenuation processes, including sorption and degradation by sunlight radiation or microbial activity. The decrease of the number of positive detects might also be attributed to the limitations of analytical methods in analyzing PhACs at sub-ng/L concentrations. This may lead to the question of the relevance of such trace levels, which is often discussed with regard to the ng/L-concentrations of PhACs that are currently instrumentally detectable. Most of the findings of PhACs in ground or drinking water can be assigned to pharmaceuticals summarized as analgesic and anti-inflammatory drugs. These compounds are frequently applied at high individual doses as prescription and/or over-the-counter (OTC) drugs in human medical care. The following sections compile reported findings of individual PhACs in sewage, surface, ground, and drinking water. Additionally, first observations and investigations of natural attenuation processes or potential techniques for the removal of the individual PhAC will also be mentioned.

All sections have been arranged according to the different prescription classes found in the aquatic environment.

A. Analgesics and Anti-Inflammatory Drugs

This section compiles data from PhACs primarily used as painkillers. Most analgesics also have anti-inflammatory and antipyretic properties. Large amounts of painkillers are prescribed in human medical care, but often, much higher quantities are sold without prescription as OTC drugs. In Germany, prescription data is accessible via the health insurance companies, but only rough estimates are possible for the amount of drugs donated in hospitals or sold as OTC drugs (Stan and Heberer, 1997).

Acetaminophen (paracetamol) and acetylsalicylic acid (ASA) are the two most popular painkillers, mainly sold as OTC drugs. In Germany, the total quantity of ASA sold each year has been estimated to be over 500 tons (Ternes, 2001). Nevertheless, other analgesics such as diclofenac or ibuprofen, sold in Germany at annual quantities of approximately 75 and 180 tons, respectively (Ternes, 2001), have been recognized as being more important for the water cycle. ASA was detected at a median concentration of only 0.22 μg/L in sewage effluents in Germany (Ternes, 1998). In the same study, the median concentration of ASA in surface water samples was below the detection limits.

As a prodrug, ASA is easily degraded by deacetylation into its more active form, salicylic acid, and into two other metabolites, namely ortho-hydroxyhippuric acid and the hydroxylated metabolite gentisic acid. Ternes (1998) and Ternes *et al.* (1998) detected salicylic acid, ortho-hydroxyhippuric acid, and gentisic acid in sewage influent samples at concentrations up to 54, 6.8, and 4.6 μg/L, respectively. Ternes (1998) and Ternes *et al.* (1998) observed that all three compounds were efficiently removed by the municipal STPs, and only salicylic acid was detected at very low concentrations in sewage effluents and rivers. Similar sewage effluent concentrations near the 100 ng/L level were also reported by Flaherty *et al.* (2002). Heberer (2002b) found average concentrations of only 0.04 μg/L for salicylic acid in sewage effluents. In this study, the average influent concentrations of 0.34 μg/L were also relatively low. On the other hand, much higher concentrations of salicylic acid, up to 13 μg/L, were detected in sewage effluents in Greece and Spain (Farré *et al.*, 2001; Heberer *et al.*, 2001a). Residues of salicylic acid do not necessarily have to derive from ASA. Other sources, such as the use of salicylic acid as keratolytic, dermatice, and food preservative, or its natural formation, are even more likely to be responsible for the occurrence of this compound in the aquatic environment (Heberer, 2002b).

The other prominent painkiller, acetaminophen, is also easily degraded and removed by the STPs. In investigations of sewage effluents and rivers in Germany, acetaminophen was only detected in less than 10% of all sewage effluents, and not detected in river water at all (Ternes, 1998). In investigations of 142 streams in the U.S. susceptible for contaminations by municipal sewage effluents, Kolpin *et al.* (2002a) detected acetaminophen in 17% of all samples at maximum concentrations up to 10 μg/L. In Czech surface water samples collected from the Elbe river, acetaminophen was found at concentrations up to 106 ng/L (ARGE, 2003). However, in surface water samples collected from the Saale and the Elbe rivers in Germany, it was only found with or below 20 ng/L (ARGE, 2003). Andreozzi *et al.* (2003b) studied the oxidation of acetaminophen from aqueous solutions by means of ozonation and H_2O_2 photolysis, and observed mineralization degrees up to 30–40%.

Approximately 75 tons of the prescription drug diclofenac are annually sold in Germany (Ternes, 2001). In long-term monitoring investigations of sewage and surface water samples from Berlin, Germany, Heberer *et al.* (2002b) identified diclofenac as one of the most important PhACs present in the water cycle. Average concentrations of 3.02 and 2.51 μg/L were detected in the influents and effluents of municipal STPs, respectively. The low removal rate of only 17% demonstrates the persistence of diclofenac in the STPs, as reported by Buser *et al.* (1998b), Stumpf *et al.* (1999), Zwiener *et al.* (2000), and Zwiener and Frimmel (2003). Ternes (1998) reported a removal rate of 69% for diclofenac in the STPs. Diclofenac was also frequently detected at concentrations up to the μg/L-level in investigations of sewage effluents and surface waters in Austria, Brazil, Canada, Czech Republic, France, Germany, Greece, Italy, Spain, Sweden, Switzerland, and the U.S. (Ahrer *et al.*, 2001; Andreozzi *et al.*, 2003a; ARGE, 2003; Buser *et al.*, 1998b; Deng *et al.*, 2003; Drewes *et al.*, 2002, 2003; Farré *et al.*, 2001; Heberer, 2002b; Heberer *et al.*, 2001a; Koutsouba *et al.*, 2003; Miao *et al.*, 2002; Möhle *et al.*, 1999; Öllers *et al.*, 2001; Sedlak and Pinkston, 2001; Soulet *et al.*, 2002; Stumpf *et al.*, 1999; Ternes, 1998; Tixier *et al.*, 2003; Werres *et al.*, 2000). Reddersen and Heberer (2003) observed a matrix-dependent formation of an artifact of diclofenac during sample preparation resulting in an up to 40% underestimation of diclofenac concentrations, especially in matrix-prone samples such as sewage effluents or surface water. The artifact identified as 1-(2,6-dichlorophenyl) indolin-2-one is formed during the acidification ($pH < 2$) of the samples, a sample preparation step inevitable when using solid phase extraction with reversed-phase adsorbents for the extraction of diclofenac.

Buser *et al.* (1998b) also observed a significant elimination of diclofenac in the water of a natural lake in Switzerland presuming a possible photolytic degradation of the residues. In laboratory experiments with spiked lake water, Buser *et al.* (1998b) confirmed a rapid and extensive photodegradation of diclofenac by sunlight. They also characterized several photoproducts, but these could not be detected under natural conditions. Photodegradation of diclofenac was, however, also observed in a recent study in Switzerland reported by Tixier *et al.* (2003) and in irradiation experiments conducted by

Huber *et al.* (2003) and Andreozzi *et al.* (2003a). Moreover, results from surface water monitoring in Berlin, Germany also indicate a possible photodegradation of diclofenac (Heberer *et al.*, 2002b). In general, the reduction of diclofenac by natural photolytic degradation also depends on some additional key parameters such as eutrophic conditions, degree of solid particulate matter, and the depth of the water. In addition to photodegradation, seasonal differences of diclofenac concentrations may also be due to more extensive application of the drug during the winter (Heberer *et al.*, 2002b), because cold and humid weather causes an increase of rheumatic diseases.

Under recharge conditions, diclofenac has also been detected in groundwater samples (Heberer *et al.*, 1997; Sacher *et al.*, 2001). Preliminary results from laboratory experiments (Mersmann *et al.*, 2002), observations (Brauch *et al.*, 2000; Kühn and Müller, 2000) and field experiments on bank filtration (Heberer and Mechlinski, 2003; Heberer *et al.*, 2001b, 2002b, 2004), slow-sand filtration (Preuss *et al.*, 2001) and SAT (Drewes *et al.*, 2002, 2003) indicate significant sorption and an attenuation of diclofenac residues in the subsoil (Verstraeten *et al*, 2002). To date, diclofenac was only sporadically found at trace-level concentrations in raw or treated drinking water (Brauch *et al.*, 2000; Heberer, 2002b; Heberer *et al.*, 2001a, b; Kühn and Müller, 2000; Ternes, 2001). Several studies have shown that diclofenac can be removed from drinking water by ozonation or filtration with granular activated carbon (GAC) (Huber *et al.*, 2003; Ternes *et al.*, 2002; Zwiener and Frimmel, 2000). Together with several other PhACs, diclofenac was also efficiently removed from surface water and municipal sewage effluents using membrane filtration (Drewes *et al.*, 2002; Heberer and Feldmann, 2004; Heberer *et al.*, 2002b; Kimura *et al.*, 2003; Sedlak and Pinkston, 2001).

In Austria, Brazil, Canada, France, Germany, Greece, Italy, and Switzerland, ibuprofen is found in sewage effluents and rivers, usually at concentrations lower than those determined for diclofenac (Andreozzi *et al.*, 2003a; ARGE, 2003; Buser *et al.*, 1999; Gans *et al.*, 2002; Heberer *et al.*, 2002b; Miao *et al.*, 2002; Öllers *et al.*, 2001; Stumpf *et al.*, 1999; Ternes, 1998; Winkler *et al.*, 2001), due to ibuprofen's better degradability during sewage treatment (Zwiener and Frimmel, 2003). Soulet *et al.* (2002) measured ibuprofen in sewage influents in Switzerland at concentrations up to more than 3 μg/L, whereas only up to 0.5 μg/L of ibuprofen were found in the respective effluent samples. Tixier *et al.* (2003) also detected ibuprofen in Swiss STP effluents at concentrations up to 1.3 μg/L, but only at low concentrations (<100 ng/L) in surface water. In Sweden, 7.11 μg/L of ibuprofen, but no diclofenac, was detected in sewage effluents by Andreozzi *et al.* (2003a). In U.S. sewage effluent samples, ibuprofen was found at concentrations equal to ($\leq$300 ng/L) or even much higher (up to 3.38 μg/L) than those measured for diclofenac (Drewes *et al.*, 2002, 2003; Sedlak and Pinkston, 2001). Kolpin *et al.* (2002a) also detected up to 1.0 μg/L of ibuprofen in samples collected from U.S. streams. In Spain, Farré *et al.* (2001) found up to 85 μg/L of ibuprofen in sewage effluent samples. In the same study, ibuprofen was also found at relatively high concentrations (up to 2.7 μg/L) in Spanish surface waters. Rodriguez *et al.* (2003) found ibuprofen at concentrations of 2.81 and 5.77 μg/L in influent and 0.91 and 2.10 μg/L in effluent samples from municipal STPs in Spain. In Czech surface water samples, ibuprofen was found at concentrations up to 146 ng/L (ARGE, 2003).

Ibuprofen is degraded in the human body to its principal metabolites hydroxy- and carboxy-ibuprofen and to carboxy-hydratropic acid (Buser *et al.*, 1999; Stumpf *et al.*, 1998), which are found together with ibuprofen in raw sewage. Stumpf *et al.* (1998) observed a significant removal of ibuprofen, especially carboxy-ibuprofen, during sewage treatment, the concentrations of hydroxy-ibuprofen in the sewage effluents (median: 0.92 μg/L) were almost similar to those in the influents. Thus, hydroxy-ibuprofen was found in 12 German surface waters at much higher concentrations (median: 0.34 μg/L) than ibuprofen or carboxy-ibuprofen (median: 0.02 μg/L, respectively) (Stumpf *et al.*, 1998). In contrast to Stumpf *et al.* (1998), Buser *et al.* (1999) observed an efficient elimination (96–99.9%) of these compounds (including hydroxy-ibuprofen) in the municipal STPs. Buser *et al.* (1999) also studied the enantioselectivity of the degradation of the parent optical isomers and their corresponding metabolites. Laboratory studies carried out by Winkler *et al.* (2001) indicated that ibuprofen and its pharmacologically inactive stereoisomer were readily degraded in a river biofilm reactor. Hydroxy- and carboxy-ibuprofen, detected as metabolites, were further degraded in the biofilm reactors. Winkler *et al.* (2001) also concluded that abiotic losses and adsorption could only play a minor role because no loss of ibuprofen was observed in a sterile reactor. Heberer and Feldmann (2004) observed an influence of external conditions on the removal efficiency of the sewage treatment process for several selected PhACs, including the analgesics ibuprofen, ketoprofen, and naproxen. During heavy rain, the concentrations of ibuprofen detected in the secondary effluent of a STP (including nitrification and denitrification) increased from less than 1 ng/L to 630 ng/L (Heberer and Feldmann, 2004). Soil column experiments indicated a significant removal of ibuprofen, most likely caused by microbial degradation (Mersmann *et al.*, 2002; Preuss *et al.*, 2001).

Several other analgesics (namely 4-aminoantiyrine, aminophenazone, codeine, fenoprofen, flurbiprofen, hydrocodone, indometacin, ketoprofen, mefenamic acid, naproxen, phenazone, phenylbutazone, propyphenazone, and the drug metabolites 1-Acetyl-1-Methyl-2-Dimethyl-Oxamoyl-2-PhenylHydrazide [AMDOPH], 4-hydroxy-antipyrine, N-Acetyl-4-Aminoantipyrine [AAA], and N-Formyl-4-AminoAntipyrin [FAA]) have also been detected in sewage and/or surface water samples (Ahrer *et al.*, 2001; Andreozzi *et al.*, 2003a; ARGE, 2003; Boyd and Grimm, 2001; Drewes *et al.*, 2002, 2003; Farré *et al.*, 2001; Gans *et al.*, 2002; Heberer, 2002b; Heberer *et al.*, 2001a, 2002b; Jelicic and Ahel, 2003; Kolpin *et al.*, 2002a; Miao *et al.*, 2002; Möhle *et al.*, 1999; Öllers *et al.*, 2001; Rodriguez *et al.*, 2003; Sedlak and Pinkston, 2001; Schmidt and Brockmeyer, 2002; Soulet *et al.*, 2002; Stumpf *et al.*, 1999, 1998; Ternes, 1998; Ternes *et al.*, 2001; Tixier *et al.*, 2003). For ketoprofen and naproxen, Tixier *et al.* (2003) proposed biodegradation and phototransformation as possible elimination processes in surface waters.

Under recharge conditions or at landfill leachates, several analgesics (namely diclofenac, ibuprofen, ketoprofen, aminophenazone, phenazone, propyphenazone, and the drug metabolites gentisic acid, N-methylphenacetin, 1-acetyl-1-methyl-2-phenylhydrazide [AMPH], dimethyloxalamide acid-[N′-methyl-N-phenyl]-hydrazide [DMOAS], N-methyl-4-aminoantiyrine, 4-aminoantiyrine, FAA, AAA, and AMDOPH) have been detected in groundwater samples in Croatia, Denmark or Germany (Ahel and Jelicic, 2001; Brauch *et al.*, 2000; Heberer and Mechlinski, 2003; Heberer *et al.*, 1997, 2001b, 2002b; Holm *et al.*, 1995; Kühn and Müller, 2000; Reddersen *et al.*, 2002; Sacher *et al.*, 2001; Schmidt and Brockmeyer, 2002). In Germany, residues of diclofenac, ibuprofen, phenazone, propyphenazone, AMPH, DMOAS, and AMDOPH have also been found at trace-level concentrations in a few drinking water samples (Heberer, 2002b; Heberer *et al.*, 2001a; Reddersen *et al.*, 2002; Ternes, 2001). In laboratory experiments, propyphenazone was adsorped at sediments, but there is also some evidence that it might be remobilized by particle transport (Mersmann *et al.*, 2002). In field experiments on bank filtration, propyphenazone was not totally removed. It was detected together with AMDOPH in shallow wells and also reached the water supply wells (Heberer and Mechlinski, 2003; Heberer *et al.*, 2001b, 2002b, 2004). Drewes *et al.* (2002, 2003) investigated the fate of several PhACs during SAT of sewage effluents after secondary treatment, and they observed a high efficacy for the removal of diclofenac, ibuprofen, ketoprofen, naproxen, and fenoprofen, but not for propyphenazone. Antiphlogistic drugs might also be removed from municipal sewage effluents by ozonation (Ternes *et al.*, 2003).

B. Antibiotics/Bacteriostatics (Antibacterial Drugs)

Several studies have been carried out in Germany (ARGE, 2003; Christian *et al.*, 2003; Hirsch *et al.*, 1999; Steger-Hartmann *et al.*, 1997; Ternes *et al.*, 2003), France, Greece, Italy, Sweden (Andreozzi *et al.*, 2003a), Austria (Gans *et al.*, 2002), Switzerland (Alder *et al.*, 2001; Golet *et al.*, 2001, 2002), and the U.S. (Lindsey *et al.*, 2001; Kolpin *et al.*, 2002a) to investigate the occurrence and fate of antibacterial drugs in STPs or surface waters. Macrolide antibiotics (azithromycin, clarithromycin, clindamycin, dehydroerythromycin [metabolite of erythromycin], roxithromycin, lincomycin, tylosin), sulfonamides (sulfamethoxazole, sulfadimethoxine, sulfadimidine, sulfamethazine, and sulfathiazole), fluoroquinolones (ciprofloxacin, norfloxacin, ofloxacin, lomefloxacin, enoxacin, and enrofloxacin), chloramphenicol, tylosin, and trimethoprim have all been found up to the low μg/L-level in sewage and surface water samples. Antibiotics have also been identified at high concentrations in hospital effluents (Alder *et al.*, 2001; Hartmann *et al.*, 1998). Hartmann *et al.* (1998) detected 3–87 μg/L of the fluoroquinolone antibiotic ciprofloxacin in hospital effluents.

In monitoring investigations of various sewage, surface, and groundwater samples in Germany, Hirsch *et al.* (1999) did not detect penicillins or tetracyclines. This result is no surprise, as penicillins are easily hydrolyzed and tetracyclines readily precipitate with cations such as calcium, and accumulate in sewage sludge or sediment (Daughton and Ternes, 1999; Stuer-Lauridsen *et al.*, 2000). Nevertheless, Lindsey *et al.* (2001) and Kolpin *et al.* (2002a) detected tetracycline drugs (chlortetracycline, oxytetracycline, and tetracycline) in U.S. surface water samples, whereas Christian *et al.* (2003) did not detect any tetracycline antibiotics in riverwater samples from North Rhine-Westphalia, Germany. In the same samples, they found five -lactam antibiotics (namely piperacillin, amoxicillin, ampicillin, mezlocillin, and flucloxacillin) but only at very low concentrations with average levels of less than 10 ng/L (Christian *et al.*, 2003).

Golet *et al.* (2001) analyzed fluoroquinolone antibiotics in primary and tertiary wastewater effluents in Switzerland. In these samples, ciprofloxacin and norfloxacin occurred at concentrations of 249–405 ng/L and 45–120 ng/L, respectively. Golet *et al.* (2002) also detected ciprofloxacin and norfloxacin in surface water at concentrations below 19 ng/L. These results were confirmed by Christian *et al.* (2003), who only sporadically found ciprofloxacin and ofloxacin in German riverwater samples at concentrations $\leq$20 ng/L. In the same study, the macrolide antibiotics azithromycin, clarithromycin, clindamycin, and roxithromycin were frequently detected at similar concentrations (maximum value of 37 ng/L for clarithromycin) whereas tylosin, which is used solely in animal medicine in Germany, was detected in only one sample at a concentration of 90 ng/L. The most prevalent antibiotic was dehydro-erythromycin, which occurred at peak concentrations up to 300 ng/L in the surface water samples (Christian *et al.*, 2003). The relatively high concentrations of erythromycin residues compared to those of the other macrolide antibiotics were explained by its main use against acne and skin diseases (Christian *et al.*, 2003), because dermal application results in lower reabsorption and metabolism of this drug. In the same study, frequent findings ($\leq$100 ng/L) were also reported for sulfamethoxazole and trimethoprim, which are often administered in combination as anti-acne drugs. The antibacterial and antiprotozoal drugs metronidazole and ronidazole were sporadically detected at concentrations up to 44 and 16 ng/L, respectively, in surface water samples in the Czech Republic (ARGE, 2003).

Sacher *et al.* (2001) reported the occurrence of sulfamethoxazole (up to 410 ng/L) and dehydro-erythromycin (up to 49 ng/L) in groundwater samples in Baden-Württemberg, Germany. Sulfamethoxazole and sulfamethazine have also been detected at low concentrations in a few groundwater samples in the U.S. and Germany (Hartig *et al.*, 1999; Hirsch *et al.*, 1999; Lindsey *et al.*, 2001). Holm *et al.* (1995) found residues of different sulfonamides at high concentrations in groundwater samples collected downgradient of a landfill in Grinsted, Denmark. However, Heberer *et al.* (in preparation) observed an efficient removal of various antibiotic and bacteriostatic drugs during bank filtration at a field site (Lake Wannsee transect) in Berlin, Germany. Antibacterial drugs can also be removed from drinking water or municipal sewage effluents by ozonation (Huber *et al.*, 2003; Ternes *et al.*, 2003).

C. Antiepileptic Drugs

The antiepileptic drug carbamazepine has frequently been detected in municipal sewage and surface water samples (Ahrer *et al.*, 2001; Andreozzi *et al.*, 2003a; ARGE, 2003; Drewes *et al.*, 2002, 2003; Gans *et al.*, 2002; Heberer, 2002b; Heberer *et al.*, 2001a, 2002b; Möhle *et al.*, 1999; Öllers *et al.*, 2001; Patterson *et al.*, 2002; Ternes, 1998; Tixier *et al.*, 2003). Investigations of influent and effluent samples from different municipal STPs have shown that it is not significantly removed (less than 10%) during sewage treatment (Gans *et al.*, 2002; Heberer, 2002b; Ternes, 1998). It was also found as a fairly persistent contaminant in surface waters in Germany and Switzerland (Heberer, 2002b; Heberer *et al.*, 2002b; Tixier *et al.*, 2003) and was detected at concentrations up to 1075 ng/L in Berlin, Germany (Heberer *et al.*, 2002b). Andreozzi *et al.* (2002) studied the abiotic transformation of carbamazepine by solar photodegradation. Their results indicated that carbamazepine is capable of photolyzing in distilled waters and riverwaters with nitrate promoting and humic acids inhibiting its degradation. Primidone, another antiepileptic drug, has also been detected in samples from municipal sewage influents and effluents and in surface waters (up to 635 ng/L) in Germany (Heberer, 2002b; Heberer *et al.*, 2001a, 2002b; Möhle *et al.*, 1999). In the U.S., primidone has been detected in secondary and tertiary treated wastewater at concentrations between 100 and 200 ng/L (Drewes *et al.*, 2002, 2003).

In soil column experiments, carbamazepine was significantly retarded compared to lithium, used as tracer compound (Mersmann *et al.*, 2002). On the other hand, no significant elimination of carbamazepine was observed under aerobic or anaerobic groundwater conditions (Mersmann *et al.*, 2002; Preuss *et al.*, 2001). Carbamazepine and primidone were not removed by using SAT for the treatment of sewage effluents (Drewes *et al.*, 2002, 2003). Additionally, different field studies have shown that carbamazepine and primidone are only slightly or not attenuated during bank infiltration (Brauch *et al.*, 2000; Heberer and Mechlinski, 2003; Heberer *et al.*,

2001b, 2002b, 2004; Kühn and Müller, 2000). Both compounds have been detected in the shallow wells and water-supply wells of different transects built to study the behavior of PhACs during bank filtration (Heberer *et al.*, 2001b, 2002b, 2004). This also explains why carbamazepine has been detected in a number of groundwater samples at a maximum concentration of 1.1 μg/L (Sacher, 2001; Seiler *et al.*, 1999; Ternes, 2001) and was also found with a concentration of 30 ng/L in drinking water (Ternes, 2001). Drinking water treatment by ozonation or granular activated charcoal (GAC) filtration can successfully remove PhACs (Andreozzi *et al.*, 2002; Huber *et al.*, 2003; Ternes *et al.*, 2002). Carbamazepine can be removed from municipal sewage effluents by ozonation (Ternes *et al.*, 2003); primidone and carbamazepine can also efficiently be removed by membrane filtration with nanofiltration or reverse osmosis (RO) membranes (Drewes *et al.*, 2002; Heberer and Feldmann, 2004; Kimura *et al.*, 2003). Even when highly contaminated surface waters or municipal sewage effluents were used as raw water sources, the concentrations of both compounds in treated water can be decreased below the detection limits (Drewes *et al.*, 2002; Heberer and Feldmann, 2004).

D. Beta-Blockers

Several beta-blockers (metoprolol, acebutolol, oxprenolol, propanolol, betaxolol, bisoprolol, and nadolol) have been detected in municipal sewage effluents up to the μg/L-level (Andreozzi *et al.*, 2003a; ARGE, 2003; Hirsch *et al.*, 1996; Huggett *et al.*, 2003; Sedlak and Pinkston, 2001; Ternes, 1998; Ternes *et al.*, 2003). Only metoprolol, propanolol, and bisoprolol have also been found in surface water samples, at relatively low concentrations (Hirsch *et al.*, 1998; Ternes, 1998). As far as beta-blockers are concerned, Hirsch *et al.* (1998) did not find any relevance for groundwater recharge or drinking water supply. However, Sacher *et al.* (2001) also reported the detection of sotalol at maximum concentrations of 560 ng/L in three groundwater samples from Baden-Württemberg, Germany. Sedlak and Pinkston (2001) have demonstrated that metoprolol is efficiently removed from sewage effluents applying RO or SAT. Beta-blockers might also be removed from municipal sewage effluents by ozonation (Ternes *et al.*, 2003).

E. Blood Lipid Regulators

The first incidental findings of the drug metabolite clofibric acid (2-(4)-chlorophen-oxy-2-methyl propionic acid) in the aquatic environment of Germany and Switzerland (Buser *et al.*, 1998a; Heberer and Stan, 1997; Heberer *et al.*, 1996, 1997; Stan *et al.*, 1992, 1994) probably initiated most of the investigations on PhACs mentioned in this chapter. However, the first detections of clofibric acid—the active metabolite of the blood lipid regulators clofibrate, etofyllin clofibrate, and etofibrate—in samples from STPs in the U.S. were reported in the 1970s (Garrison *et al.*, 1976; Hignite and Azarnoff, 1977). In Germany, clofibric acid was found at concentrations up to 4 μg/L in groundwater samples collected from former sewage irrigation fields near Berlin (Heberer and Stan, 1997). Underneath the sewage farm areas, it could even be found in samples from the fourth and fifth groundwater aquifers, down to a depth of 125 m. Up to 270 ng/L of clofibric acid has been detected in Berlin drinking water samples (Heberer and Stan, 1997; Stan *et al.*, 1994). Buser *et al.* (1998a) detected clofibric acid at the low ng/L-range in Swiss lakes from populated areas, and also in the North Sea. Clofibric acid was identified as a refractory contaminant in several investigations of municipal sewage influents and effluents (Heberer, 2002b; Heberer *et al.*, 2002b; Patterson *et al.*, 2002; Ternes, 1998; Tixier *et al.*, 2003; Soulet *et al.*, 2002; Stumpf *et al.*, 1999). Zwiener *et al.* (2000) and Zwiener and Frimmel (2003) carried out biodegradation studies using a pilot sewage plant and biofilm reactors operated under oxic or anoxic conditions. In spiking experiments with synthetic sewage water, they confirmed the persistence of clofibric acid under anoxic and oxic conditions as well.

Meanwhile, a number of findings in sewage, surface, and groundwater have been reported for clofibric acid from Austria, Brazil, Canada, Germany, Italy, Sweden, Switzerland, and the U.S. (Ahrer *et al.*, 2001; Andreozzi *et al.*, 2003a; ARGE, 2003; Boyd and Grimm, 2001; Heberer and Feldmann, 2004; Heberer *et al.*, 1997, 1998, 2001a,b; Öllers *et al.*, 2001; Patterson *et al.*, 2002; Soulet *et al.*, 2002; Stumpf *et al.*, 1999; Ternes, 1998, 2001; Tixier *et al.*, 2003; Werres *et al.*, 2000; Winkler *et al.*, 2001).

Bezafibrate, gemfibrozil, clofibrate, fenofibrate, and its metabolite fenofibric acid, have been detected up to the μg/L-level in sewage effluents and surface water samples (Ahrer *et al.*, 2001; Andreozzi *et al.*, 2003a; ARGE, 2003; Drewes *et al.*, 2002, 2003; Farré *et al.*, 2001; Heberer *et al.*, 2001b, 2002b; Heberer and Feldmann, 2004; Miao *et al.*, 2002; Sedlak and Pinkston, 2001; Stumpf *et al.*, 1999; Ternes, 1998; Werres *et al.*, 2000). Recently, atorvastatin has also been detected in sewage effluent samples (Miao and Metcalfe, 2003). Usually, gemfibrozil is detected in sewage effluents at the ng/L-level, but in several samples of secondary and tertiary treated wastewater, it was also detected at much higher levels, up to 6 μg/L (Drewes *et al.*, 2002, 2003; Sedlak and Pinkston, 2001). Heberer and Feldmann (2004) observed a 10-fold increase of the sewage effluent concentrations for bezafibrate, gemfibrozil, and fenofibric acid during a heavy rain event.

In laboratory experiments using soil columns (Mersmann *et al.*, 2002; Preuss *et al.*, 2001), bezafibrate, clofibric acid, and gemfibrozil did not show any significant sorption or elimination under aerobic or anaerobic groundwater conditions. It leached almost tracer-like through the soil columns without retardation. This observation was also confirmed in several studies on bank filtration, where clofibric acid was reaching the water-supply wells without being completely removed in the subsoil (Heberer *et al.*, 2001b, 2002b, 2004; Verstraeten *et al.*, 2002). However, Preuss *et al.* (2001) observed a moderate elimination of clofibric acid and gemfibrozil during slow-sand filtration. On the other hand, bezafibrate was found to be easily attenuated during bank filtration (Heberer *et al.*, 2001b, 2002b, 2004) or slow-sand filtration (Preuss *et al.*, 2001). Nevertheless, bezafibrate has also been reported to occur in groundwater samples at a maximum concentration of 190 ng/L (Ternes, 2001). Gemfibrozil was found in groundwater samples at maximum concentrations of 340 ng/L (Heberer, 2002b). Besides several findings of clofibric acid (Heberer, 2002b; Heberer and Stan, 1997; Stan *et al.*, 1994; Ternes, 2001), single detections have also been reported for fenofibric acid (42 ng/L) and bezafibrate (27 ng/L) in drinking water (Ternes, 2001).

Drinking water treatment by ozonation or GAC filtration decreases the concentrations of bezafibrate and clofibric acid (Boyd and Grimm, 2001; Huber *et al.*, 2003; Ternes *et al.*, 2002; Zwiener and Frimmel, 2000), but clofibric acid might not totally be removed even when applying a combination of both techniques (Ternes *et al.*, 2002). Clofibric acid, as well as bezafibrate, gemfibrozil, and fenofibric acid, can efficiently be removed by RO from various raw water sources, including highly contaminated surface waters or municipal sewage effluents (Heberer and Feldmann, 2004; Heberer *et al.*, 2002a).

F. Contrast Media

Iodinated X-ray contrast media applied at high amounts, mostly in hospitals but also in practical surgeries, have been identified by Gartiser *et al.* (1996) as the main contributors to the loads of total adsorbable organic halogens (AOX) in clinical wastewaters. Oleksy-Frenzel *et al.* (2000) used adsorbable organic iodine (AOI) detection and measured high concentrations (up to 130 μg I/L) of organic iodine compounds in the influent and effluent of a municipal treatment plant in Berlin, and up to 10 mg I/L in a hospital wastewater, observing no degradation or only minor attenuation during sewage purification. They assumed that AOI contamination of the aquatic environment was primarily due to the presence of iodinated X-ray contrast media. This has, however, not yet been fully confirmed by individual component identification. Putschew *et al.* (2001) identified 39% of the AOI detected in sewage effluents as contrast agents. In Berlin, Germany, high AOI values of more than 10 μg I/L have not only been measured in sewage and surface waters, but also in bank filtrate and raw drinking water samples (Putschew *et al.*, 2000). In surface waters, between 18 and 33% of the AOI could be identified as being iodinated contrast media; in bank filtrate and raw drinking water samples, only between 3.4 and 25% of the AOI were identified (Putschew and Jekel, 2001; Putschew *et al.*, 2000). Although many of the postulated metabolites have been analyzed, only one of these compounds was also identified in the samples (Putschew *et al.*, 2001). Thus, it was assumed that the majority of the AOI consists of several other unknown metabolites of iodinated contrast media (Putschew *et al.*, 2000).

The five X-ray contrast agents diatrizoate, iohexol, iopamidol, iopromide, and iomeprol were found up to μg/L-concentrations in municipal sewage effluents and in surface water samples (Putschew *et al.*, 2001; Ternes and Hirsch, 2000; Ternes *et al.*, 2003). Iothalamic acid and ioxithalamic acid have sometimes also been detected at ng/L-concentrations in influents and effluents of STPs and in surface waters (Ternes and Hirsch, 2000).

Generally, the loads of the X-ray contrast media are significantly increased on weekdays, because X-ray examinations are performed in hospitals and radiological practices predominantly from Monday to Friday (Ternes and Hirsch, 2000). Ternes and Hirsch (2000) stated that, compared to the other drug residues, the iodinated X-ray contrast media exhibited generally higher maximum levels in STP effluents. Nevertheless, considering their high maximum levels, the average contamination was not as great as expected. The median concentrations of the X-ray contrast media of $\leq$0.75 μg/L measured by Ternes and Hirsch (2000) are at least one order of magnitude less than the corresponding maximum concentration levels. In all countries with a developed medical care system, it can be expected that X-ray contrast media are present at appreciable quantities in the sewage effluents, leading to a contamination of receiving waters.

Iodinated contrast agents are very persistent in the aquatic environment and also easily leach into the groundwater aquifers. Thus, diatrizoate, iopromide, iopamidol, and amidotrizoic acid were detected up to the μg/L-level in groundwater and bank filtrate samples (Putschew *et al.*, 2000; Sacher *et al.*, 2001; Ternes and Hirsch, 2000). Iothalamic acid and ioxithalamic acid have been detected in a few samples at low ng/L-concentrations by Ternes and Hirsch (2000). Ternes (2001) and Putschew *et al.* (2000) also reported positive findings of diatrizoate, iopromide, and iopamidol in drinking water and raw water used for drinking water production.

Iodinated X-ray contrast media are not readily biodegradable. Nevertheless, Steger-Hartmann *et al.* (2002) observed that iopromide was amenable to primary degradation when using a test system to simulate sewage treatment. The resulting degradation product (5-amino-N,N′-bis (2,3-dihydroxypropyl)-2,4,6-triiodo-N-methyliso-phthalamide) showed a faster photolysis than the parent compound, and was also further degraded in a test system simulating surface water conditions (Steger-Hartmann *et al.*, 2002). Additionally, oxidation of iopromide is possible when using conventional ozonation or advanced oxidation processes (AOPs) applied in drinking water treatment (Huber *et al.*, 2003). However, Ternes *et al.* (2003) still detected the iodinated X-ray contrast media diatrizoate, iopamidol, iopromide, and iomeprol at appreciable concentrations, even when applying 10–15 mg/L ozone for further treatment of municipal sewage effluents. Thus, the ionic diatrizoate only exhibited removal efficiencies of ≤14%, while the non-ionic iodinated X-ray contrast media (ICM) were more than 80% removed (Ternes *et al.*, 2003).

The rare earth element gadolinium (Gd), used in the form of organic complexes in magnetic resonance imaging (MRI), is also consecutively discharged via hospital effluents and public sewage systems into the receiving surface waters (Kümmerer, 2001). It was detected in hospital effluents at concentrations up to 100 μg/L (Kümmerer and Helmers, 2000). In rivers influenced by STP discharges, Gd has been found at concentrations of about 0.2 μg/L, significantly higher than the natural background value of approximately 0.001 μg/L (Bau and Dulski, 1996; Kümmerer and Helmers, 2000).

G. Cytostatic Drugs

Cytostatics are frequently used in chemotherapy. Thus, residues of cytostatic drugs almost exclusively originate from hospital applications and may occur in hospital sewage at concentrations up to the low μg/L-level (Steger-Hartmann *et al.*, 1997). In effluents from those municipal STPs receiving and purifying hospital effluents, cytostatic drugs have been found at trace concentrations, mostly at the low ng/L-level (Kümmerer *et al.*, 1997; Steger-Hartmann *et al.*, 1996; Ternes, 1998). Steger-Hartmann *et al.* (1996) detected ifosfamide and cyclophosphamide in sewage samples from a university hospital at concentrations of 24 and 146 ng/L, respectively. Kümmerer *et al.* (1997) found ifosfamide at mean concentrations of 109 ng/L in effluents from a oncologic hospital. The influents and effluents of the receiving municipal STP contained were measured at mean concentrations between 6.2 and 9.3 ng/L, without observing any significant reduction during sewage treatment. In 4 out of 16 effluent samples from German STPs, Ternes (1998) detected cyclophosphamide at maximum concentrations of 20 ng/L. Ifosfamide was only detected in two samples, but in one of the samples at a concentration of 2.9 μg/L. Until now, cytostatics have not been detected in surface waters, but Kümmerer *et al.* (1997) calculated a predicted environmental concentration (PEC) of 0.8 ng/L for ifosfamide in German surface waters. Due to their high pharmacologic potency, such compounds often exhibit carcinogenic, mutagenic, or embryotoxic properties. Thus, further investigations on their occurrence and fate may provide insight on their risk potential for humans and the environment (Kümmerer, 2001).

H. Oral Contraceptives

Synthetic steroids are frequently prescribed as oral contraceptives, but because of their high pharmacological potency, the total amounts sold annually are relatively low. Ternes *et al.* (1999b) estimated the annual prescriptions of 17α-ethinylestradiol in Germany at only 50 kg per year. Thus synthetic steroid hormones, such as the estrogens 17α-ethinylestradiol (EE2) and mestranol, can only appear at trace-level concentrations (low ng/L-range) in sewage effluents. This presumption was confirmed by results from several investigations of STPs in Brazil, Canada, Germany, France, England, Italy, The Netherlands, and the U.S. (Adler *et al.*, 2001; Baronti *et al.*, 2000; Belfroid *et al.*, 1999; Bruchet *et al.*, 2002; Desbrow *et al.*, 1998; Heberer, 2002b; Huang and Sedlak, 2001; Johnson *et al.*, 2000; Kuch and Ballschmiter, 2001; Spengler *et al.*, 1999; Ternes *et al.*, 1999a; Verstraeten *et al.*, 2003; Xiao *et al.*, 2001; Zühlke *et al.*, 2004). Mestranol has only sporadically been detected in sewage effluents at concentrations up to 4 ng/L (Spengler *et al.*, 1999; Ternes *et al.*, 1999a). In investigations of 20 German sewage effluent samples, Stumpf *et al.* (1996) reported a median concentration of 17 ng/L for EE2, a result that has, however, never been confirmed in any of the following studies. Generally, the median concentration of EE2 in sewage effluents in Germany, England, France, The Netherlands, and the U.S. as reported by several authors (Adler *et al.*, 2001; Belfroid *et al.*, 1999; Bruchet *et al.*, 2002; Desbrow *et al.*, 1998; Heberer, 2002b; Huang and Sedlak, 2001; Johnson *et al.*, 2000; Kuch and Ballschmiter, 2001; Spengler *et al.*, 1999; Ternes *et al.*, 1999a; Verstraeten *et al.*, 2003; Xiao *et al.*, 2001; Zühlke *et al.*, 2004) is approximately 1–3 ng/L or even lower, below the analytical detection limit. Canadian sewage effluent samples contained EE2 at a median concentration of 9 ng/L (Ternes *et al.*, 1999a). Much higher concentrations of EE2 (up to 831 ng/L; median value; 73 ng/L excluding non-detects) were reported by Kolpin *et al.* (2002a) in samples collected from various streams in the U.S. These findings have extensively been debated (Ericson *et al.*, 2002; Kolpin *et al.*, 2002b). In an investigation of six activated sludge STPs near Rome, Italy, Baronti *et al.* (2000) determined average concentrations of 3.0 ng/L for EE2 in sewage influent samples. The median sewage effluent concentration of EE2 was 0.45 μg/L. Baronti *et al.* (2000) calculated a removal rate of 85% for EE2 and concluded that activated sludge treatment efficiently removed EE2. Similar but slightly lower removal rates (75.7%; n = 31) were also observed by Zühlke *et al.* (2004). They detected EE2 at mean concentrations of 2.0 ng/L in the effluents (mean influent concentration: 8.6 ng/L) of a municipal STP in Berlin, Germany, after secondary treatment including nitrification and denitrification. EE2 could not be detected at levels above the analytical limit of quantitation of 0.4 ng/L in effluents from another municipal STP in Berlin, Germany, which uses the same technology (Verstraeten *et al.*, 2003).

However, Ternes *et al.* (1999b) did not observe a significant reduction of EE2 concentrations in aerobic batch experiments containing a diluted slurry of activated sludge from a STP near Frankfurt, Germany. In the same experiments, mestranol was rapidly eliminated. On the basis of the daily human excretion of conjugated estrogens, Baronti *et al.* (2000) presumed that deconjugation of estrogens preferentially occurs in sewers. However, in investigations of influents and effluents from German STPs, Adler *et al.* (2001) observed that conjugated steroids constituted up to 50% of the total steroid concentration. Andersen *et al.* (2003) investigated the fate of estrogenic steroids in the treatment train of a German municipal STP with an activated sludge system for nitrification and denitrification, including sludge recirculation. In these investigations, removal rates of more than 90% were observed for EE2 with a mean concentration of <1 ng/L (limit of detection) in the secondary effluent (mean influent concentration: 8.2 ng/L). In contrast to the natural estrogens, which were to a large extent degraded in the denitrifying and aerated nitrifying tanks, EE2 was only degraded in the nitrifying tank of the activated sludge system (Andersen *et al.*, 2003).

Analysis of a water sample from the Tiber river, Italy, downstream of small towns whose sewages are treated by percolating filter STPs or directly discharged into the river, revealed the presence of EE2 at 0.04 ng/L (Baronti *et al.*, 2000). In general, EE2 was only detected in a few surface water samples at maximum concentrations up to 4.3 ng/L (Adler *et al.*, 2001; Belfroid *et al.*, 1999; Stumpf *et al.*, 1996) but most of the samples were below the limits of detection (Adler *et al.*, 2001; Belfroid *et al.*, 1999; Huang and Sedlak, 2001; Stumpf *et al.*, 1996; Verstraeten *et al.*, 2003; Xiao *et al.*, 2001; Zühlke *et al.*, 2004). Although the detected concentrations are very low, they may still be important for the aquatic environment because *in vitro* studies have shown that exposure of fish to only 0.1 ng/L of EE2 (Purdom *et al.*, 1994) may provoke feminization in some species of male wild fishes.

Due to their physico-chemical properties, and on the basis of their hydrophobicity (log K_{ow} ~4) and potential for biotransformation (Holthaus *et al.*, 2002; Jurgens *et al.*, 2002; Ternes *et al.*, 1999b), the hormones would not be expected to persist in the subsurface. Nevertheless, Adler *et al.* (2001) and Kuch and Ballschmitter (2001) recently reported several positive detections of EE2 in ground and drinking water in Germany. Adler *et al.* (2001) found EE2 in groundwater and in raw and purified drinking water up to concentrations of 2.4 ng/L. They used high-performance liquid chromatography (HPLC) with single mass spectrometry (MS), applying single ion monitoring (SIM) detection for the analysis of EE2. Kuch and Ballschmitter (2001) detected mean concentrations of 0.80 and 0.35 ng/L of EE2 in surface and drinking water samples, respectively. In their study, they used solid-phase extraction, pentafluorobenzoylation, and gas chromatography with MS (GC-MS) detection, applying negative chemical ionization (NCI) for the analysis of EE2. The results from both studies have not been confirmed by any other analytical method, such as GC-tandem MS (GC-MS/MS) or HPLC-MS/MS detection. However, further confirmation of these unexpected results seems to be necessary; both analyses by HPLC-MS or GC-NCI/MS are not as selective as GC with electron impact ionization MS or tandem MS (MS/MS), or HPLC with MS/MS detection, and might therefore be interfered with by matrix compounds. Investigations of surface and bank filtrate samples by HPLC-MS/MS in Berlin, Germany did not show any evidence of the infiltration of estrogenic steroids during groundwater recharge of surface water under the influence of sewage effluents (Verstraeten *et al.*, 2003; Zühlke *et al.*, 2004). Although several other PhACs were detected at concentrations up to the μg/L-level (Heberer *et al.*, 2004), EE2 was neither detected in surface water nor in bank-filtered water above the detection limit of 0.1 ng/L (Verstraeten *et al.*, 2003; Zühlke *et al.*, 2004). However, it was shown that even short distances between the river or lake banks and monitoring wells lead to dramatic decreases of estrogen concentrations, demonstrating the potential of groundwater recharge systems for the retention of estrogenic steroids.

I. Other PhACs

The bronchodilator drugs (β_2-sympathomimetics) salbutamol (albuterol in the U.S.) and terbutaline, and in a few cases clenbuterol and fenoterol, were reported by Ternes (1998) to occur at concentrations less than 20 ng/L in municipal sewage effluents. For all four compounds, the median sewage effluent concentrations were below the detection limits. In surface waters, only sporadic detections have been reported (Hirsch *et al.*, 1996; Ternes, 1998). The bronchodilator drug ambroxole and the bronchodilator drug metabolites 6,8-dibromo-3-(trans-4-hydroxycyclohexyl)-1,2,3,4-tetrahydroquinazoline (Na873) and 3,5-dibromoanthranilic acid were detected in surface water samples in Germany (Schmidt and Brockmeyer, 2002; ARGE, 2003). Schmidt and Brockmeyer (2002) also detected the sympathomimetic drug xylometazoline in surface water samples from Berlin, Germany.

In investigations of STP effluents and surface waters, Ternes *et al.* (2001) detected the tranquilizer diazepam, the antidiabetic drug glibenclamide, and the calcium influx inhibitor nifedipine. All three compounds were only found in a few samples, at maximum concentrations clearly below 100 ng/L. Gans *et al.* (2002) detected the calcium ion influx inhibitor drug verapamil at low concentrations (<70 ng/L) in influent and effluent samples collected from STPs in Austria. Möhle *et al.* (1999) detected the drug pheneturide and the hemorrheologic agent pentoxifylline in sewage influents and effluents in Germany. Pentoxifylline was also found at low concentrations up to 30 ng/L in samples collected from the Elbe river in 1998, but not in 1999 and 2000 (ARGE, 2003).

In surface water investigations in the U.S. commissioned by the U.S. Geological Survey, Kolpin *et al.* (2002a) detected low ng/L-concentrations of several other drugs, such as the histamine H_2-receptor antagonists cimetidine and ramitidine, the calcium ion influx inhibitor diltiazem, the angiotensin converting enzyme inhibitor enalaprilat, the nifedipine metabolite dehydronifedipine, the antidiabetic drug metformin, and the antidepressant fluoxetine.

Eckel *et al.* (1993) detected pentobarbital at a concentration of 1 μg/L in groundwater from a landfill in Florida. In groundwater samples near Reno, Nevada, Seiler *et al.* (1999) identified residues of the antidiabetic drug chlorpropamide and the anticonvulsant phensuximide. 5,5-diallylbarbituric acid was found together with several other pharmaceuticals and drug intermediates in groundwater from a landfill in Grinsted, Denmark (Holm *et al.*, 1995).

III. Conclusions

The studies show that some PhACs originating from human therapy are not eliminated completely in the municipal STPs and are therefore discharged as contaminants into the receiving waters. More than 100 compounds, pharmaceuticals and several drug metabolites, have been detected up to the μg/L-level in municipal sewage and surface waters located downstream from municipal sewage treatment plants. Under recharge conditions, several polar PhACs such as clofibric acid, carbamazepine, primidone, and iodinated contrast agents can leach through the subsoil into the groundwater aquifers. Positive findings of PhACs have also been reported in groundwater contaminated by landfill leachates or manufacturing residues. Laboratory experiments and field studies on bank filtration have, however, also shown that some PhACs are naturally attenuated in the subsoil. To date, only in a few cases have PhACs also been detected at trace-level concentrations in drinking water samples. Purification or removal techniques such as ozonation and membrane filtration are able to remove contaminations of PhACs efficiently from drinking water, surface water, and even from municipal sewage effluents.

References

Adler, P., Steger-Hartmann, T., and Kalbfus, W. (2001). Distribution of natural and synthetic estrogenic steroid hormones in water samples from southern and middle Germany. *Acta Hydrochim. Hydrobiol.* **29**, 227–241 (German).

Ahel, M., and Jelicic, I. (2001). Phenazone analgesics in soil and groundwater below a municipal solid waste landfill. *In* "Pharmaceuticals and Personal Care Products in the Environment: Scientific and Regulatory Issues" (C. G. Daughton and T. Jones-Lepp, Eds.), pp. 100–115. Symposium Series 791, American Chemical Society, Washington, D.C.

Ahrer, W., Scherwenk, E., and Buchberger, W. (2001). Determination of drug residues in water by the combination of liquid chromatography or capillary electrophoresis with electrospray mass spectrometry. *J. Chromatogr. A* **919**, 69–78.

Alder, A. C., McArdell, C. S., Golet, E. M., Ibric, S., Molnar, E., Nipales, N. S., and Giger, W. (2001). Occurrence and fate of fluoroquinolone, macrolide, and sulfonamide antibiotics during wastewater treatment and in ambient waters in Switzerland. *In* "Pharmaceuticals and Personal Care Products in the Environment: Scientific and Regulatory Issues" (C. G. Daughton and T. Jones-Lepp, Eds.), pp. 56–69. Symposium Series 791. American Chemical Society, Washington, D.C.

Andersen, H., Siegrist, H., Halling-Sørensen, B., and Ternes, T. A. (2003). Fate of estrogens in a municipal sewage treatment plant. *Environ. Sci. Technol.* **37,** 4021–4026.

Andreozzi, R., Marotta, R., Pinto, G., and Pollio, A. (2002). Carbamazepine in water: Persistence in the environment, ozonation treatment and preliminary assessment on algal toxicity. *Wat. Res.* **36,** 2869–2877.

Andreozzi, R., Marotta, R., and Paxéus, N. (2003a). Pharmaceuticals in STP effluents and their solar photodegradation in aquatic environment. *Chemosphere* **50,** 1319–1330.

Andreozzi, R., Caprio, V., Marotta, R., and Paxéus, N. (2003b). Paracetamol oxidation from aqueous solutions by means of ozonation and H_2O_2/UV system. *Wat. Res.* **37,** 993–1004.

ARGE (Arbeitsgemeinschaft für die Reinhaltung der Elbe) (2003). Pharmaceuticals in the Rivers Elbe and Saale. ARGE, Hamburg, Germany. Report available online at: http.// www.arge-elbe.de/wge/Download/Berichte/03Arzn.pdf

Baronti, C., Curini, R., d'Ascenzo, G., DiCorcia, A., Gentili, A., and Samperi, R. (2000). Monitoring natural and synthetic estrogens at activated sludge sewage treatment plants and in a receiving river water. *Environ. Sci. Technol.* **34,** 5059–5066.

Bau, M., and Dulski, P. (1996). Anthropogenic origin of positive gadolinium anomalies in river waters. *Earth Planet Sci. Lett.* **143,** 245–255.

Belfroid, A. C., Van der Horst, A., Vethaak, A. D., Schäfer, A. J., Rijs, G. B. J., Wegener, J., and Cofino, W. P. (1999). Analysis and occurrence of estrogenic hormones and their glucuronides in surface water and waste water in The Netherlands. *Sci. Total Environ.* **225,** 101–108.

Boyd, G. R., and Grimm, D. A. (2001). Occurrence of pharmaceutical contaminants and screening of treatment alternatives for southeastern Louisiana. *Ann. N.Y. Acad. Sci.* **948,** 80–89.

Brauch, H. J., Sacher, F., Denecke, E., and Tacke, T. (2000). Efficiency of bank filtration for the removal of polar organic tracer compounds. *Wasser Abwasser* **14,** 226–234 (German).

Bruchet, A., Prompsy, C., Filippi, G., and Souali, A. (2002). A broad spectrum analytical scheme for the screening of endocrine disruptors (EDs), pharmaceuticals and personal care products in wastewaters and natural waters. *Wat. Sci. Technol.* **46,** 97–104.

Buser, H.-R., Müller, M. D., and Theobald, N. (1998a). Occurrence of the pharmaceutical drug clofibric acid and the herbicide mecoprop in various Swiss lakes and in the North Sea. *Environ. Sci. Technol.* **32,** 188–192.

Buser, H.-R., Poiger, T., and Müller, M. D. (1998b). Occurrence and fate of the pharmaceutical drug diclofenac in surface waters: Rapid photodegradation in a lake. *Environ. Sci. Technol.* **32,** 3449–3456.

Buser, H.-R., Poiger, T., and Müller, M. D. (1999). Occurrence and environmental behavior of the pharmaceutical drug ibuprofen in surface waters and in wastewater. *Environ. Sci. Technol.* **33,** 2529–2535.

Christian, T., Schneider, R. J., Färber, H. A., Skutlarek, D., Meyer, M. T., and Goldbach, H. E. (2003). Determination of antibiotic residues in manure, soil, and surface waters. *Acta Hydrochim. Hydrobiol.* **31,** 36–44.

Daughton, C. G., and Jones-Lepp, T. (2001). Pharmaceuticals and personal care products in the environment: Scientific and regulatory issues. Symposium Series 791. American Chemical Society, Washington, D.C.

Daughton, C. G., and Ternes, T. A. (1999). Pharmaceuticals and personal care products in the environment: Agents of subtle change? *Environ. Health Perspect.* **107,** 907–938.

Deng, A., Himmelsbach, M., Zhu, Q.-Z., Frey, S., Sengl, M., Buchberger, W., Niessner, R., and Knopp, D. (2003). Residue analysis of the pharmaceutical diclofenac in different water types using ELISA and GC-MS. *Environ. Sci. Technol.* **37,** 3422–3429.

Desbrow, C., Routledge, E. J., Brighty, G. C., Sumpter, J. P., and Waldock, M. (1998). Identification of estrogenic chemicals in STP effluent. 1. Chemical fractionation and *in vitro* biological screening. *Environ. Sci. Technol.* **32,** 1549–1558.

Drewes, J., Heberer, Th., and Reddersen, K. (2002). Fate of pharmaceuticals during indirect potable reuse. *Water Sci. Technol.* **46,** 73–80.

Drewes, J., Heberer, Th., Rauch, T., and Reddersen, K. (2003). Fate of pharmaceuticals during groundwater recharge. *Ground Wat. Monit. Remed.* **23,** 64–73.

Eckel, W. P., Ross, B., and Isensee, R. K. (1993). Pentobarbital found in ground water. *Ground Wat.* **31,** 801–804.

Ericson, J. F., Laenge, R., and Sullivan, D. E. (2002). Comment on "pharmaceuticals, hormones, and other organic wastewater contaminants in U.S. streams, 1999–2000: A national reconnaissance." *Environ. Sci. Technol.* **36,** 4005–4006.

Farré, M., Ferrer, I., Ginebreda, A., Figueras, M., Olivella, L., Tirapu, L., Vilanova, M., and Barcelo, D. (2001). Determination of drugs in surface water and wastewater samples by liquid chromatography-mass spectrometry: Methods and preliminary results including toxicity studies with *Vibrio fischeri*. *J. Chromatogr. A* **938,** 187–197.

Flaherty, S., Wark, S., Street, G., Farley, J. W., and Brumley, W. C. (2002). Investigation of capillary electrophoresis-laser induced fluorescence as a tool in the characterization of sewage effluent for fluorescent acids: Determination of salicylic acid. *Electrophoresis* **23,** 2327–2332.

Gans, O., Sattelberger, R., and Scharf, S. (2002). Analysis of selected pharmaceuticals effluents of municipal sewage treatment plants in Austria. *Vom Wasser* **98,** 165–176 (German).

Garrison, A. W., Pope, J. D., and Allen, F. R. (1976). GC/MS analysis of organic compounds in domestic wastewaters. *In* "Identification and Analysis of Organic Pollutants in Water." (C. H. Keith, Ed.), pp. 517–556. Ann Arbor Science Publishers, Ann Arbor, Michigan.

Gartiser, S., Brinker, L., Erbe, T., Kümmerer, K., and Willmund, R. (1996). Contamination of hospital wastewater with hazardous compounds as defined by §7a WHG. *Acta Hydrochim. Hydrobiol.* **24,** 90–97.

Golet, E. M., Alder, A. C., Hartmann, A., Ternes, T. A., and Giger, W. (2001). Trace determination of fluoroquinolone antibacterial agents in urban wastewater by solid-phase extraction and liquid chromatography with fluorescence detection. *Anal. Chem.* **73,** 3632–3638.

Golet, E. M., Alder, A. C., and Giger, W. (2002). Environmental exposure and risk assessment of fluoroquinolone antibacterial agents in wastewater and river water of the Glatt Valley Watershed, Switzerland. *Environ. Sci. Technol.* **36,** 3645–3651.

Halling-Sørensen, B., Nielsen, N., Lansky, P. F., Ingerslev, F., Hansen, L., Lützhøft, H. C., and Jørgensen, S. E. (1998). Occurrence, fate and effects of pharmaceutical substances in the environment—a review. *Chemosphere* **36,** 357–394.

Hartig, C., Storm, T., and Jekel, M. (1999). Detection and identification of sulphonamide drugs in municipal waste water by liquid chromatography coupled with electrospray ionisation tandem mass spectrometry. *J. Chromatogr. A* **854,** 163–173.

Hartmann, A., Alder, A. C., Koller, T., and Widmer, R. M. (1998). Identification of fluoroquinolone antibiotics as the main source of umuC genotoxicity in native hospital wastewater. *Environ. Toxicol. Chem.* **17,** 377–382.

Heberer, Th. (2002a). Occurrence, fate, and removal of pharmaceutical residues in the aquatic environment: A review of recent research data. *Toxicol. Lett.* **131,** 5–17.

Heberer, Th. (2002b). Tracking down persistent pharmaceutical residues from municipal sewage to drinking water. *In* "Attenuation of Groundwater Pollution by Bank Filtration" (Th. Grischek and K. Hiscock, Eds.). *J. Hydrol.* **266,** 175–189.

Heberer, Th., and Feldmann, D. (2004). Removal of pharmaceutical residues from contaminated raw water sources by membrane filtration. *In* "Pharmaceuticals in the

Environment, Sources, Fate, Effects and Risks, 2nd edition" (K. Kümmerer, Ed.), pp. 391–410. Springer-Verlag, Berlin.

Heberer, Th., and Mechlinski, A. (2003). Fate and transport of pharmaceutical residues during bank filtration. *Hyroplus* **12,** 53–60.

Heberer, Th., Dünnbier, U., Reilich, Ch., and Stan, H. J. (1997). Detection of drugs and drug metabolites in groundwater samples of a drinking water treatment plant. *Fresenius' Environ. Bull.* **6,** 438–443.

Heberer, Th., Feldmann, D., Reddersen, K., Altmann, H., and Zimmermann, Th. (2002a). Production of drinking water from highly contaminated surface waters: Removal of organic, inorganic, and microbial contaminants applying mobile membrane filtration units. *Acta Hydrochim. Hydrobiol.* **30,** 24–33.

Heberer, Th., Fuhrmann, B., Schmidt-Bäumler, K., Tsipi, D., Koutsouba, V., and Hiskia, A. (2001a). Occurrence of pharmaceutical residues in sewage, river, ground and drinking water in Greece and Germany. *In* "Pharmaceuticals and Personal Care Products in the Environment: Scientific and Regulatory Issues" (C. G. Daughton and T. Jones-Lepp, Eds.), pp. 70–83. Symposium Series 791, American Chemical Society, Washington, D.C.

Heberer, Th., Mechlinski, A., Knappe, A., Pekdeger, A., Skutlarek, D., and Färber, H., in preparation. Natural attenuation of antibiotic and bacteriostatic drug residues during bank filtration. *Chemosphere.*

Heberer, Th., Reddersen, K., and Mechlinski, A. (2002b). From municipal sewage to drinking water: Fate and removal of pharmaceutical residues in the aquatic environment in urban areas. *Water Sci. Technol.* **46,** 81–88.

Heberer, Th., and Stan, H. J. (1996). Occurrence of polar organic contaminants in Berlin drinking water. *Vom Wasser* **86,** 19–31 (German).

Heberer, Th., and Stan, H. J. (1997). Determination of clofibric acid and N-(Phenylsulfonyl)-sarcosine in sewage, river and drinking water. *Int. J. Environ. Anal. Chem.* **67,** 113–124.

Heberer, Th., Verstraeten, I. M., Meyer, M. T., Mechlinski, A., and Reddersen, K. (2001b). Occurrence and fate of pharmaceuticals during bank filtration—preliminary results from investigations in Germany and the United States. *Water Res. Update* **120,** 4–17. Also available online at: www.ucowr.siu.edu/updates/pdfn/V120_A2.pdf

Heberer, Th., Mechlinski, A., Fanck, B., Knappe, A., Massmann, G., Pekdeger, A., and Fritz, B. (2004). Field-studies on the fate and transport of Pharmaceutical residues in bank filtration. *In* "Fate and Transport of Pharmaceuticals and Endocrine Disrupting Compounds (EDCs) During Ground Water Recharge" (Th. Heberer and I. M. Verstraeten, Eds.). *J. Ground Wat. Monit. Remed.* **24,** 70–77.

Hignite, C., and Azarnoff, D. L. (1977). Drugs and drug metabolites as environmental contaminants: Chlorophenxoyisobutyrate and salicylic acid in sewage water effluent. *Life Sci.* **20,** 337–342.

Hirsch, R., Ternes, T.A., Haberer, K., Mehlich, A., Ballwanz, F., and Kratz, K.-L. (1996). Nachweis von Betablockern and Bronchospasmolytika in der aquatischen Umwelt. *Vom Wasser* **87,** 263–274.

Hirsch, R., Ternes, T., Haberer, K., and Kratz, K.-L. (1999). Occurrence of antibiotics in the aquatic environment. *Sci. Total Environ.* **225,** 109–118.

Holm, J. V., Rügge, K., Bjerg, P. L., and Christensen, T. H. (1995). Occurrence and distribution of pharmaceutical organic compounds in the groundwater downgradient of a landfill (Grindsted, Denmark). *Environ. Sci. Technol.* **29,** 1415–1420.

Holthaus, K. I. E., Johnson, A. C., Jurgens, M. D., Williams, R. J., Smith, J. J. L., and Carter, J. E. (2002). The potential for estradiol and ethinylestradiol to sorb to suspended and bed sediments in some English rivers. *Environ. Toxicol. Chem.* **21,** 2526–2535.

Huang, C. H., and Sedlak, D. L. (2001). Analysis of estrogenic hormones in municipal wastewater effluent and surface water using enzyme-linked immunoadsorbent assay and gas chromatography/tandem mass spectrometry. *Environ. Toxicol. Chem.* **20,** 133–139.

Huber, M. M., Canonica, S., Park, G.Y., and von Gunten, U. (2003). Oxidation of pharmaceuticals during ozonation and advanced oxidation processes. *Environ. Sci. Technol.* **37,** 1016–1024.

Huggett, D. B., Khan, I. A., Foran, C. M., and Schlenk, D. (2003). Determination of betaadrenergic receptor blocking pharmaceuticals in United States wastewater effluent. *Environ. Pollut.* **121,** 199–205.

Jelicic, I., and Ahel, M. (2003). Occurrence of phenazone analgesics and caffeine in Croatian municipal wastewaters. *Fresen. Environ. Bull.* **12,** 46–50.

Johnson, A., Belfroid, A. C., and di Corcia, A. (2000). Estimating steroid oestrogen inputs into activated sludge treatment works and observations on their removal from the effluent. *Sci. Total Environ.* **256,** 163–173.

Jurgens, M. D., Holthaus, K. I. E., Johnson, A. C., Smith, J. J. L., Hetheridge, M., and Williams, R. J. (2002). The potential for estradiol and ethinylestradiol degradation in English rivers. *Environ. Toxicol. Chem.* **21,** 480–488.

Khan, S. J., and Ongerth, J. E. (2002). Estimation of pharmaceutical residues in primary and secondary sewage sludge based on quantities of use and fugacity modelling. *Wat. Sci. Technol.* **46,** 105–113.

Kimura, K., Amy, G., Drewes, J., Heberer, Th., Kim, T.-U., and Watanabe, Y. (2003). Rejection of organic micro-pollutants (disinfection by-products, endocrine disrupting compounds, and pharmaceutically active compounds) by NF/RO membranes. *J. Membrane Sci.* **227,** 113–121.

Kolpin, D. W., Furlong, E. T., Meyer, M. T., Thurman, E. M., Zaugg, S. D., Barber, L. B., and Buxton, H. T. (2002a). Pharmaceuticals, hormones, and other organic wastewater contaminants in U.S. streams, 1999–2000: methods, development and national reconnaissance. *Environ. Sci. Technol.* **36,** 1202–1211.

Kolpin, D. W., Furlong, E. T., Meyer, M. T., Thurman, E. M., Zaugg, S. D., Barber, L. B., and Buxton, H. T. (2002b). Response to comment on "Pharmaceuticals, hormones, and other organic wastewater contaminants in U.S. streams, 1999–2000: A national reconnaissance." *Environ. Sci. Technol.* **36,** 4007–4008.

Koutsouba, V., Heberer, Th., Fuhrmann, B., Schmidt-Bäumler, K., Tsipi, D., and Hiskia, A. (2003). Determination of polar pharmaceuticals in sewage water of Greece by gas chromatography–mass spectrometry. *Chemosphere* **51,** 69–75.

Kuch, H. M., and Ballschmiter, K. (2001). Determination of endocrine-disrupting phenolic compounds and estrogens in surface and drinking water by HRGC-(NCI)-MS in the picogram per liter range. *Environ. Sci. Technol.* **35,** 3201–3206.

Kühn, W., and Müller, U. (2000). Riverbank filtration—an overview. *J. AWWA* **92,** 60–69.

Kümmerer, K. (2001). Drugs in the environment: Emission of drugs, diagnostic aids and disinfectants into wastewater by hospitals in relation to other sources—a review. *Chemosphere* **45,** 957–969.

Kümmerer, K., Steger-Hartmann, T., and Meyer, M. (1997). Biodegradability of the antitumor agent ifosfamide and its occurrence in hospital effluents and communal sewage. *Wat. Res.* **31,** 2705–2710.

Kümmerer, K., and Helmers, E. (2000). Hospitals as a source of gadolinium in the aquatic environment. *Environ. Sci. Technol.* **34,** 573–577.

Lindsey, M. E., Meyer, M., and Thurman, E. M. (2001). Analysis of trace levels of sulfonamide and tetracycline antimicrobials in groundwater and surface water using solid-phase extraction and liquid chromatography/mass spectrometry. *Anal. Chem.* **73,** 4640–4646.

Mersmann, P., Scheytt, T., and Heberer, Th. (2002). Column experiments on the transport behavior of pharmaceutically active compounds in the saturated zone. *Acta Hydrochim. Hydrobiol.* **30,** 275–284 (German).

Miao, X. S., and Metcalfe, C. D. (2003). Determination of pharmaceuticals in aqueous samples using positive and negative voltage switching microbore liquid chromatography/electrospray ionization tandem mass spectrometry. *J. Mass. Spectrom.* **38,** 27–34.

Miao, X. S., Koenig, B. G., and Metcalfe, C. D. (2002). Analysis of acidic drugs in the effluents of sewage treatment plants using liquid chromatography-electrospray ionization tandem mass spectrometry. *J. Chromatogr. A* **952,** 139–147.

Möhle, E., Horvath, S., Merz, W., and Metzger, J. W. (1999). Determination of hardly degradable organic compounds in sewage water—identification of pharmaceutical residues. *Vom Wasser* **92,** 207–223 (German).

Oleksy-Frenzel, J., Wischnack, S., and Jekel, M. (2000). Application of ion-chromatography for the determination of the organic-group parameters AOCl, AOBr and AOI in water. *Fresenius J. Anal. Chem.* **366,** 89–94.

Öllers, S., Singer, H. P., Fässler, P., and Müller, R. S. (2001). Simultaneous quantification of neutral and acidic pharmaceuticals and pesticides at the low-ng/l level in surface and waste water. *J. Chromatogr. A* **911,** 225–234.

Patterson, D. B., Brumley, W. C., Kelliher, V., and Ferguson, P. L. (2002). Application of US EPA methods to the analysis of pharmaceuticals and personal care products in the environment: Determination of clofibric acid in sewage effluent by GC-MS. *Am. Lab.* **34,** 20–28.

Preuss, G., Willme, U., and Zullei-Seibert, N. (2001). Behavior of some pharmaceuticals during artificial groundwater recharge—Elimination and effects on microbiology. *Acta Hydrochim. Hydrobiol.* **29,** 269–277 (German).

Purdom, C. E., Hardiman, P. A., Bye, V. J., Eno, N. C., Tyler, C. R., and Sumpter, J. P. (1994). Estrogenic effects of effluents from sewage treatment works. *Chem. Ecol.* **8,** 275–285.

Putschew, A., and Jekel, M. (2001). Iodinated X-ray contrast media in the anthropogenic influenced water cycle. *Vom Wasser* **97,** 103–114 (German).

Putschew, A., Schittko, S., and Jekel, M. (2001). Quantification of triiodinated benzene derivatives and X-ray contrast media in water samples by liquid chromatography-electrospray tandem mass spectrometry. *J. Chromatogr. A* **930,** 127–134.

Putschew, A., Wischnack, S., and Jekel, M. (2000). Occurrence of triiodinated X-ray contrast agents in the aquatic environment. *Sci. Total Environ.* **255,** 129–134.

Reddersen, K., Heberer, Th., and Dünnbier, U. (2002). Occurrence and identification of phenazone drugs and their metabolites in ground and drinking water. *Chemosphere* **49,** 539–545.

Reddersen, K., and Heberer, Th. (2003). Formation of an artifact of diclofenac during acidic extraction of environmental water samples. *J. Chromatogr. A* **1011,** 221–226.

Rodriguez, I., Quintana, J. B., Carpinteiro, J., Carro, A. M., Lorenzo, R. A., and Cela, R. (2003). Determination of acidic drugs in sewage water by gas chromatography-mass spectrometry as tert.-butyldimethylsilyl derivatives. *J. Chromatogr. A* **985,** 265–274.

Sacher, F., Lange, F. Th., Brauch, H.-J., and Blankenhorn, I. (2001). Pharmaceuticals in groundwaters. Analytical methods and results of a monitoring program in Baden-Württemberg, Germany. *J. Chromatogr. A* **938,** 199–210.

Schmidt, R., and Brockmeyer, R. (2002). Vorkommen und Verhalten von Expektorantien, Analgetika und Xylometazolin und deren Metaboliten in Gewässern und bei der Uferfiltration. *Vom Wasser* **98,** 37–54.

Sedlak, D. L., and Pinkston, K. E. (2001). Factors affecting the concentrations of pharmaceuticals released to the aquatic environment. *Wat. Res. Upd.* **120,** 56–64.

Seiler, R. L., Zaugg, S. D., Thomas, J. M., and Howcroft, D. L. (1999). Caffeine and pharmaceuticals as indicators of waste water contamination in wells. *Ground Wat.* **37,** 405–410.

Soulet, B., Tauxe, A., and Tarradellas, J. (2002). Analysis of acidic drugs in Swiss waste-waters. *Int. J. Environ. Anal. Chem.* **82,** 659–667.

Spengler, P., Körner, W., and Metzger, J. W. (1999). Hardly degradable substances with estrogenic activity in effluents of municipal and industrial sewage plants. *Vom Wasser* **93,** 141–157 (German).

Stan, H. J., and Heberer, Th. (1997). Pharmaceuticals in the aquatic environment. *In* "Dossier Water Analysis" (M. J. F. Suter, Ed.), *Analusis* 25, pp. M20–M23.

Stan, H. J., Heberer, Th., and Linkerhägner, M. (1994). Occurrence of clofibric acid in the aquatic system—is their therapeutic use responsible for the loads found in surface, ground and drinking water? *Vom Wasser* **83,** 57–68.

Stan, H. J., and Linkerhägner, M. (1992). Identification of 2-(4-chlorophenoxy)-2-methylpropionic acid in ground water using capillary-gas chromatography with atomic emission detection and mass spectrometry. *Vom Wasser* **79,** 75–88 (German).

Steger-Hartmann, T., Kümmerer, K., and Schecker, J. (1996). Trace analysis of the antineoplastics ifosfamide and cyclophosphamide in sewage water by two-step solid-phase extraction and gas chromatography-mass spectrometry. *J. Chromatogr. A* **726,** 179–184.

Steger-Hartmann, T., Kümmerer, K., and Hartmann, A. (1997). Biological degradation of cyclophosphamide and its occurrence in sewage water. *Ecotox. Environ. Safety* **36,** 174–179.

Steger-Hartmann, T., Länge, R., Schweinfurth, H., Tschampel, M., and Rehmann, I. (2002). Investigations into the environmental fate and effects of iopromide (ultravist), a widely used iodinated X-ray contrast medium. *Water Res.* **36,** 266–274.

Stuer-Lauridsen, F., Birkved, M., Hansen, L. P., Holten Lützhøft, H.-C., and Halling-Sørensen, B. (2000). Environmental risk assessment of human pharmaceuticals in Denmark after normal therapeutic use. *Chemosphere* **40,** 783–793.

Stumpf, M., Ternes, Th., Haberer, K., and Baumann, W. (1996). Determination of natural and synthetic estrogens in sewage plants and river water. *Vom Wasser* **87,** 251–261.

Stumpf, M., Ternes, Th., Haberer, K., and Baumann, W. (1998). Isolation of ibuprofen-metabolites and their importance as pollutants of the aquatic environment. *Vom Wasser* **91,** 291–303.

Stumpf, M., Ternes, T. A., Wilken, R.-D., Rodrigues, S. V., and Baumann, W. (1999). Polar drug residues in sewage and natural waters in the state of Rio de Janeiro, Brazil. *Sci. Total Environ.* **225,** 135–141.

Ternes, T. A. (1998). Occurrence of drugs in German sewage treatment plants and rivers. *Water Res.* **32,** 3245–3260.

Ternes, T. A. (2001). Pharmaceuticals and metabolites as contaminants of the aquatic environment. *In* "Pharmaceuticals and Personal Care Products in the Environment: Scientific and Regulatory Issues" (C. G. Daughton and T. Jones-Lepp, Eds.), pp. 39–54. Symposium Series 791, American Chemical Society, Washington, D.C.

Ternes, T. A., and Hirsch, R. (2000). Occurrence and behavior of X-ray contrast media in sewage facilities and the aquatic environment. *Environ. Sci. Technol.* **34,** 2741–2748.

Ternes, T. A., Stumpf, M., Schuppert, B., and Haberer, K. (1998). Simultaneous determination of antiseptics and acidic drugs in sewage and river water. *Vom Wasser* **90,** 295–309.

Ternes, T. A., Stumpf, M., Mueller, J., Haberer, K., Wilken, R.-D., and Servos, M. (1999a). Behavior and occurrence of estrogens in municipal sewage treatment plants—I. Investigations in Germany, Canada and Brazil. *Sci. Total Environ.* **225,** 81–90.

Ternes, T. A., Kreckel, P., and Mueller, J. (1999b). Behavior and occurrence of estrogens in municipal sewage treatment plants—II. Aerobic batch experiments with activated sludge. *Sci. Total Environ.* **225,** 91–99.

Ternes, T. A., Bonerz, M., and Schmidt, T. (2001). Determination of neutral pharmaceuticals in wastewater and rivers by liquid chromatography-electrospray tandem mass spectrometry. *J. Chromatogr. A* **938,** 175–185.

Ternes, T. A., Meisenheimer, M., Mcdowell, D., Sacher, F., Brauch, H. J., Haist-Gulde, B., Preuss, G., Wilme, U., and Zulei-Seibert, N. (2002). Removal of pharmaceuticals during drinking water treatment. *Environ. Sci. Technol.* **36,** 3855–3863.

Ternes, T. A., Stuber, J., Herrmann, N., McDowell, D., Ried, A., Kampmann, M., and Teiser, B. (2003). Ozonation: A tool for removal of pharmaceuticals, contrast media and musk fragrances from wastewater? *Water Res.* **37**, 1976–1982.

Tixier, C., Singer, H. P., Oellers, S., and Müller, S. R. (2003). Occurrence and fate of carbamazepine, clofibric acid, diclofenac, ibuprofen, ketoprofen, and naproxen in surface waters. *Environ. Sci. Technol.* **37**, 1061–1068.

Verstraeten, I. M., Heberer, Th., and Scheytt, T. (2002). Occurrence, characteristics, and transport and fate of pesticides, pharmaceutical active compounds, and industrial and personal care products at bank-filtration sites. *In* "Riverbank Filtration: Improving Source-Water Quality" (C. Ray, G. Melin, and R. B. Linsky, Eds.), pp. 175–227. Kluwer Academic Publishers, Dordrecht.

Verstraeten, I. M., Heberer, Th., Vogel, J., Speth, Th., Zühlke, S., and Dünnbier, U. (2003). Overview of occurrence of endocrine-disrupting and other wastewater compounds during water treatment with case studies from Lincoln, Nebraska (USA), and Berlin, Germany. *In* "Endocrine Disrupting Compounds in the Environment. Practice Periodical of Hazardous, Toxic and Radioactive Waste Management" (C. Adam, Ed.), October 2003, Volume 7, Issue 4, pp. 253–263.

Werres, F., Stien, J., Balsaa, P., Schneider, A., Winterhalter, P., and Overath, H. (2000). Automatisierte Bestimmung polarer Arzneimittelrückstände in Wässern mittels Festphasenmikroextraktion (SPME) und Derivatisierung. *Vom Wasser* **94**, 135–147.

Winkler, M., Lawrence, J. R., and Neu, T. R. (2001). Selective degradation of ibuprofen and clofibric acid in two model river biofilm systems. *Water Res.* **35**, 3197–3205.

Xiao, X.-Y., McCalley, D. V., and McEvoy, J. (2001). Analysis of estrogens in river water and effluents using solid-phase extraction and gas chromatography—negative chemical ionisation mass spectrometry of the pentafluorobenzoyl derivatives. *J. Chromatogr. A* **923**, 195–204.

Zühlke, S., Dünnbier, U., Heberer, Th., and Fritz, B. (2004). Analysis of endocrine disrupting steroids: Investigation of their release into the environment and their behavior during bank filtration. *In* "Fate and Transport of Pharmaceuticals and Endocrine Disrupting Compounds (EDCs) During Ground Water Recharge" (Th. Heberer and I. M. Verstraeten, Eds.). *J. Ground Water Monitoring & Remediation (GWMR)*, in press.

Zwiener, C., and Frimmel, F. H. (2000). Oxidative treatment of pharmaceuticals in water. *Wat. Res.* **34**, 1881–1885.

Zwiener, C., and Frimmel, F. H. (2003). Short-term tests with a pilot sewage plant and biofilm reactors for the biological degradation of the pharmaceutical compounds clofibric acid, ibuprofen, and diclofenac. *Sci. Total Environ.* **309**, 201–211.

Zwiener, C., Glauner, T., and Frimmel, F. H. (2000). Biodegradation of pharmaceutical residues investigated by SPE-GC/ITD-MS and on-line derivatization. *High Res. Chromatogr.* **23**, 474–478.

Alistair B. A. Boxall*
Paul Blackwell*
Romina Cavallo[†]
Paul Kay*
Johannes Tolls[†]

*Cranfield Centre for EcoChemistry
Cranfield University
Shardlow, Derby, DE72 2GN, United Kingdom

[†]Institute for Risk Assessment Science
Faculty of Veterinary Medicine
Utrecht University
3584 CL Utrecht, The Netherlands

The Sorption and Transport of a Sulphonamide Antibiotic in Soil Systems

I. Introduction

Livestock are given veterinary medicines to treat disease and protect their health. After application of the drug, the substance may be metabolized; a mixture of the parent compound and metabolites will then be excreted in the urine and feces. Releases of veterinary medicines into the environment occur both directly (e.g., through the treatment of animals on pasture) and indirectly, via the application of animal manure (containing excreted products) to land. Once released to land, the veterinary medicines may leach to groundwater, or be transported to surface waters through drainage waters and overland flow. Because of historic, measurable impacts in the environment, a number of groups of veterinary medicines have been extensively studied, primarily sheep dip chemicals and anthelmintics (McKellar, 1997; Strong, 1993). However, with the exception of a few studies and reviews (Baguer *et al.*, 2000; Halling Sørensen, 2000; Loke *et al.*, 2000; Rabolle and Spliid, 2000; Tolls, 2001; Wollenberger *et al.*, 2000), limited information is available on the fate, behavior, and effects of other major classes of veterinary medicines used to treat livestock, and very little data are available on concentrations in the environment (Hamscher *et al.*, 2000; Nessel *et al.*, 1989).

In order to assess the risks posed to the environment by veterinary medicines used to treat livestock, a number of models and guidelines have been developed for predicting concentrations of veterinary medicines in soil, groundwater, and surface waters (Committee for Veterinary Medicinal Products [CVMP], 1996; Montforts, 1999; Spaepen *et al.*, 1997). These models are based on assumptions derived from experience with pesticides and industrial chemicals. However, the mode of application to soil (in manure/slurry) and the properties of the veterinary medicines differ from most pesticides and industrial chemicals, so the use of assumptions may be inappropriate. For example, besides being a potential vector for veterinary pharmaceutical into soils, manure contains high levels of ammonia that will increase the pH of the soil solution, thereby altering the speciation of the veterinary medicines and thus affecting the sorption of the compound. Transport of manure-associated and dissolved organic carbon-associated veterinary medicines through the soil profile to surface waters and groundwaters may also be important pathways.

The overall aim of this study was therefore to investigate the sorption behavior of one of the major groups of veterinary medicines, the sulphonamides, in soils and soil/manure mixtures and to determine the potential movement of the sulphonamides to surface and groundwaters in order to test current risk assessment models and assumptions. The specific objectives were to: (1) investigate the sorption behavior in different soil types; (2) evaluate the effects of pH changes caused by manure amendment on sorption; (3) determine concentrations in drainage waters and soil pore water at the field scale; and (4) use the information obtained from the sorption and field studies to assess current models and guidelines used in the risk assessment process for veterinary medicines.

The sulfonamides are the fifth most widely used group of veterinary antibiotics within the European Union (EU), accounting for 2% of sales in 1997 (Ungemach, 1999). In the UK and the Netherlands, they are the second most widely used veterinary antibiotic, accounting for approximately 82 tons (21% of total sales of veterinary antibiotics) and 75 tons (25% of total sales) per year, respectively (Jongbloed *et al.*, 2001; VMD, 2001). After application, treated animals excrete the sulfonamides as unaltered parent compound or as acetic acid conjugates. In the manure tank, the conjugates are reconverted to the parent compound (Langhammer, 1989). The sulphonamide sulfachloroyridazine was selected as a model compound; like all sulphonamides, sulfachloropyridazine is an N-substituted derivative of sulphanilamide. At high pH values, the sulfachloropyridazine will dissociate to the deprotonated sulfachloropyridazine anion (Fig. 1). Experimental work was performed on two contrasting agricultural soils.

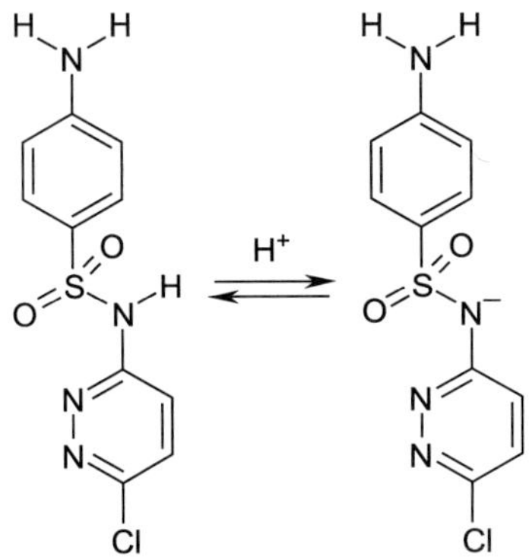

FIGURE 1 Structure of sulfachloropyridazine and its deprotonated anion.

II. Methods

Both laboratory and field investigations were performed to determine the sorption behavior and environmental fate of sulfachloropyridazine. In addition, concentrations of the study compound in soil and soil water were predicted using the model developed by Spaepen *et al.* (1997) and current risk assessment guidelines (CVMP, 1996). Each of these is described in detail below.

A. Sorption Experiments

The sorption behavior in two soil types, namely a clay loam and a sandy loam, was investigated. The soils were obtained from field sites in the UK and were dried and sieved prior to use in the experiments. Pig manure from untreated animals was obtained from the demonstration farm of the veterinary department of Utrecht University and frozen until used in the experiments. Batch sorption experiments according to the OECD Technical Guideline 106 (OECD, 2000) were performed with 2.5–5 g aliquots of soil, with 10 ml of 10 mM $CaCl_2$ as the aqueous phase. In order to inhibit microbial activity, the soil suspension was adjusted to 10 mM NaN_3. Different amounts of sulfachloropyridazine, dissolved in MeOH:H_2O, were added (such that the solution contained less than 0.5% MeOH) to give a range of sulfachloropyridazine concentrations. The soil suspension was then equilibrated with the sulfachloropyridazine by shaking for 48 h. After this time, the solid and liquid phases were separated by centrifugation; the liquid phase was filtered over a 0.22 mm Teflon filter and analyzed.

Given that sulfonamides are weak organic acids with pK_A-values (range, 4.5–7.5) in the same range as pH-values of soil solutions, it can be expected that sorption is pH-dependent. In order to investigate the effect of pH on sorption of sulfachloropyridazine in the two soils, the soils were equilibrated with different concentrations of HCl and NaOH, in addition to 10 mM $CaCl_2$. In that manner, the pH varied from 4.6 to 7.8. The sorption coefficients for the sulfachloropyridazine at different pH values were then determined using the approach described above.

The effect of manure amendment on sulfachloropyridazine sorption was examined by performing sorption experiments in the presence of pig manure. Since 10 mL of pig manure per kg of soil represents the typical UK practice, manure additions were centered about this value.

B. Prediction of Environmental Concentrations of Sulfachloropyridazine

The likely amount of sulfachloropyridazine released to agricultural fields, and the resulting concentrations of sulfachloropyridazine in soil water and soil, were predicted using the modeling approach developed by Spaepen *et al.* (1997) and guidelines developed by the Committee for Veterinary Medicinal Products (CVMP, 1996). The input values for the model are shown in Table I. Due to a lack of experimental data on metabolism and degradation during storage, it was assumed that sulfachloropyridazine was not extensively metabolized and is persistent in manure. The treatment regime used in the calculations corresponded to the treatment of pigs and was selected to represent a realistic worst case scenario.

C. Field Studies

The behavior of sulfachloropyridazine was investigated at the field scale using two study sites with contrasting soil types. The first study site comprised a field with an underdrained clay soil (clay site), and the second site

TABLE I Model Input Data Used to Predict Concentrations of Study Compounds in Soil and Water Following Treatment of Pigs

Parameter	*Input value*
Dosage (mg/kg/d)	20
Length of treatment (d)	5
Number of treatments/animal	2
Fraction excreted unchanged	0.95
Mixing depth (cm)	5
Average body weight (kg)	95
Number of animals raised per place per year	2.5
Yearly output of excreta per place (kg/place/yr)	1764
Yearly production of phosphorous (kg P_2O_5/place/yr)	8.32
Yearly production of nitrogen (kg N/place/yr)	9.59
K_D (liters/kg)	0.9, 1.8

comprised a field with a sandy loam soil (sandy site). Sulfachloropyridazine was applied to both study fields, in spiked pig slurry at an application rate of 1.2 kg/ha. The slurry was obtained from a working pig farm in Leicestershire, UK from pigs that had not received any treatment with sulphonamides. The slurry was applied to the clay site using a slurry spreader on October 15, 2000 following normal agricultural practice, and was incorporated into the soil 10 days later. Study plots on the sandy site were treated manually on January 17, 2001; the application rates were estimated using the UK scenario described by Spaepen *et al.* (1997). The sampling methodology varied between the two sites.

The clay site was instrumented in order to obtain samples of drainflow. The outlet from the field drains was instrumented with a weir box and pressure transducer to record discharge, and an automatic water sampler to take samples during periods of flow. During periods of drainflow, water samples were obtained every hour. After collection, all samples were stored at −20°C prior to chemical analysis. Rainfall and soil hydrology measurements were also made at the site.

At the sandy site, there were three study plots: one control and two treatment. Each plot was instrumented with 12 suction samplers, four each at 40, 80, and 120 cm depth. In the treatment plots, six tensiometers, two each at 40, 80, and 120 cm depth, and one neutron probe access tube to 150 cm depth were installed; in the control plot, three tensiometers and three equitensiometers, one at each depth, were installed. These were connected to a DL2000 datalogger along with a raingauge in order to provide continuous rainfall and soil moisture data. The sampling trigger criteria were rainfall events of >10 mm over 24 hrs, >15 mm over 48 hrs, or a less intense but more prolonged event. After collection, all samples were stored at −20°C prior to chemical analysis.

Rainfall was recorded at both sites, supplemented with meteorological data from Nottingham University's Sutton Bonnington site.

D. Chemical Analyses

The filtrate samples from the sorption studies were analyzed by isocratic reversed phase high performance liquid chromatography (HPLC) using a Supelco Discovery column (150 × 4.6 mm, 5 μm particle size) in combination with a solvent composition of 70% 10 mM PO_4 buffer (pH 3.1) and 30% acetonitrile. Sulfachloropyridazine was detected with a UV-absorbance detector at a wavelength of 260 nm.

Water samples from the two field sites were analyzed, after extraction, by HPLC with UV detection. Prior to extraction, buffer (100 ml 0.1 M Na_2EDTA; 60 ml 0.2 M citric acid; 40 ml 0.4 M Na_2HPO_4; and 2 ml H_3PO_4) and methanol were added to each sample (5 ml buffer and 2 ml methanol per 100 ml of sample). Sulfachloropyridazine was then extracted from the buffer solution using preconditioned Isolute SAX (IST, Hengoed, Wales) and Oasis HLB (Waters, Elstree, England) solid phase extraction cartridges in tandem at a flow rate of 10 ml/min. The SAX cartridge, which removed interfering humic material, was then removed. The HLB cartridge was washed, and sulfachloropyridazine was then eluted from the cartridge using 2×1 ml methanol. Concentrations of sulfachloropyridazine in each extract were determined using a Dionex Summit system equipped with a GENESIS 4 μm C_{18} endcapped column. Analysis was performed using gradient elution over 25 min with a tetrahydofuran (THF), acetonitrile (ACN) and 0.05% trifluoroacetic acid in water (TFA) mobile phase. The mobile phase contained 5% THF throughout the analysis. At the start of the analysis, the mobile phase contained 2.5% AC and 92.5% TFA, and this composition was maintained for 4 min. The composition then changed linearly between 4 and 18 min to 75% AC and 20% TFA. Between 18 and 20 min, the composition changed linearly to 2.5% AC and 92.5% TFA. These conditions were then maintained for the remainder of the run. The injection volume was 20 μl and the detection wavelength used was 285 nm.

The analytical method used on the samples obtained from the field site was validated using spiked water obtained from the River Trent, UK. Six samples of water were spiked to give a concentration of 1 μg/liter sulfachloropyridazine, and six samples were spiked to give a sulfachloropyridazine concentration of 10 μg/liter. The spiked samples were then extracted and analyzed using the methodology described above. Measured concentrations were compared with spiked concentrations to provide a measure of recovery.

III. Results

A. Analytical Detection Limits and Recoveries

Recoveries of sulfachloropyridazine from spiked Trent water ranged from 99.9% at a concentration of 10 μg/liter to 105% at a concentration of 1 μg/liter. The limit of detection was 250 ng/liter.

B. Sorption Experiments

Logarithmic isotherms obtained for sulfachloropyridazine in the two study soils are shown in Fig. 2. Fitting a Freundlich isotherm ($\log C_s = 1/n \times \log C_w + \log K_F$) to the data resulted in Freundlich coefficients (1/n) of 0.97 and 0.91 for the clay loam and the sandy loam, respectively. As the values of 1/n were close to unity, sorption can be reasonably approximated with a linear sorption coefficient K_D. Fitting a linear model to the data ($C_s = K_D \times \log C_w$) resulted in K_D values of 1.8 and 0.9 liter/kg for the clay loam and for the sandy loam, respectively.

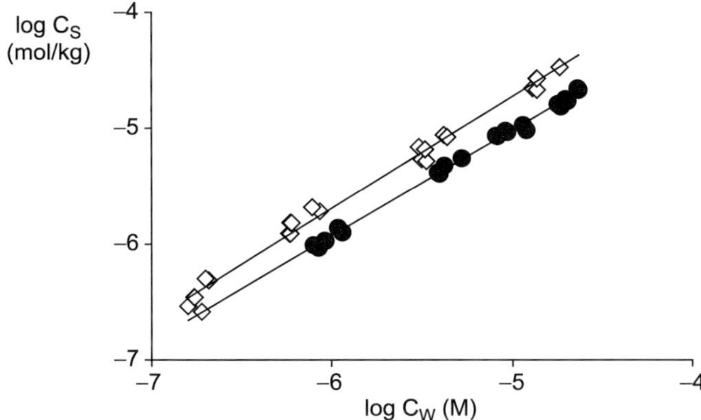

FIGURE 2 Logarithmic isotherms obtained for sulfachloropyridazine in the clay loam (diamonds) and the sandy loam (circles).

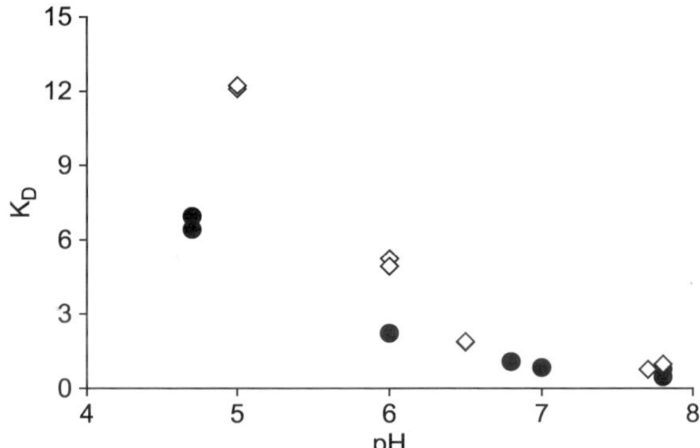

FIGURE 3 Dependence of the sorption coefficient K_D on the pH in the clay loam (diamonds) and the sandy loam (circles).

Sorption coefficients in both soil types increased as pH decreased (Fig. 3). Sorption of the deprotonated sulfachloropyridazine anion (prevalent at high pH) therefore appears to be significantly weaker than that of the neutral sulfachloropyridazine (concentration increases with decreasing pH).

Addition of manure resulted in an increase in the pH of the soil (Fig. 4). As predicted on the basis of the pH-dependence investigations, the apparent value of K_D decreased with the increase in pH. Comparison of the pH-K_D relationship resulting from amending the soil with manure to that obtained by manipulating the pH with NaOH (Fig. 5) indicated that the relationships were similar; the reduced sorption in manure amended soil is primarily due

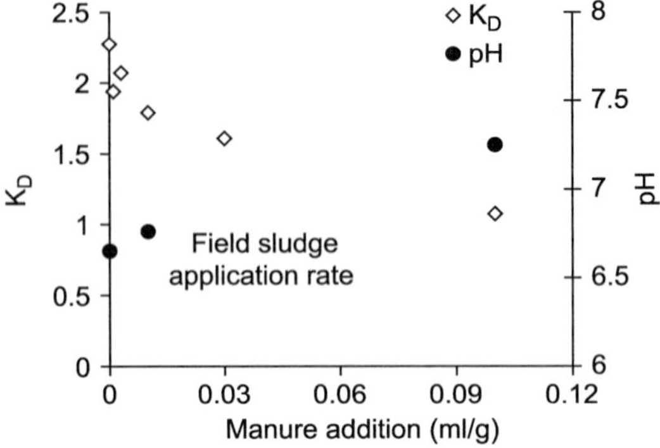

FIGURE 4 Effect of manure amendment on the soil solution pH and K_D in the clay loam.

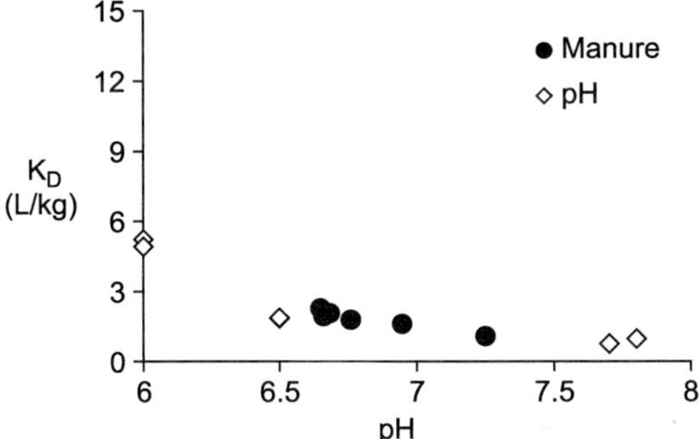

FIGURE 5 Comparison of the effect of acid/base induced pH-manipulation (diamonds) and of manure addition (circles) on the sorption of sulfachloropyridazine to the clay loam.

to the increased pH. Conversely, an effect of dissolved organic matter (DOM) leading to an additional decrease of K_D values was not observed at typical manure application rates.

C. Predicted Amounts Released to Fields and Resulting Environmental Concentrations

Using the uniform approach developed by Spaepen *et al.* (1997) and the CVMP guidelines, and assuming no metabolism or degradation during storage, it was estimated that 1.2 kg of sulfachloropyridazine would be applied per hectare. The predicted concentration in soil was 1.6 mg/kg. Because of its low K_D value in the test soils, predicted concentrations of sulfachloropyridazine in soil water were high, from 860 (clay site) to 1800 (sandy site) μg/liter.

TABLE II Peak Antibiotic Concentrations Measured in Either Drainflow (Clay Site) or Soil Pore Water (Sandy Site)

	Maximum sulfachloropyridazine concentration (μg/liter)	
Days after application	*Clay site*	*Sandy site*
6	590	ND
14	64	ND
20	NA	0.78
21	1	NA
28	NA	0.58
37	0.85	NA
78	NA	ND
120	0.63	NA
121	NA	0.15
196	0.60	NA

ND, not detected; NA, sampling timepoint not relevant to this site.

D. Field Studies

The clay site received 849 mm of rainfall between application of the spiked slurry and the last sampling event (annual precipitation was 931 mm). Drainflow samples were collected from the clay site 6, 14, 21, 37, 120, and 196 days after treatment (DAT). Peak concentrations measured during each storm event are shown in Table II. During the first drainflow event following application, sulfachloropyridazine was present at a maximum concentration of 590 μg/liter. The peak concentration of sulfachloropyridazine was also high in the second period of drainflow, 64 μg/liter. In subsequent events, peak concentrations were around 1 μg/liter or less; concentrations had fallen below the limit of detection nine months after application. Peak concentrations were generally recorded near the beginning of each period of drainflow, when discharge was at its lowest (Fig. 6).

The sandy site received a total of 313 mm of rainfall between application of the spiked slurry and the last sampling event. Samples of soil pore water were obtained on 6 occasions—6, 9, 20, 28, 78, and 121 DAT. Concentration of sulfachloropyridazine in all samples was below 1 μg/liter (Table II).

IV. Discussion

The present data for the sorption behavior of sulfachloropyridazine are in strong agreement with the sorption coefficients reported for other sulphonamides—K_D for sulfathiazole equals 4.9 liters/kg (Thurman and Lindsey, 2000), and K_D for sulfamethazine ranges from 0.6 to 3.2 liters/kg

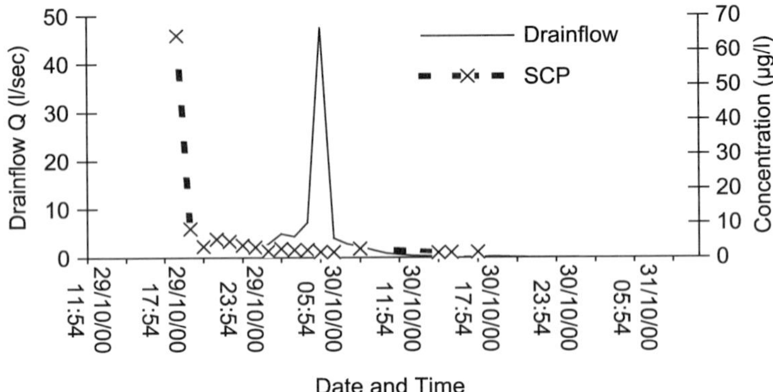

FIGURE 6 Typical chemo- and hydro-graphs for a drainflow event at the clay site.

for different soils (Langhammer, 1989). The K_D values obtained in this study were low, particularly under basic conditions, and indicate that sulfachloropyridazine will be mobile in soil. Therefore, facilitated transport of sulfachloropyridazine with manure-born DOM is unlikely to contribute significantly to sulfachloropyridazine mobility.

Results of investigations into the effects of soil pH on the sorption behavior of sulfachloropyridazine indicate that sorption increases with a decrease in pH. However, fertilization of agricultural soils in temperate climates maintains the soil pH values in quite a narrow range, between 6 and 7.5 (Scheffer and Schachtschabel, 1989); within this window, the variation of K_D values amounts to a factor of 5.9 and 3.6 for the clay loam and the sandy loam, respectively. From a risk assessment perspective, this degree of uncertainty is small. It therefore appears appropriate to simplify the description of sulfachloropyridazine sorption by neglecting pH-dependence. The difference in the sorptive capacity between the two soils (approximately a factor of 2) can be viewed in the same way. Nevertheless, both issues deserve some attention, because they may be highly relevant for other veterinary medicines, and because little is known about the mechanisms and soil constituents involved in sorption of sulfachloropyridazine to soils.

The results obtained from the clay field site supported the previous sorption studies and demonstrated that sulfachloropyridazine is highly mobile in soils; after application, it will be rapidly transported to field drains and ultimately enter surface waters. Predicted soil pore water concentrations, obtained using the model and current risk assessment guidelines provided by the CVMP (1996), were within a factor of two of each other—860 (predicted) and 590 (measured) μg/liter. It therefore appears that the simple approach recommended for estimating predicted concentrations in surface waters is suitable for estimating likely concentrations of hydophilic substances, such as sulphonamides. However, other monitoring results not presented here indicate that these guidelines may not be appropriate for highly sorptive substances such as oxytetracyclines, where other transport routes (e.g., transport via particulates and DOM) may be important.

In contrast to the clay site, concentrations of sulfachloropyridazine in soil pore water of the sandy loam site were low and all less than 1 μg/liter. Concentrations in groundwater are likely to be lower still, due to further attenuation from dilution and possible degradation. The calculations obtained using the model and CVMP guidelines therefore greatly overestimate the likely concentrations of sulfachloropyridazine in groundwater. One possible explanation for the differences between the results obtained at the sandy site and the clay site is that the sulfachloropyridazine is rapidly degraded in the sandy soil. While data are currently not available on the persistence of sulfonamides in soils, a number of studies have investigated the degradation of sulfonamides in other media (Hektoen *et al.*, 1995; Ingerslev, 2000; Lunestad *et al.*, 1995; Samuelsen *et al.*, 1994; Van Dijk and Keukens, 2000). These studies indicate that sulphonamides are impersistent to moderately persistent in manure, moderately to very persistent in sediments, moderately to very persistent in water, and impersistent in activated sludge. The results of degradation studies are variable, so it is difficult to say whether sulfachloropyridazine is likely to be rapidly degraded in the sandy soil. The degradation of sulfachloropyridazine in both the sandy and clay soils is therefore being investigated in the laboratory.

A limited amount of data were available on the ecotoxicity of sulfachloropyridazine (Novartis, 1999). The 48 hr EC50 (median effect concentration) for daphnids was 250 mg/liter, whereas the 96 hr LC50 (median lethal concentration) for fish was over 1000 mg/liter. Comparison of these data with the maximum concentrations observed in this study indicates that sulfachloropyridazine is unlikely to pose a risk to fish and invertebrates, as the maximum concentrations are more than two orders of magnitude less than measured effects concentrations. Effects on fish and invertebrates in the "real" environment are even more unlikely, due to a number of factors including: (1) the amount of sulfachloropyridazine applied to the field sites was a "worst case"; (2) further dilution of the drainflow will occur in the receiving water; and (3) maximum concentrations during a drainflow event are likely to be short lived <1 hr. However, due to a lack of experimental data, effects on other groups of aquatic organisms (e.g., plants, algae, fungal species) and terrestrial organisms cannot be ruled out.

V. Conclusions

This study has investigated the sorption behavior of the sulphonamide sulfachloropyridazine, and the potential for this compound to be transported to surface and groundwaters. Results from the sorption studies indicate that, like other sulphonamides, sulfachloropyridazine has a low sorption potential and is therefore likely to be highly mobile. While changes in soil pH affect the sorption of sulfachloropyridazine, the effects are small and probably do not need to be considered in the risk assessment process. However, these effects may be more important for other sulphonamides and other classes of veterinary medicines, so an understanding of the underlying mechanisms affecting the sorption of veterinary medicines is desirable. Results from the sorption studies were supported by the results from the clay site demonstrating that sulfachloropyridazine is highly mobile and will be rapidly transported to surface waters. Comparison of these monitoring results with predictions obtained using current risk assessment guidelines demonstrated that, even though they are simplistic, the guidelines can be usefully applied to hydrophilic substances such as the sulphonamides. The sorption studies and predictions of soil water concentrations indicated that concentrations in soil water at the sandy site would be high. However, concentrations at this site were all below 1 μg/liter. One possible explanation for this is that the sulfachloropyridazine was rapidly degraded at the sandy site. Even though high concentrations of sulfachloropyridazine were observed in drainflow, the available ecotoxicological data indicate that it is unlikely that the compound will affect aquatic organisms.

Acknowledgments

This project was funded under the EU Framework V program, project number EVK1-CT-1999-00003. The authors would like to thank Novartis Animal Health for providing the sulfachloropyridazine used in the studies.

References

Baguer, A. J., Jensen, J., and Henning Krogh, P. (2000). Effects of the antibiotics oxytetracycline and tylosin on soil fauna. *Chemosphere* **40**, 751–757.

Committee for Veterinary Medicinal Products (1996). Note for guidance: Environmental risk assessment for veterinary medicinal products other than GMO-containing and immunological products EMEA/CVMP/055/96-Final. European Agency for Evaluation of Medicinal Products, London, UK.

Halling Sørensen, B. (2000). Algal toxicity of antibacterial agents used in intensive farming. *Chemosphere* **40**, 731–739.

Hamscher, G., Sczesny, S., Abu-Quare, A., Hoper, H., and Nau, H. (2000). Substances with pharmacological effects including hormonally active substances in the environment: Identification of tetracyclines in soil fertilised with animal slurry. *Deutsch Tieraarztl. Wschr.* **107**, 293–348.

Hektoen, H., Berge, J. A., Hormazabal, V., and Yndestad, M. (1995). Persistence of antibacterial agents in marine sediments. *Aquaculture* **133**, 175–184.

Ingerslev, F., and Halling-Sørensen, B. (2000). Biodegradation properties of sulfonamides in activated sludge. *Environ. Toxicol. Chem.* **19**, 2467–2473.

Jongbloed, R. H., Kan, C. A., Blankendaal, V. G., and Bernhard, R. (2001). Milieurisico's van diergenesmiddelen en veevoeradditivien in Nederlands oppervlaktewateren 31635, TNO-MEP. Apeldorn, The Netherlands.

Langhammer, J.-P. (1989). Untersuchungen zum Verbleib antimikrobiell wirksamer Arzneistoffe als Rückstände in Gülle und im landwirtschaflichen Umfeld (PhD thesis). Rheinische Friedrich-Wilhelms-Universität, Bonn, Germany.

Loke, M.-L., Ingerslev, F., Halling Sørensen, B., and Tjornelund, J. (2000). Stability of tylosin A in manure containing test systems determined by high performance liquid chromatography. *Chemosphere* **40,** 759–765.

Lunestad, B. T., Samuelsen, O. B., Fjelde, S., and Ervik, A. (1995). Photostability of eight antibacterial agents in seawater. *Aquaculture* **134,** 217–225.

McKellar, Q. A. (1997). Ecotoxicology and residues of anthelmintic compounds. *Vet. Parasitol.* **72,** 413–435.

Montforts, M. H. H. M. (1999). Environmental risk assessment for veterinary medicinal products Part 1. Other than GMO-containing and immunological products. RIVM report 601300 001, April 1999. National Institute for Public Health and Environment, Bilthoven, The Netherlands.

Nessel, R. J., Wallace, D. H., Wehner, T. A., Tait, W. E., and Gomez, L. (1989). Environmental fate of ivermectin in a cattle feed lot. *Chemosphere* **18,** 1531–1541.

Novartis (1999). Safety data sheet: Sulfachloropyridazine-Na. Release date 18 October 1999. Novartis Animal Health, Basel, Switzerland.

OECD (2000). Adsorption-Desorption Using a Batch Equilibrium Method, Technical Guideline 106. OECD-Organisation for Economic Cooperation and Development, Paris.

Rabolle, M., and Spliid, N. H. (2000). Sorption and mobility of metronidazole, olaquindox, oxytetracycline and tylosin in soil. *Chemosphere* **40,** 715–722.

Samuelsen, O. B., Lunestad, B. T., Ervik, A., and Fjelde, S. (1994). Stability of antibacterial agents in an artificial marine aquaculture sediment studied under laboratory conditions. *Aquaculture* **126,** 283–290.

Scheffer, F., and Schachtschabel, P. (1989). Lehrbuch der Bodenkunde. Enke, Stuttgart, Germany.

Spaepen, K. R. I., Leemput, L. J. J., Wislocki, P. G., and Verscheuren, C. (1997). A uniform approach to estimate predicted environmental concentrations of the residues of veterinary medicines in soils. *Envir. Toxicol. Chem.* **16,** 1977–1982.

Strong, L. (1993). Overview: The impact of avermectins on pastureland ecology. *Vet. Parasitol.* **48,** 3–17.

Thurman, E. M., and Lindsey, M. E. (2000). Transport of antibiotics in soil and their potential for groundwater contamination. Poster presented at the 3rd SETAC World Congress, May 22–25, 2000, Brighton, UK.

Tolls, J. (2001). Sorption of veterinary pharmaceuticals—a review. *Envir. Sci. Tech.* **35,** 3397–3406.

Ungemach, F. R. (2000). Figures on the quantities of antibacterials used for different purposes in the EU countries and interpretation. *Acta. Vet. Scand. Suppl.* **93,** 89–98.

Van Dijk, J., and Keukens, H. J. (2000). The stability of some veterinary drugs and coccidiostats during composting and storage of laying hen and broiler faeces. *In* "Residues of veterinary drugs in food. Proceedings of the Euroresidue IV conference." (L. A. Van Ginkel and A. Ruiter, Eds.). Veldhoven, The Netherlands.

VMD (2001). Sales of antimicrobial products used as veterinary medicines and growth promoters in the UK in 1999. Report obtained from Veterinary Medicines Directorate website at: http://www.vmd.gov.uk/general/publications/antisales01.pdf.

Wollenberger, L., Halling Sorensen, B., and Kusk, K. O. (2000). Acute and chronic toxicity of veterinary antibiotics to Daphnia magna. *Chemosphere* **40,** 723–730.

Jürg Oliver Straub
EurProBiol CBiol MIBiol
Corporate Safety & Environmental Protection CSE
F. Hoffmann-La Roche Ltd
CH–4070 Basel, Switzerland

Concentrations of the UV Filter Ethylhexyl Methoxycinnamate in the Aquatic Compartment: A Comparison of Modeled Concentrations for Swiss Surface Waters with Empirical Monitoring Data

I. Introduction

Ultraviolet (UV) filters are substances that absorb UV radiation of wavelengths of <200–400 nm. This range is commonly divided into UV-A (400–320 nm), UV-B (320–280 nm), and UV-C (280–<200 nm) wavebands (Diffey, 1991; IARC, 1992). Absorbed UV energy is re-emitted as thermal radiation. A few UV filters are inorganic, but most are organic substances, characterized by single or multiple aromatic structures, often with attached hydrophobic groups to improve formulation properties. In general, single UV filters have a relatively small absorption bandwidth, necessitating the combination of several UV filters, depending on the desired function. The widespread use of UV filters is due to the potential of high-energy UV radiation to damage organic compounds. UV radiation may damage synthetic compounds exposed to sunlight (e.g., paints, plastics, and fibers), interfering with their function and shortening their useful lifespan, but in addition, UV radiation has been shown to damage biological systems (Diffey, 1991). Consequently, sunlight has been recognized as a confirmed human carcinogen, with both UV-A and UV-B explicitly listed as probable human carcinogens by the International Agency for Research on Cancer (IARC, 1992).

UV filters used for sunscreens and cosmetic products may, in certain respects, be compared to pharmaceuticals. They are used for their health-protective effects and, also similar to OTC drugs, are widely used. Indeed, UV filters are regulated as OTC drugs in the U.S. (CFR, 2001), which translates into relatively high amounts being marketed. Suncare products are applied superficially to the skin, and the UV filters are designed to remain on the uppermost layers of the skin; penetration through the skin is low. Consequently, UV filters are generally washed off the skin, either during bathing or swimming, while the rest may be transferred to towels or clothes (Balk *et al.*, 2001), which will be washed. Therefore, the main release into the environment is by way of household wastewater and sewage works, with some direct emission through swimming in summer; the aquatic compartment is the primary compartment of concern. Additionally, the season of environmental exposure must be considered; a greater amount of products containing UV filters are used during summer, which may cause increased environmental exposure during that time of year.

Again, similar to some pharmaceuticals, certain UV filters have been detected in surface waters in low concentrations (Gewässerschutzamt, unpublished data; Nagtegaal *et al.*, 1997; Poiger *et al.*, 2004). While the environmental significance and possible risks of such low concentrations of synthetic substances remain unclear, these issues are being increasingly debated (Daughton and Ternes, 1999; Kümmerer, 2001; Ternes, 1998). Concerns regarding the potential estrogenic effects of some UV filters (Schlumpf *et al.*, 2001) have also directed attention to environmental exposure of these substances (Balk *et al.*, 2001; Poiger *et al.*, 2004). In this paper, specifically the aquatic exposure of the common UV-B filter ethylhexyl methoxycinnamate (EHMC) will be discussed. Predicted environmental concentrations (PECs) of EHMC in surface waters of Switzerland from an Environmental Compatibility Assessment for Swiss authorities will be compared with available monitoring data.

II. Methods

A. Basic Data

EHMC (CAS 5466–77–3; formerly known as Octyl Methoxycinnamate or OMC) is an organic UV-B filter that was originally developed in the 1930s and has been one of the most widely used sunscreens for decades

(Gonzenbach, personal communication). EHMC is a cinnamic acid derivative that has a methoxy group on the *para* position and is esterified with 2-ethylhexanol (Fig. 1); the chemical name is 2-ethylhexyl-3-(4-methoxyphenyl)-2-propenoate. Physico-chemical and environmental fate data for EHMC were collated from Roche internal documentation and available external sources; in addition, some tests were commissioned at contract laboratories. Basic data of EHMC are listed in Table I.

Dependable use estimates of EHMC for Switzerland are difficult to determine. There are several producers and vendors, both within and outside

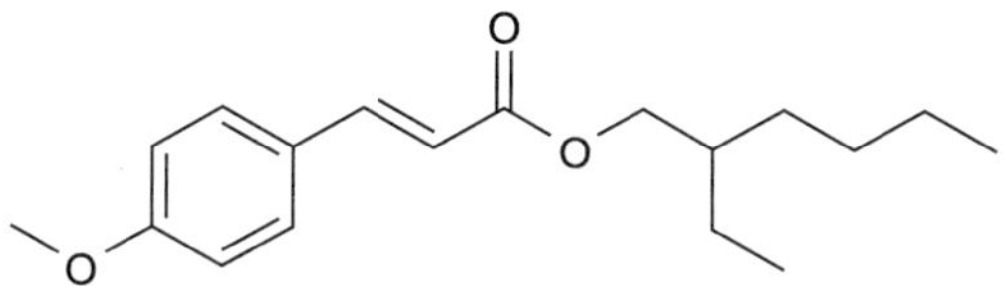

FIGURE 1 EHMC structure.

TABLE I EHMC Physico-Chemical and Environmental Fate Basic Data

Parameter	*Value*	*Source*
Physico-chemical parameters		
Empirical formula	$C_{18}H_{26}O_3$	Roche
Molecular mass	290.4 g/mol	Roche
Melting point	$<-25°C$	Roche
Boiling point	310°C	Roche
Characterization	Oily liquid	Roche
Water solubility	<750–41 μg/l	BASF, 1999; Roche; Sabaliunas *et al.*, 2001
Vapor pressure	$\sim 1.5 \times 10^{-6}$ Pa	Roche
Density	1.007–1.013 kg/l	Roche
Environmental fate parameters		
$\log K_{ow}$	≥6 (6.0–6.3)	Merck, 2000; Roche
K_{oc}	~1230 (QSAR)	SRC, 2000
Adsorption rate to activated sludge	~91% (QSAR)	SRC, 2000
K_H	$<1.8 \times 10^{-6}$ atm × m^3/mol (QSAR)	SRC, 2000
Atmospheric degradation	t ~ 2.4–26.2 h, $OH^•$ resp. O_3 (QSAR)	SRC, 2000
Aquatic photolysis	DT_{50} ~ 5–9 d (summer, 40 degrees N)	NOTOX, 2002
Hydrolysis	Negligible; very slow (QSAR)	NOTOX, 2002; SPARC, 2001
Ready biodegradability	78% (28 d, readily, OECD 301F); 71% (28 d, readily, OECD 301C)	H&R, 1998; Roche
Ultimate anaerobic biodegradation	67% (79 d, ultimately, ISO 11734)	BMG, 2001

of Switzerland. There are also many cosmetics companies that formulate suncare products and operate in Europe, including Switzerland. This means that both pure EHMC and formulated products are imported and exported. In order to estimate the amount actually used in Switzerland in the form of final products, which is needed for a reasonable derivation of PECs, the European Cosmetics trade organization (COLIPA) was contacted. An annual EHMC amount for Switzerland was back-calculated from the overall sales of suncare products in Europe, the average content of EHMC in suncare products, and the relative share for Switzerland based on population numbers. This calculation resulted in use of up to 20 t EHMC per year for Switzerland. For the subsequent modelling, this upper limit was used.

Geographical basic data for Switzerland (Table II) were found on the Internet from websites of Swiss ministries, the Hydrological Office (BWG), and in particular, *The Environment in Switzerland* on the website of the Swiss Agency for the Environment, Forests, and Landscape (BUWAL, 2001).

B. PECs

Prior to the actual calculation of PECs, the pathways and estimated amounts for EHMC from consumer products to the environment were modeled (Fig. 2). First, it was assumed that wastes from production and formulation do not need to be considered, as both processes should be closely supervised and waste streams fully treated. From suncare products containing a total of 20 t EHMC/yr, an appreciable amount of 10% (2 t/yr) was estimated to end up in household waste, both from expired products and from residual amounts in packagings. In Switzerland, 85% of municipal waste is incinerated while 15% is landfilled (BUWAL, 2001). The remaining EHMC (90%, 18 t/yr) is used while participating in outdoor activities in

TABLE II Basic Swiss Geographic Parameters

Inhabitants	7.1×10^6
Area	4.13×10^{10} m^2
Natural surface waters	~1.52×10^9 m^2
Natural soil	~2.69×10^{10} m^2
Agriculturally used soil	~4.92×10^9 m^2
Water dynamics	
Average annual precipitation	1.456 m/year
Approx. discharge of 4 major Swiss rivers (Rhine, Rhone, Ticino, Inn)	~7.02×10^6 m^3/h (high estimate) ~6.11×10^6 m^3/h (low estimate)
Average sewage flow	~0.4 m^3/inh × d

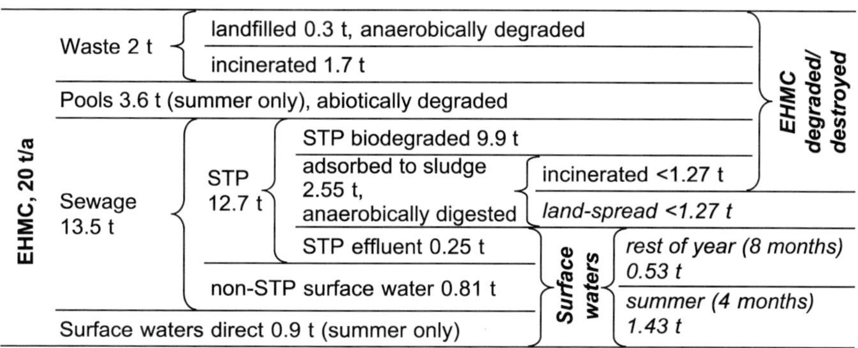

FIGURE 2 Modeled release pathways of EHMC into the environment in Switzerland.

summer, roughly four months. Swimming and certain summer activities result in almost whole-body applications, while for the rest of the year application is restricted to face, neck, hands, and arms, resulting in less product use. In summer, swimming is assumed to take place mainly (approximately 80%) in public or private pools, and only to a lesser extent (20%) in surface waters like lakes and rivers. All of these uses will release EHMC indirectly or directly into water. Direct loss of EHMC to water is estimated at 4.5 t/yr (25%), of which the percentages listed above will end up in confined pools or directly in surface waters. In confined pools, EHMC is expected to be abiotically degraded due to treatment of pool waters with highly reactive chlorine compounds or ozone and, additionally, through direct insolation. The remaining 13.5 t/yr (75%) will end up in sewage, either through washing off the skin, or by adsorption onto bathing towels or clothes which in turn are washed. Due to higher use in summer, it is estimated that half of the 13.5 t will be used during the four summer months, and the other half during the remaining eight months.

In Switzerland, 94% of household wastewater is treated in sewage treatment plants (STP)(BUWAL, 2001), while for simplicity, the remainder is considered to be untreated direct effluent, as there is no information on septic tanks. For a simple, crude calculation, it is assumed that 78% EHMC (Roche ready biodegradation test) will be degraded in the STP, while 91% of the remainder will adsorb into activated sludge (QSAR; SRC, 2000); thus only 9% of the undegraded 22% will be released from STPs to surface waters. Superfluous activated sludge is anaerobically digested for several weeks, then (in Switzerland) half is incinerated and the other half is spread on agricultural land or landfilled (BUWAL, 2001). Due to ultimate anaerobic degradability, only upper limits for the residual concentration in digested sludge can be given. Figure 2 presents a flow diagram of modeled EHMC amounts and release pathways into the environment.

C. Models

The first aquatic PEC for EHMC is a rather crude "mechanical" model using the pathways and amounts described above, with release per hour determined for summer and rest-of-year months, divided by the water flow of the four major Swiss rivers. This simple model accounts only for a minimal degradation in STPs, but not for further degradation, adsorption, or subsequent partitioning in the environment.

For more realistic modelling of EHMC fate and aquatic PECs, two software models were used, both configured for Swiss basic data as per Table II. The models used were the Mackay Level III model (Mackay *et al.*, 1996) and the Dutch Uniform System for the Evaluation of Substances (USES, 1999), which is based on the EU Technical Guidance Document for Risk Assessment (ECB, 1999) with advanced modeling added. Both models use basic substance data to derive fugacity, the pairwise distribution coefficients between environmental compartments (Mackay *et al.*, 1996), and degradation parameters or half-lives to allow for environmental fate processes. The Level III model was run with inputs into surface waters and soil (from sludge spreading) for summer and rest-of-year months as per Fig. 2, while the USES model was run with whole-year amounts. However, as USES has no exits for incinerated or landfilled wastes, these amounts were suppressed. In the case of activated sludge for disposal, the amount from STPs for land-spreading as computed by USES was halved in order to reflect the 50% sludge incineration rate for Switzerland.

D. Monitoring Data

The first recorded analyses of UV filters in the aquatic environment were published by Nagtegaal and colleagues in 1997 for the Meerfelder Maar, a bathing lake in Germany, based on sampling in the years 1991 and 1993. With a reported limit of detection (LoD) of 2 ng/l, no EHMC was detected in either year. For the purpose of the present work, it will be assumed that Nagtegaal's group took at least two samples per season, four in all. However, measured environmental concentrations (MECs) do exist. These have not yet been published but were made available for this paper. The Gewässerschutzamt (Water Protection Board of the Swiss Canton of) Basel-Stadt monitored EHMC at their automated sampling station on the Palmrain bridge across the river Rhine below Basel during the summer season of 1997 (GSA, unpublished data). A paper copy of the respective MEC graph was graciously made available to Roche. It shows 47 data points from mid-May to August from 1 ng/l (probably below LoD, not stated); 33 data points are less than or equal to 5 ng/l, and another 12 were between 6 and 10 ng/l, one was 13 ng/l, and one was 28 ng/l. The average from these 47 data points is 5.5 ng/l. Further, a series of water samplings was taken and analyzed for

UV filters in the relatively large Lake Zurich and the much smaller Lake Hüttner in Switzerland during the summer of 1998, on behalf of the Swiss Environment Office (Poiger *et al.*, 2004); the LoD for EHMC was also 2 ng/l in this study. Dr. Poiger graciously made the 39 single analytical data available for use in the present paper. Overall, this contributes nearly 90 data points, including some below a comparable LoD. Data from Sabaliunas and colleagues (2001) from England were not used because of their indirect sampling technique and relatively high LoD (Sabaliunas, personal communication).

E. Probabilistic Environmental Concentration

Detections below LoD are valuable data that can be usefully incorporated into probabilistic analyses (Campbell *et al.*, 1999; Solomon *et al.*, 1996). All analytical data available were pooled, ordered by magnitude, and percent-ranked. The percent ranks were plotted against the logarithm of the concentration on a log-probit graph. Finally, 50th, 90th, 95th, and 99th percentile concentration values were determined using the regression line. This procedure assumes stochastic distribution of the values.

III. Results

A. Environmental Fate of EHMC

EHMC is a substance that enters the environment through water, mostly through sewage. In outdoor swimming pools, EHMC is modeled to be abiotically degraded; however, there are no analytical data for this. In STPs, most EHMC is degraded due to its ready biodegradability. Most of the remainder is adsorbed to activated sludge and anaerobically degraded during sludge digestion; non-incinerated sludge is land-spread, allowing further aerobic or anaerobic degradation of EHMC. Only a relatively minor part is released to surface waters with STP effluent. In summer, an appreciable additional amount of EHMC enters the surface waters directly through swimming. Based on its low vapor pressure and QSAR-modeled Henry constant, EHMC does not appreciably partition to air; even if it did, it would be short-lived based on predicted half-lives for OH-radical- or ozone-mediated degradation. In the aquatic environment, the low water solubility, high octanol-water partition coefficient, and relatively high QSAR-modeled organic carbon partition coefficient make partitioning from water to undissolved solids and sediment likely, where further biological degradation is expected. Moreover, as a substance specifically designed to absorb high-energy UV radiation, comparatively direct aquatic photolysis with a summer DT_{50} of 5–9 days (NOTOX, 2002) is a further degradation pathway for EHMC in surface waters and air. Overall, high degradation rates and short half-lives for EHMC in all environmental compartments are assumed, based on experimental results and modeled data. EHMC is not a persistent substance in the environment, however, due to the relatively high amounts released over all seasons; measurable steady-state concentrations are still expected, reaching their heights during summer.

B. Aquatic PECs

By crude "mechanical" reckoning, summer surface water PECs range from 70.8 ng/l (high flux) to 81.3 ng/l (low flux), while for the rest of the year PECs are 13.1 ng/l or 15.1 ng/l, respectively. The Level III model calculates a PEC for water of 2.4 ng/l in summer and 0.44 ng/l for the remaining eight months. Lastly, USES 3.0 gives a whole-year average PEC of 7.6 ng/l for the "region" configured to Swiss basic data.

C. Environmental Concentrations

Log-probit plotting of the available 90 data points resulted in a graph with a good fit of the percent-rank data to the straight regression line ($r^2 = 0.989$; Fig. 3). This good fit confirms the prior assumption of

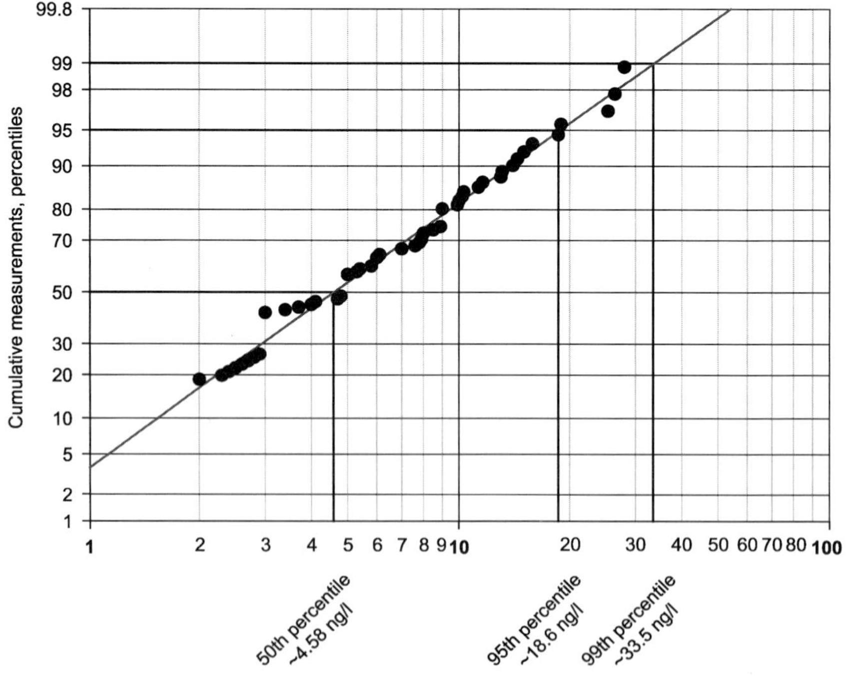

FIGURE 3 Log-probit graph of 90 EHMC monitoring data points. (See Color Insert.)

genuinely stochastic (non-biased and non-skewed) distribution of all measurement data. The 50th percentile or median value, based on the regression line, is 4.58 ng/l, while the 90th percentile corresponds to 13.2 ng/l, the 95th percentile to 18.6 ng/l, and the 99th percentile to 33.5 ng/l.

IV. Discussion

The Environmental Compatibility Assessment, containing the prediction of the environmental fate and the three PECs of EHMC, was completed and sent to the BUWAL in early 2001, before some of the analytical data became available and a probabilistic analysis could be performed. Still, the prediction agrees well with the environmental fate evidenced by Sabaliunas and colleagues (2001). This study is based on analytical measurements of EHMC before and after treatment at two STPs in England, with additional quantitative GIS-supported modeling of the subsequent fate in the river and qualitative and indirectly quantitative confirmation of the latter using semipermeable membrane devices (SPMD). While the modeled and SPMD-confirmed rapid disappearance of EHMC from the water may primarily reflect sorption to sediment, genuine degradation undoubtedly takes place. Sabaliunas and coworkers (2001) determined 91% removal in the STP with trickling filters, whereas the rate was 98% in the STP with activated sludge. This removal closely corresponds to the crude reckoning using the bare ready degradation rate plus the adsorption to sludge.

Removal may also be expected to occur in surface waters, as shown indirectly in aerobic and anaerobic tests and in the physico-chemical photodegradation test. In comparison with biological degradation, the importance abiotic processes is assumed to increase with further removal and dilution of EHMC in surface waters, as there must be a concentration threshold where the energy invested in the synthesis of specific catabolic enzyme cascades exceeds the energy gained from this degradation. However, higher local concentrations of EHMC adsorbed to undissolved solids or sediment particles, due to high $\log K_{ow}$ and extrapolated K_{oc}, may render biodegradation profitable again in such microenvironments. Alternatively, co-metabolism through enzymes for other, structurally similar substrates may explain further removal (Zwiener, personal communication). In any case, a high degree of supplementary degradation in the water column or the sediment is likely when the results from the present MEC analysis corroborate further removal, compared to the worst-case static PEC.

The high $\log K_{ow}$, on the other hand, suggests strong bioaccumulation, or partitioning into living organisms. EHMC was reported in fish by Nagtegaal and colleagues (1997) in concentrations of 0.27 μg/kg total fish for roach and 2.71 μg/kg for perch. However, these two data are derived from different years (1991 *vs.* 1993), and the EHMC in water was below the

LoD of 2 ng/l in both years. Hence, strictly speaking, no bioaccumulation factor (BAF) can be derived from these data and due to the different sampling times, possible biomagnification from roach to perch can only be a matter of speculation. But even the theoretical minimal BAF based on the data of Nagtegaal and colleagues (2.71 μg/kg for perch divided by the LoD, 2 ng/l, resulting in 1355) is not confirmed by a recent OECD bioaccumulation test performed on behalf of Roche using radio-labelled EHMC (NOTOX, 2000). This test resulted in a five-day BAF for the two concentrations tested of 60–430, which is far lower than predicted from the $\log K_{ow}$ or QSAR but which does include the theoretical minimal BAF of 135 derived for roach. In addition, the Roche test also showed rapid depuration with a DT_{50} below 2 days and a DT_{90} below 6 days after returning the fish to clean, non-spiked water. The high $\log K_{ow}$ means fast depuration through active excretion, in agreement with the rapid excretion of EHMC observed in human volunteers (Roche, unpublished data). Thus, even allowing for limited seasonal bioaccumulation, it is unlikely that living systems constitute a major environmental partitioning compartment for EHMC.

Probabilistic analysis of a large set of monitoring data results in a median MEC of 4.6 ng/l during the summer months. The extended sampling period from April to October in the dataset of Poiger and colleagues (2004), may cause the media to be slightly lower due to influence from non-summer measurements; data from the Rhine (GSA, unpublished data), which reflect only the summer period from mid-May to August 1997, result in a median of 3.2 ng/l and an average value of 5.5 ng/l, showing a small range around the overall value. Therefore, 4.6 ng/l is considered to be a reasonable, realistic summer concentration for EHMC. In addition, 90% of the measurements are expected to be smaller than 13.5 ng/l, 95% less than 19 ng/l, and 99% below 34 ng/l, showing a relatively narrow distribution on the high side of the median. With the exception of the marginal April and October MECs of Poiger and colleagues (2004), there is no empirical data for the non-summer months with clearly lower PECs.

Comparing the median MEC with the six-fold higher maximum values raises the concern of how published environmental concentrations should be interpreted. Obviously, maximum concentrations do not always reflect typical, realistic concentrations. Hence, caution is advised when adopting published MECs without qualification of what they represent. Only clear declaration of the quality of MECs, by characterization of the values as 95th or 90th percentiles such as in Ternes (1998), will allow readers to form a more realistic picture of actual concentrations.

With the above MECs for comparison, the PECs can now be better qualified. Even the crude mechanistic PEC between 70 and 80 ng/l is only about one factor away from realistic values. This is most certainly due to the ready biodegradability of EHMC that, in combination with a high rate of sludge adsorption, removes most indirect EHMC. On the other hand, without taking into account any further elimination, direct EHMC in surface waters from swimming and non-STP sewage accounts for a considerable load. Both computerized models give highly realistic PECs with summer or whole-year values, within the same dimension as the MEC. Any attempt to perfect the half-lives or partitioning coefficients is pointless, as the former are more system-related than substance-specific values (Balk, personal communication; Blok, 2001). The object of this comparison was to juxtapose PECs and MECs of a substance with both "easy" (biodegradability) and "difficult" (solubility, $logK_{ow}$) characteristics. The result is that, at least in this case, even a crude, rule-of-thumb PEC is fairly accurate; the computerized environmental fate models used here are even better. They are certainly useful in deriving realistic PECs for environmental assessments.

Based on a cohort of experimental lab and field data, EHMC is not a persistent compound in the environment. Realistic summer MECs for Switzerland are around 4.6 ng/l and, if the manner of deriving the environmental release and PECs of EHMC is reasonable, winter concentrations are below 1 ng/l. This prediction, and the whole seasonal behavior of EHMC, can only be tested through dependable analyses over the whole year.

Acknowledgments

My sincere appreciation to the Gewässerschutzamt Basel-Stadt and to Dr. Thomas Poiger for kindly making their unpublished original analytical data available. Thanks to Ing. Froukje Balk for her comments and for Dr. Han Blok's October 2001 Ph.D. thesis. Further thanks to Drs. Hans Gonzenbach, Darius Sabaliunas, and Christian Zwiener for discussions and important inputs.

References

Balk, F., Schwegler, A., and Rijs, G. (2001). Gebruik, emissie en aanwezigheid van UV-filters in zonnebrandcrèmes in watersystemen. Een verkennende studie. RIZA, The Netherlands Ministry for Traffic and Water. RIZA-Werk document 2001.206x.

BASF (1999). Safety data sheet Uvinul MC80. ME00615 1999. BASF, Ludwigshafen.

Blok, H. J. (2001). A quest for the right order. Biodegradation rates in the scope of environmental risk assessment of chemicals. Proefschrift (PhD thesis), Univ. Utrecht, Utrecht.

BMG (2001). Häner, A. MCX, Ultimate biodegradability. Evaluation of the anaerobic biodegradability in an aqueous medium: ISO 11734, 1995. BMG report no. 334/c-01 on behalf of Roche. BMG Engineering, Zürich-Schlieren.

BUWAL (2001). The Environment in Switzerland. BUWAL, available at http://www.buwal.ch/e/themen/grundl/raum/ek01u00.pdf

Campbell, P. J., Arnold, D. J. S., Brock, T. C. M., Grandy, N. J., Heger, W., Heimbach, F., Maund, S. J., and Streloke, M. (Eds.) (1999). Guidance document in higher tier aquatic risk assessment for pesticides (HARAP). SETAC-Europe, Brussels.

CFR (2001). US Code of Federal Regulations, 21CFR352; Sunscreen products for over-the-counter human use. U.S. Govt. Printing Office, Washington, D.C.

Daughton, C. G., and Ternes, T. A. (1999). Pharmaceuticals and personal care products in the environment: Agents of subtle change? *Environ. Health Perspect.* **107,** 907–938.

Diffey, B. L. (1991). Solar ultraviolet radiation effects on biological systems. *Phys. Med. Biol.* **36,** 299–328.

ECB (1999). European Chemicals Bureau. Technical guidance document in support of commission directive 93/67/EEC on risk assessment for new notified substances and commission regulation (EC) No. 1488/94 on risk assessment for existing substances. Office for Official Publications of the European Communities, Luxemburg.

GSA (Gewässerschutzamt Basel-Stadt), unpublished data. Parsol MCX im Rhein 1997. Excel graph.

H&R (1998). Chemical safety data sheet Neo heliopan AV, 25.03.1998. Haarman & Reimer, Holzminden.

IARC (1992). *Solar and ultraviolet radiation. IARC Monographs on the Evaluation of Carcinogenic Risks to Humans,* **vol. 55,** IARC, Lyon.

Kümmerer, K. (Ed.) (2001) *Pharmaceuticals in the environment; sources, fate, effects and risks.* Springer, Berlin.

Mackay, D., DiGuardo, A., Paterson, S., and Gowan, C. E. (1996). Evaluating the environmental fate of variety of types of chemicals using the EQC model. *Envir. Toxicol. Chem.* **15,** 1627–1637. http://www.trentu.ca/academic/aminss/envmodel/models.html

Merck (2000). Sicherheitsdatenblatt Eusolex 2292. 04/2000, CD-ROM 2000/1. Merck, Darmstadt.

Nagtegaal, M., Ternes, T. A., Baumann, W., and Nagel, R. (1997). UV-Filtersubstanzen in Wasser und Fischen. UWSF Z. *Umweltchem. Ökotox.* **9,** 79–86.

NOTOX (2000). Bioconcentration: 14-day flow-through fish test in rainbow trout with MCX and [phenyl ring U-^{14}C]MCX. NOTOX report 285569 on behalf of F. Hoffmann-La Roche Ltd, Basel.

NOTOX (2002). Photodegradation of MCX in water. Notox report 285558 on behalf of F. Hoffman-La Roche Ltd, Basel.

Poiger, T., Buser, H.-R., Balmer, M. E., Bergqvist, P. A., and Müller, M. D. (2004). Occurrence of UV filter compounds from sunscreens in surface waters: Regional mass balance in two Swiss lakes. *Chemosphere* **55,** 951–963.

Roche (updated). Internal substance documentation. F. Hoffmann-La Roche Ltd, Basel.

Sabaliunas, D., Webb, S. F., Peeters, S., and Eckhoff, W. S. (2004). Environmental concentrations, fate and safety of the organic UV filter Octyl Methoxycinnamate in surface waters. Poster, 11th. SETAC Europe Annual Congress, Madrid.

Schlumpf, M., Cotton, B., Conscience, M., Haller, V., Steinmann, B., and Lichtensteiger, W. (2004). *In vitro* and *in vivo* estrogenicity of UV screens. *Environ. Health Perspect.* **109,** 239–244.

Solomon, K. R., Baker, D. B., Richards, P., Dixon, K. R., Klaine, S. J., La Point, T. W., Kendall, R. J., Giddings, J. M., Giesy, J. P., Hall, L. W. J., Weisskopf, C., and Williams, M. (2004). Ecological risk assessment of atrazine in North American surface waters. *Envir. Toxicol. Chem.* **15,** 31–76.

SPARC (2004). SPARC on-line calculator. The University of Georgia, Athens, GA. Available at: http://ibmlc2.chem.uga.edu/sparc/index.cfm

SRC (2004). EPISuite, version 3.10. Syracuse Research Corporation, Syracuse, NY., on behalf of the US Environmental Protection Agency.

Ternes, T. A. (2004). Occurrence of drugs in German sewage treatment plants and rivers. *Wat. Res.* **32,** 3245–3260.

USES (1999). Anonymous/Rijksinstituut voor Volksgezondheid en Milieuhygiëne, 1999. Uniform System for the Evaluation of Substances (USES), v. 3.0. RIVM, Bilthoven.

Diederik Schowanek and Simon Webb
Procter & Gamble Eurocor
B-1853 Strombeek-Bever, Belgium

Exposure Simulation for Pharmaceuticals in European Surface Waters with GREAT-ER

I. Introduction

The assessment of whether a substance presents a risk to organisms in the environment is based on a comparison of the predicted environmental concentration (PEC) of the substance with its predicted no-effect concentration (PNEC) to organisms in ecosystems. This assessment can be performed for different compartments (air, water, and soil) and on different spatial scales (local, regional, and continental). The European Union (EU) practice related to risk assessment of industrial chemicals is described in the Technical Guidance Documents (TGD) supporting the Commission Directive on Risk Assessment of New Chemicals (93/67/EEC) and Commission Regulation on Risk Assessment of Existing Substances (1488/94/EEC) in support of Existing Substances Regulation (CEC, 1996), and is applied in the computerized calculation model EUSES (European Union System for the Evaluation of Substances, European Chemicals Bureau, 1997). Directive

93/39/EEC effectively extends the requirement for environmental risk assessments (ERA) to new pharmaceutical actives as part of the authorization process. Similarly, the reported occurrence of existing pharmaceutical actives in surface waters has necessitated a consideration of the risk to aquatic biota associated with such observations. The environmental risk assessment of existing pharmaceuticals has been reported by Halling-Sørenson *et al.* (1998), Stuer-Lauridsen *et al.* (2000), and Webb (2001b). The objective of this study is to evaluate the potential of the Geo-Referenced Regional Exposure Assessment Tool for European Rivers (GREAT-ER) model as an exposure assessment tool in the environmental risk assessment of pharmaceuticals. Such a use of GREAT-ER has previously been proposed for pharmaceuticals by the EU Scientific Committee on Toxicity, Ecotoxicity, and the Environment (CSTEE, 2000).

II. GREAT-ER System Description

The GREAT-ER model was developed as an aquatic chemical exposure prediction tool for use within environmental risk assessment schemes and river basin management. The GREAT-ER software calculates the distribution of predicted environmental concentrations (PECs) of consumer chemicals in surface waters. Compared to the generic EUSES model, realism is increased within GREAT-ER by incorporating spatial and temporal characteristics of the receiving environment in the models and underlying databases (Fig. 1). The design of the GREAT-ER system has been approached in a modular way, as described in detail in Feijtel *et al.* (1997). A variety of river catchments in the EU are available to the user or are under development. More details and regular updates can be found on www.great-er.org.

A. GIS Data Manipulation

In the data manipulation module, input data sourced from several databases and from the hydrology module (see below) are transformed into appropriate GIS formats (Wagner *et al.*, 1998).

B. Hydrology

The hydrology module combines several hydrological databases with a hydrological model. It provides the GREAT-ER system with the required river flow distributions, flow velocities, and river characteristics. The Micro Low Flows model, developed by the Institute of Hydrology to predict natural river flows at ungauged sites, has been augmented with

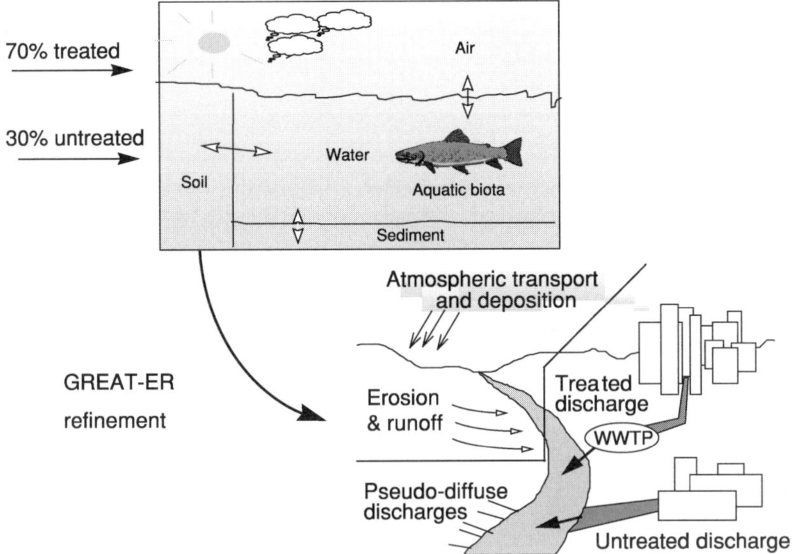

FIGURE 1 Refinement of generic regional exposure models (EUSES) by taking actual discharge pathway, treatment, and river flow data into account (GREAT-ER). WWTP, wastewater treatment plant. (See Color Insert.)

artificial influence data (abstractions, reservoirs, and discharges) to give reliable predictions of flow distributions in the Yorkshire River (Young *et al.*, 1998). For the Lambro and the Rur Rivers, simpler hydraulic models based on accumulated river length have been used (Schulze, 2001).

C. Waste Pathway and River Modelling

This module is used for the prediction of chemical emission, chemical removal/transformation during conveyance and treatment, and chemical fate in rivers (Boeije *et al.*, 2000). Chemical fate in wastewater treatment plants and rivers is described deterministically, with several levels of complexity available to reflect the current information concerning both the chemical and the environment. For example, chemical removal during sewage treatment can be either on a simple percentage removal basis or predicted using the Simple Treat model, which is currently also used in EUSES (European Chemicals Bureau, 1997). Additionally, GREAT-ER applies a stochastic technique (Monte Carlo simulation), which allows most input parameters to be described that in terms of a distribution, normal, lognormal, or uniform distributions that can be specified. The Monte Carlo approach generally requires about 1000 runs for sufficient convergence to be obtained. Thus, GREAT-ER can produce a statistical distribution of predicted environmental concentrations, as required for probabilistic risk assessment.

D. End-User Desktop GIS

In this module, access to and visualization of the databanks and model results is achieved; linking of the models with the data banks is also achieved. The GIS databanks, waste pathway models, and river models are integrated into one coherent simulation system. This integration process results in an operational end-user system, which runs on a PC platform. The hydrological models and the ARC/INFO spatial data processing steps are not integrated into the end-user software system. The user interface is the front end between the user and the software system. It allows the selection of catchments and chemicals, as well as the input of model and scenario parameters. The user interface also handles filtering and visualization of model results by the GIS. Avenue (ESRI) has been used for the development of this interface in an ArcView (ESRI) environment. ArcView 3.0a or 3.1 software is required to run GREAT-ER.

III. Output of GREAT-ER

GREAT-ER offers a set of possibilities for analysis of the simulation results (Schowanek *et al.*, 2001).

A. Color-Coded River Maps

GREAT-ER's direct output provides predicted chemical concentrations linked to a river network, which are visualized as color-coded digital GIS river maps (Fig. 2). To capture the spatial variability, the predicted concentrations are represented as quartiles of the distributions of all concentrations in the catchment. PECs can also be shown as absolute concentration classes as defined by the user. The GIS analysis tools and color-coded maps allow identification of any locations (hot spots) within a region where site-specific PEC values may exceed the PNEC. General water quality maps may be overlain onto the simulation output to compare chemical presence with physico-chemistry or biology-based water quality indices.

B. Concentration Profiles

Profiles of predicted concentrations through the studied catchment can be generated and exported (Fig. 3). Such simulated profiles clearly illustrate chemical fate from a river's headwater down to its mouth, and can be used to directly compare model predictions with monitoring data, where available.

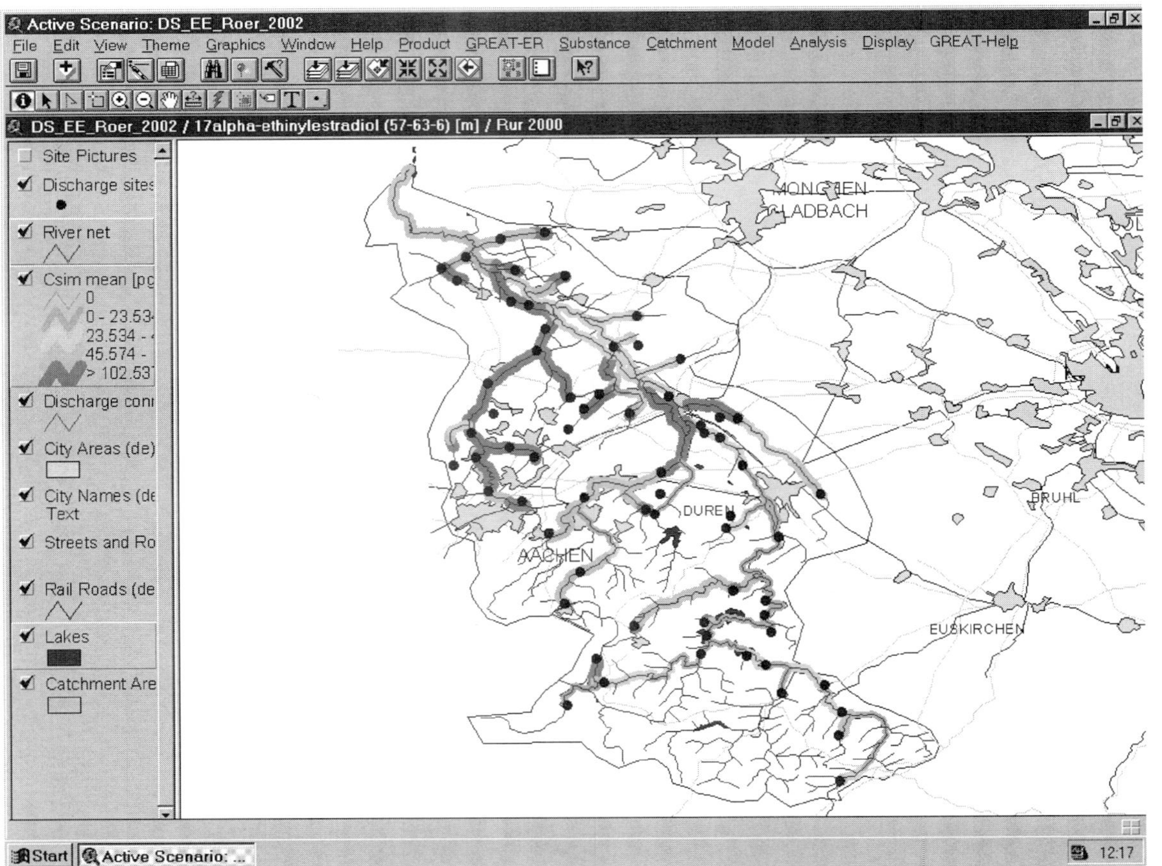

FIGURE 2 GREAT-ER 1.03 user interface and color coded GIS map with the simulation of ethinyl oestradiol in the Rur basin, Germany. (See Color Insert.)

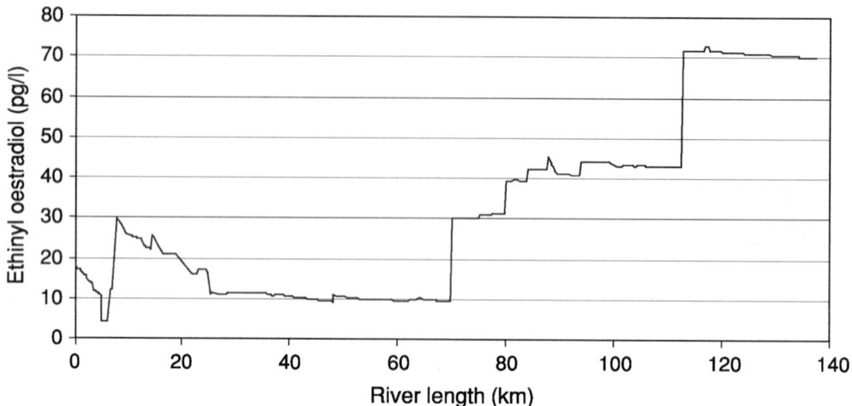

FIGURE 3 Simulated PEC of ethinyl oestradiol under mean flow conditions in the Rur river, Germany. River stretch length, 300 m; km0, Kalterherberg; step changes at 70 km and 120 km represent the respective inputs from Duren and Aachen. (See Color Insert.)

C. Aggregated PECs

Geo-referenced model results can be aggregated to obtain a spatially averaged PEC (Fig. 4), which is representative of the river basin under study (Boeije *et al.*, 2000). GREAT-ER can generate a $PEC_{initial}$, which comes from the distribution of concentrations in the river stretch below each emission point source under mean flow conditions as default. This can be considered a GIS-analog of the PEC-local concept used in the EU TGD. GREAT-ER can also generate a $PEC_{catchment}$ by incorporating the concentration distributions in each river stretch in the catchment. This involves a weighting procedure that can be based on stretch flow increment, length, or volume. This concept can be considered a GIS analog of the EU TGD PEC-regional.

IV. Methods

Six common human pharmaceuticals (ethinyl oestradiol, paracetamol, aspirin, dextropropoxyphene, clofibrate, and oxytetracycline) were selected to illustrate the functioning of GREAT-ER for the Aire, Lambro, and Rur river catchments where tonnage data were available. Details of the compounds and their usage, metabolism, and environmental fate as employed in the GREAT-ER version 1.03 simulations are given in Table I.

An audit of UK compound usage data was commissioned from IMS (Intercontinental Medical Statistics, UK and Ireland) and reflects all sales of products containing these compounds into retail pharmacies, dispensing general practitioners, and hospital pharmacies in 1995. Data relating to

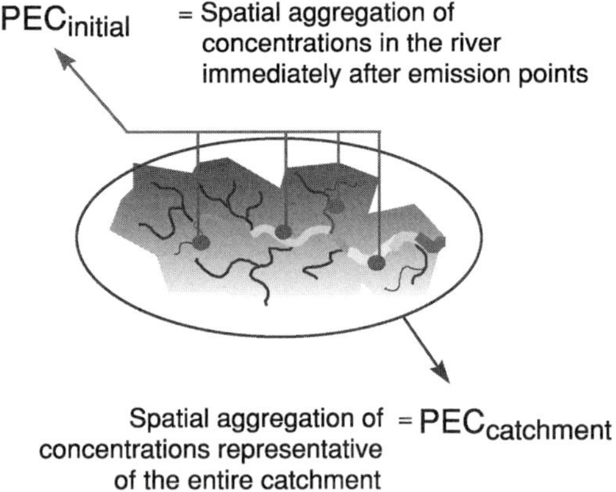

FIGURE 4 Schematic illustration of the $PEC_{initial}$ and $PEC_{catchment}$ concepts, as developed within GREAT-ER for GIS-assisted regional risk assessment. (See Color Insert.)

the UK usage of over-the-counter (OTC) analgesics for the same period were obtained from the Paracetamol Information Centre and the European Aspirin Foundation. The therapeutic category is as detailed in the Merck Index (1989). Per-capita use and worst-case sewage influent concentrations (assuming no human metabolism) are calculated on the basis of a UK population of 57.6 million and a specific water consumption of 259 liters/capita/day (WSA, 1994).

Effective post-metabolism exposure as factored into the simulations is based on a consideration of Dollery (1991). In some cases, such as ethinyl oestradiol or clofibrate, a significant proportion of an administered compound may be excreted unoxidized as conjugates in urine or bile/feces. In these cases, it has been assumed that biological activity can be restored following microbially mediated biotransformation (de-conjugation). This has been reported for non-estrogenic steroid metabolites, such as oestradiol-3-glucuronide (Panter *et al.*, 1999). Henschel *et al.* (1997) similarly speculate on the microbial reactivation of paracetamol conjugates. In the case of oxytetracycline, no metabolism occurs and 100% availability is assumed. For dextropropoxyphene, extensive N-demethylation means that only about 1% is excreted from the body unchanged, although the major metabolite, norpropoxyphene, is known to retain some pharmacological activity. For aspirin, near-complete metabolism to salicylic acid is assumed.

Removal during wastewater treatment is based upon actual observations (Stumpf *et al.*, 1999; Ternes, 1998; Ternes *et al.*, 1999) or a worst-case assumption of 0% for oxytetracycline and dextropropoxyphene. Biodegradation data are from Richardson and Bowron (1995) and Schweinfurth *et al.* (1996). River water dieaway rates employed are the default values from the EU Technical Guidance Document (CEC, 1996): a half life of 15 days for a

TABLE I Overview of Compound Usage, Metabolism, and Environmental Fate

	Ethinyl oestradiol	*Paracetamol*	*Aspirin*	*Dextropro-poxyphene*	*Clofibrate*	*Oxytetracycline*
Formula	$C_{20}H_{24}O_2$	$C_8H_9NO_2$	$C_9H_8O_4$	$C_{22}H_{29}NO_2$	$C_{10}H_{11}ClO_3$	$C_{22}H_{24}N_2O_9$
Molecular Weight	296.41	151.17	180.16	339.48	214.66	460.44
CAS #	57-63-6	103-90-2	50-78-2	469-62-5	882-09-7	79-57-2
Therapeutic category	Estrogen	Analgesic; anti-pyretic	Analgesic; anti-pyretic; anti-inflammatory	Narcotic analgesic	Anti-hyperlipoprote inemic	Anti-bacterial
Annual use (tons)	0.029	2000	770	42.5	1.5	33.7
Use (mg/cap/yr)	0.5	34,722	13,368	738	26	585
Worst case Influent (μg/l)	0.005	367.3	141.4	7.81	0.28	6.19
Excreted post-metabolism (%)	<1% unchanged, 30% as conjugates	2–5% unchanged, 85% congugates	Assume 1%	<1% unchanged	98% as conjugates of free acid	100% unchanged
Biodegradation	Non-readily biodegradable	Readily biodegradable + acclimation	Readily biodegradable	Non-readily biodegradable	Non-readily biodegradable	Non-readily biodegradable (Tetracycline)
Total removal via wastewater treatment (%)	78% AS, 64% TF	98% AS	81% AS	Assume 0%	51% AS; 34% AS 15% TF	Assume 0%
River water die-away K (d^{-1})	0	4.7×10^{-2}	4.7×10^{-2}	0	0	0

AS, activated sludge; TF, trickling filter.

readily biodegradable chemical, and infinity for compounds with no or very slow degradation.

In the case of ethinyl oestradiol, simulations were similarly conducted for the Lambro (Italy) and the Rur (Germany) Rivers. Reported consumption of ethinyl oestradiol in Italy is 16 kg/year (Zuccato *et al.*, 2001). The corresponding figure for Germany is 50 kg/year (Ternes, 2001a).

Risk characterization employing PECs from GREAT-ER, or measured environmental concentrations (MECs) and PNECs from available ecotoxicity data, was also conducted. PEC/PNEC ratios less than 1 are assumed to imply environmental safety (Webb, 2001b).

V. Results and Discussion

The results of the GREAT-ER exposure simulations in the Aire River are presented in Table II. PEC initial values ranged from 6.8 μg/l for paracetamol down to 0.13 ng/l for ethinyl oestradiol. PEC catchment concentrations were correspondingly lower (approximately a 0.55 factor for the Aire). When available, some examples of actual measurements from surface waters are also provided for comparative purposes, although it should be noted that most relate to observations in German surface waters. Simulations of ethinyl oestradiol for the Lambro and Rur Rivers yielded PEC initial values of 0.040 ng/l and 0.185 ng/l, respectively. The simulation for the Rur is also illustrated in Figs. 2 and 3.

Observations of actual concentrations of ethinyl oestradiol in UK sewage effluents are reported by Desbrow *et al.* (1996). They observed that ethinyl oestradiol was undetectable (less than the detection limit of 0.2 ng/l) in more than half of the effluents sampled, and where detectable, it was usually below 1 mg/l. Allowing for dilution, these observations are not inconsistent with the surface water PECs reported here. Williams *et al.* (1999) have also estimated likely ethinyl oestradiol concentrations in the Aire River using the EXAMS model. Under conditions of low flow (summer), they predict concentrations of 0.15 ng/l and 0.10 ng/l at 1 km and 10 km downstream of an effluent discharge in a modeled stretch of the river, respectively. This is very similar to concentrations predicted in both this study and the Aire catchment modeled with GREAT-ER by Williams *et al.* (2001).

It is interesting to note that oxytetracycline was not detected in German surface waters (detection limit, 0.02 μg/l) by Hirsch *et al.* (1999). This contrasts with the predicted surface water PECs of approximately 1.6 μg/l reported here. One possible explanation may be that oxytetracycline is photo-degradable in surface waters as has been reported for tetracycline (Peterson *et al.*, 1993). Similarly, the complexing properties of tetracyclines with calcium and other similar ions have also been highlighted as a possible reason explaining their absence from the water column (Hirsch *et al.*, 1999).

TABLE II Calculation of PEC Initial and PEC Catchment (Flow Increment Method) in the Aire River for Six Common Pharmaceuticals

	Ethinyl oestradiol	*Paracetamol*	*Aspirin*	*Dextropropoxyphene*	*Clofibrate*	*Oxytetracycline*
GREAT-ER PEC initial (μg/l)	0.13 ng/l	6.8	0.09	0.02	0.06	1.61
GREAT-ER PEC catchment (μg/l)	0.09 ng/l	4.3	0.05	0.01	0.03	0.89
CPMP PEC (μg/l) (Webb, 2001b with effluent surface water dilution factor 10)	0.5 ng/l	36.7	14.1	0.78	0.03	0.62
MEC (μg/l)	<0.2 ng/l (Kalbfus, 1995) <1.0 ng/l (Williams *et al.*, 2001) <0.5 ng/l (Ternes, 2001b)	Max. <0.15 (Ternes, 1998)	Median <0.02, 90%-ile 0.16, max. 0.34 (Ternes, 1998, 2001b)	1 (Richardson and Bowron, 1985)	Median 0.066, 90%-ile 0.21, Max. 0.55 clofibric acid (Ternes, 1998, 2001b)	Median <0.02, 90%-ile <0.02, Max. <0.02 (Hirsch *et al.*, 1999)

Likewise, paracetamol has not been detected in surface waters (Richardson and Bowron, 1995; Ternes, 1998). Aspirin concentration reported by Ternes (2001b) for German surface waters is 0.16 μg/l at the 90th percentile, with a maximum of 0.34 μg/l. This is in the same range as calculated by GREAT-ER. Differences between the simulated PECs for the UK and reported concentrations of clofibrate in German surface waters may to some extent reflect differing use patterns in the two countries (1.5 t/yr in the UK; 16 t/yr for Germany reported by Ternes, 1998).

To further refine the GREAT-ER predictions, more detailed information about the environmental fate and behavior of pharmaceuticals is needed. The GREAT-ER model itself is able to accommodate loss or partitioning processes such as photodegradation, adsorption, and biodegradation in sewers, provided that realistic input data can be found. Readily biodegradable substances such as paracetamol are likely to be subject to some degree of in-sewer biodegradation and as such, the surface water PECs generated here are likely to be overestimates. In addition, the surface water default half-life of 15 days can probably be reduced substantially to a few days or less, as has been observed for other readily biodegradable substances in field monitoring (Schulze, 2001).

In all cases, the PECs predicted by GREAT-ER are less than those yielded via application of the exposure model in the current draft EU guidance (CPMP, 2001) for the risk assessment of new pharmaceuticals (Table II).

PEC/PNEC ratios for the six compounds in the Aire River are presented in Table III. The risk characterization employs the PEC initial values from Table II and PNEC values derived by the application of an assessment factor to acute or chronic aquatic ecotoxicity endpoints. When available, chronic data is given primacy. A full description of the available ecotoxicity data for these compounds is given in Webb (2001a). An assessment factor of 1000 is applied to acute endpoints. Assessment factors of 100, 50, or 10 are applied to chronic endpoints, depending upon the number of taxa represented in the available database. The magnitude of these factors are as employed in the Technical Guidance Document (CEC, 1996). For ethinyl oestradiol and oxytetracycline, the PEC/PNEC ratios are greater than unity and require evaluation and possible refinement. PEC/PNEC ratios for all other compounds are less than unity, implying environmental safety.

In the case of ethinyl oestradiol, it is unlikely that a further investment in the ecotoxicity database would change the PNEC significantly (perhaps with the exception of a mesocosm study). Furthermore, the observations of Desbrow *et al.* (1996) and Williams *et al.* (2001) support the simulated PEC values reported here. However, Desbrow *et al.* (1996) also detail how the majority (90%) of oestrogenic activity in sewage effluent in the UK is accounted for by the presence of the natural estrogens, oestrone and 17β-oestradiol. The source of these natural estrogens appears to be excretion by women. Desbrow *et al.* (1996) therefore conclude that although ethinyl

TABLE III Risk Characterization Calculations (PEC/PNEC Ratios) for Six Common Pharmaceuticals

	Ethinyl oestradiol	*Paracetamol*	*Aspirin*	*Dextropro-poxyphene*	*Clofibrate*	*Oxytetracycline*
Ecotoxicity data (lowest endpoint underlined)	3 Acute (Algae, Fish, *Daphnia*) Köpf (1995); Schweinfurth *et al.* (1996) 3 Chronic (Algae, Fish, *Daphnia*) Köpf (1995); Länge *et al.* (1997, 2001)	3 Acute (Algae, Fish, *Daphnia*) Henschel *et al.* (1997); Kühn *et al.* (1989)	1 Acute (*Daphnia*) Calleja *et al.* (1994)	1 Acute (*Daphnia*) Lilius *et al.* (1994)	3 Acute (Algae, Fish, *Daphnia*) Henschel *et al.* (1997); Köpf (1995) 2 Chronic (Algae, *Daphnia*) Köpf (1995)	3 Acute (Algae, Fish, Invertebrate) Holten-Lutzhøft *et al.* (1998); Hughes (1973); Johnson (1976)
Lowest endpoint (mg/l)	Acute 0.84 mg/l Chronic 1 ng/l	9.2	168	14.6	Acute 12 mg/l Chronic 10 μg/l	0.23
Assessment factor	Chronic 10	1000	1000	1000	Chronic 50	1000
PNEC (μg/l)	0.1 ng/l	9.2	168	14.6	0.2	0.23
GREAT-ER PEC initial (μg/l)	0.13 ng/l	6.8	0.09	0.02	0.06	1.61
PEC/PNEC ratio	1.3	0.74	<0.01	<0.01	0.3	7.0

oestradiol may contribute to the overall estrogenicity of sewage effluents, it appears likely that natural estrogens are responsible for the majority of the feminized responses observed in fish populations exposed to sewage effluents, although some contribution from ethinyl oestradiol to the reported effects cannot be precluded.

For oxytetracycline, it is likely that an incorporation of potential photodegradation or complexation and subsequent loss to sediment or suspended solids (Hirsch *et al.*, 1999) would have some impact upon the PEC. There is also scope to improve the PNEC by generating chronic effects data. Comparisons of the oxytetracycline MEC from Hirsch *et al.* (1999) with the PNEC would yield a MEC/PNEC ratio <0.1. In addition, there is similarly scope for additional revision of the ecotoxicity database for oxytetracycline, as the current PNEC is based on acute data.

VI. Conclusions

A preliminary estimation of the environmental concentrations of six pharmaceuticals was made by GREAT-ER. In first instance, the usage figures needed to be corrected for metabolism by the human body. After this correction, the pharmaceuticals can be treated as other down-the-drain domestic chemicals.

Some degree of uncertainty remains on loss processes for most pharmaceuticals, such as the removal during sewage treatment, in-stream die-away rates, photodegradation, and sorption to sediments. Literature data were used for these simulations at tier 1 of the GREAT-ER model, and 0% removal was assumed if no data were available. Generation of removal data or application of the Simple Treat module in tier 2 of GREAT-ER could further refine the assessment.

The highest predicted concentrations in the Aire River are found for paracetamol (high per-capita usage, partial metabolism, high sewage treatment removal, surface water die-away) and the lowest for ethinyl oestradiol (low per-capita usage, partial metabolism, some sewage treatment removal, no surface water die-away).

Based on these initial simulations, the next step in the risk assessment process would be to refine the input data and GREAT-ER calculations for substances where measured concentrations differ substantially from the PEC, such as paracetamol or oxytetracycline.

Acknowledgments

The authors thank the European Centre for Ecotoxicology and Toxicology of Chemicals (ECETOC), the Environmental Risk Assessment Steering Management committee (ERASM) of the Association Internationale de la Savonnerie, de la Détergence et des Produits d'Entretien

(A.I.S.E.) and the Comité Européen de Agents de Surface et Intermédiares Organiques (CESIO), and the UK Environment Agency for their financial and managerial support during the development of GREAT-ER. A copy of the CD-ROM version 1.03 and user manual can be obtained free of charge at the GREAT-ER web-site (http://www.great-er.org).

References

Boeije, G. M., Vanrolleghem, P., and Matthies, M. (1997). A geo-referenced aquatic exposure prediction methodology for 'down-the-drain' chemicals. Contribution to GREAT-ER #3. *Wat. Sci. Technol.* **36,** 251–258.

Boeije, G. M., Wagner, J.-O., Koorman, F., Vanrolleghem, P. A., Schowanek, D. R., and Feijtel, T. C. J. (2000). New PEC definitions for river basins applicable to GIS-based environmental exposure assessment. Contribution to GREAT-ER #8. *Chemosphere* **40,** 255–265.

Calleja, M. C., Personne, G., and Geladi, P. (1994). Comparative acute toxicity of the first 50 multicentre evaluation of *in vitro* cytotoxicity chemicals to aquatic non-vertebrates. *Arch. Environ. Contam. Toxicol.* **26,** 69–78.

CEC (1996). Technical Guidance Document in support of the Commission Directive 93/67/EEC on risk assessment for new notified substances and Commission Regulation (EEC) No. 1488/94 on risk assessment for existing substances. European Chemicals Bureau, Ispra, Italy.

CPMP (2001). Draft discussion paper on environmental risk assessment of non-genetically modified organism (non-GMO) containing medicinal products for human use. Committee for Proprietary Medicinal Products (CPMP), European Medicines Evaluation Agency (EMEA), London.

CSTEE (2000). Scientific Committee on Toxicity, Ecotoxicity and the Environment Opinion on Draft CPMP Discussion paper on Environmental Risk assessment of Medicinal Products for Human Use. Available online at http://europa.eu.int/comm/food/fs/sc/sct/out111_en.pdf.

Desbrow, C., Routledge, E., Sheahan, D., Waldock, M., and Sumpter, J. P. (1996). The identification of oestrogenic substances in sewage treatment effluents. Report of the Environment Agency, Bristol, UK.

Dollery, C. T., Ed. (1991) .*Therapeutic Drugs—Volume 1&2.* Churchill Livingstone, Edinburgh.

EEC (1993). Commission Directive 93/39/EEC amending Directives 65/65/EEC, 75/318/EEC and 75/319/EEC in respect of medicinal products.

EEC (1993). Commission Directive 93/67/EEC laying down the principles for assessment of risks to man and the environment of substances notified in accordance with Council Directive 67/458/EEC.

EEC (1994). Commission Regulation 1488/94/EEC laying down the principles for the assessment of risks to man and the environment of existing substances in accordance with Council Regulation 793/93/EEC.

EUSES (1997). European Uniform system for the evaluation of substances (EUSES), version 1.0. European Chemical Bureau, Ispra, Italy.

Feijtel, T. C. J., Boeije, G. M., Matthies, M., Young, A., Morris, G., Gandolfi, C., Hansen, B., Fox, K., Holt, M., Koch, V., Schroder, F. R., Cassani, G., Schowanek, D., Rosenblom, J., and Niessen, H. (1997). Development of a geography-referenced regional exposure assessment tool for european rivers—GREAT-ER. Contribution to GREAT-ER #1. *Chemosphere* **34,** 2351–2374.

Halling-Sørenson, B., Nors Nielsen, S., Lanzky, P. F., Ingerslev, F., Holten-Lutzhøft, H. C., and Jørgenson, S. E. (1998). Occurrence, fate and effects of pharmaceutical substances in the environment—a review. *Chemosphere* **36,** 357–393.

Henschel, K. P., Wenzel, A., Diderick, M., and Fliedner, A. (1997). Environmental hazard assessment of pharmaceuticals. *Reg. Toxicol. Pharmacol.* **25**, 220–225.

Hirsch, R., Ternes, T., Haberer, K., and Kratz, K.-L. (1999). Occurrence of antiobiotics in the aquatic environment. *Sci. Tot. Environ.* **225**, 109–118.

Holten-Lützhøft, H.-C., Halling-Sørensen, B., and Jørgensen, S. E. (1998). Algal testing of antiobiotics applied in Danish fish farming. Proc. 8th Annual Meeting of SETAC Europe, Bordeaux 1998.

Hughes, J. S. (1973). Acute Toxicity of Thirty Chemicals to Stripped Bass (Morone saxatilis). Presented at the Western Association of State Game and Fish Commissioners, Salt Lake City, Utah, July 1973.

Johnson, S. K. (1976). Twenty-Four Hour Toxicity Tests of Six Chemicals to Mysis Larvae of Penaeus setiferus. Texas A & M University Extension Disease Laboratory, Publication No. FDDL-S8.

Kalbfus, W. (1995). Effects in Bavarian watercourses through synthetic oestrogens. Presented at the 50th Seminar of the Bavarian Association for Waters Supply: Substances with Endocrine Effects in Water.

Köpf, W. (1995). Effects of endocrine substances in bioassays with aquatic organisms. Presented at the 50th Seminar of the Bavarian Association for Waters Supply: Substances with Endocrine Effects in Water.

Kühn, R., Pattard, M., Pernak, K. D., and Winter, A. (1989). Results of the harmful effects of selected water pollutants (anilines, phenols, aliphatic compounds) to *Daphnia magna*. *Wat. Res.* **23**, 495–499.

Länge, R., Schweinfurth, H., Croudace, C., and Panther, G. (1997). Growth and reproduction of fathead minnow (Pimephales promelas) exposed to the synthetic steroid hormone Ethinylestradiol in a life cycle test (Abstract). Proc. 7th Annual Meeting of SETAC Europe, Amsterdam 1997.

Länge, R., Hutchinson, T. H., Croudace, C. P., Siegmund, F., Schweinfurth, H., Hampe, P., Panter, G. H., and Sumpter, J. P. (2001). Effects of the synthetic oestrogen 17a-Ethinylestradiol over the life-cycle of the fathead minnow (*Pimephales promelas*). *Environ. Toxicol. Chem.* **20**, 1216–1227.

Lilius, H., Isomaa, B., and Holmström, T. (1994). A comparison of the toxicity of 50 reference chemicals to freshly isolated rainbow trout hepatocytes and *Daphnia magna*. *Aquat. Toxicol.* **30**, 47–60.

Merck (1989). The Merck Index. *In* "An Encyclopaedia of Chemicals, Drugs and Biologicals" (S. Budavari, Ed.), 11th ed. Merck & Co. Inc., Rahway, NJ.

Panter, G. H., Thompson, R. S., Beresford, N., and Sumpter, J. P. (1999). Transformation of a non-oestrogenic steroid metabolite to an oestrogenically active substance by minimal bacterial activity. *Chemosphere* **38**, 3579–3596.

Peterson, S. M., Batley, G. E., and Scammell, M. S. (1993). Tetracycline in antifouling paints. *Marine Pollut. Bull.* **26**, 96–100.

Richardson, M. L., and Bowron, J. M. (1995). The fate of pharmaceutical chemicals in the aquatic environment. *J. Pharm. Pharmacol.* **37**, 1–12.

Schowanek, D., Fox, K., Holt, M., Schroeder, F. R., Koch, V., Cassani, G., Matthies, M., Boeije, G., Vanrolleghem, P., Young, A., Morris, G., Gandolfi, C., and Feijtel, T. C. J. (2001). GREAT-ER: A new tool for management and risk assessment of chemicals in river basins. Contribution to GREAT-ER #10. *Wat. Sci. Technol.* **43**, 179–185.

Schulze, C. (2001). Modelling and evaluation of the aquatic fate of detergents (PhD thesis). University of Osnabrueck, Osnabrueck, Germany.

Schweinfurth, H., Länge, R., and Gunzel, P. (1996). Environmental fate and ecological effects of steroidal estrogens. Presentation at the Oestrogenic Chemicals in the Environment conference organised by IBC Technical Services Ltd., May 9th–10th, 1996, London.

Stuer-Lauridsen, F., Birkved, M., Hansen, L. P., Holten-Lutzhøft, H. C., and Halling-Sørenson, B. (2000). Environmental risk assessment of human pharmaceuticals in Denmark after normal therapeutic use. *Chemosphere* **40**, 783–793.

Stumpf, M., Ternes, T. A., Wilken, R-D., Rodrigues, S. V., and Baumann, W. (1999). Polar drug residues in sewage and natural waters in the state of Rio de Janeiro, Brazil. *Sci. Tot. Environ.* **225**, 134–141.

Ternes, T. A. (1998). Occurrence of drugs in German sewage treatment plants and rivers. *Wat. Res.* **32**, 3245–3260.

Ternes, T. (2001a). Pharmaceuticals and metabolites as contaminants of the aquatic environment. *In* "Pharmaceuticals and Personal Care Products in the Environment—Scientific and Regulatory Issues" (C. G. Daughton and T. L. Jones-Lepp, Eds.), pp. 39–54. ACS Symposium Series 791, American Chemical Society, Washington, D.C.

Ternes, T. A. (2001b). Analytical methods for the determination of pharmaceuticals in aqueous environmental samples. *Trends Anal. Chem.* **20**, 419–434.

Ternes, T. A., Stumpf, M., Mueller, J., Haberer, K., Wilken, R.-D., and Servos, M. (1999). Behaviour and occurrence of estrogens in municipal sewage treatment plants—I. Investigations in Germany, Canada and Brazil. *Sci. Tot. Environ.* **225**, 81–90.

Wagner, J.-O., Koorman, F., and Matthies, M. (1998). GREAT-ER analysis tools and connectivity—exposure at a regional scale. Proc. 8th Annual Meeting of SETAC Europe, Bordeaux, 1998.

Webb, S. F. (2001a). A data based perspective on the environmental risk assessment of human pharmaceuticals II—collation of available ecotoxicity data. *In* "Pharmaceuticals in the Environment—Sources, Fate, Effects, and Risks" (K. Kümmerer, Ed.), pp. 175–201. Springer-Verlag, Heidelberg, Germany.

Webb, S. F. (2001b). A data based perspective on the environmental risk assessment of human pharmaceuticals II—aquatic risk characterization. *In* "Pharmaceuticals in the Environment—Sources, Fate, Effects, and Risks" (K. Kümmerer, Ed.), pp. 203–219. Springer-Verlag, Heidelberg, Germany.

Williams, R. J., Jürgens, M. D., and Johnson, A. C. (1999). Initial predictions of the concentrations and distribution of 17β-Oestradiol, Oestrone and Ethinyl Oestradiol in 3 English rivers. *Wat. Res.* **33**, 1663–1671.

Williams, R. J., Johnson, A. C., Smith, J. J. L., Jürgens, M. D., and Holthaus, K. (2001). Fate and behavior of steroid oestrogens in aquatic systems. Report to England and Wales Environment Agency/DEFRA (Technical Report P2-162/TR). Centre for Ecology & Hydrology, Wallingford, UK.

WSA (1994). *In* "Waterfacts '94" (D. Burnell, Ed.) (1994). Water Services Association, London.

Young, A. R., Gustard, A., Crocker, K. M., and Round, C. E. (1998). Hydrological models for use in regional and local exposure assessment methodologies. Proc. 8th Annual SETAC Europe meeting, Bordeaux, France.

Zuccato, E., Bagnati, R., Fioretti, F., Natangelo, M., Calamari, D., and Fanelli, R. (2001). Environmental loads and detection of pharmaceuticals in Italy. *In* "Pharmaceuticals in the Environment—Sources, Fate, Effects, and Risks" (K. Kümmerer, Ed.), pp. 19–27. Springer-Verlag, Heidelberg, Germany.

Simon Webb*
Thomas Ternes[†]
Michel Gibert[‡]
Klaus Olejniczak[§]

*CEFIC—European Chemistry Industry Council
Brussels B-1160, Belgium

[†]Bundesanstalt Gewaesserkunde
D-56068 Koblenz, Germany

[‡]VEOLIA Environment
75008 Paris, France

[§]BfArM—Federal Institute for Drugs and Medical Devices
D-53175 Bonn, Germany

Indirect Human Exposure to Pharmaceuticals via Drinking Water

I. Introduction

Over the last 25 years, there have been various attempts to determine the concentrations of pharmaceuticals and their metabolites in drinking water (Aherne and Briggs, 1989; Aherne *et al.*, 1985; James *et al.*, 1998; Kalbfus, 1995; Kuch and Ballschmitter, 2001; Stan *et al.*, 1994; Ternes, 2001a,b; Waggott, 1981; Zuccato, 2000). Currently, there is no specific regulatory guidance on how the significance of the potential presence of pharmaceuticals at trace concentrations in drinking water supplies may be assessed. Risk assessment of pharmaceuticals for marketing authorization in both the United States and European Union does not address this point (CPMP, 2001; FDA-CDER, 1998). Within the European Union, the quality of water for human consumption is determined by the Drinking Water Directive (Council Directive 98/93/EC on the quality of water intended for human consumption). Of the 48 parameters within the directive, none relate to pharmaceuticals. Within the U.S. there are over 170 Environmental Protection Agency (EPA) drinking water standards or health advisories relating to organic compounds in potable water supplies. Again, none applies to pharmaceuticals (EPA, 2002). Nevertheless, numerous evaluations of the potential for adverse human effects arising from indirect exposure to pharmaceuticals via drinking water have been made (Christensen, 1998; Richardson and Bowron, 1985; Schulman *et al.*, 2002; Webb, 2001). Christensen (1998) estimated risk associated with indirect exposure via drinking water and diet to ethinyl oestradiol, phenoxymethylpenicillin, and cyclophophamide arising from diffuse emissions. The benchmark effects were male endogenous estrogen production, tolerable food residues based on allergic reactions, and genotoxic carcinogenicity, respectively. In all cases, risk was deemed negligible.

The most comprehensive evaluation to date in terms of the number of compounds considered has been that of Webb (2001). This entailed quantitative estimates of potential indirect exposure to pharmaceuticals via drinking water in the UK for approximately 60 compounds. I_{70} values based on a lifetime (70 years or 25,550 days) ingestion of 2 liters/day of water were calculated using worst-case predictions for UK drinking water concentrations (assuming no human metabolism, no removal during sewage treatment, no surface water effluent dilution, and no removal during drinking water treatment). Calculated worst-case ingested quantities for the UK were small, and a lifetime ingestion of a pharmaceutical compound via potable water was less than the daily recommended therapeutic dose for 80% of the compounds assessed. This implied a margin of at least 25,000 between indirect exposure and efficacious therapeutic dosage. For compounds where worst-case lifetime ingestion was more than the daily therapeutic dose (ethinyl oestradiol), refinement of the exposure based on measured concentrations demonstrated the degree of conservatism associated with the worst-case assumptions. Contributing to the discrepancy between observed concentrations of pharmaceuticals in potable water and the worst-case concentrations employed in Webb (2001) are human metabolism, sewage treatment, surface water effluent dilution, and drinking water treatment. Schulman *et al.* (2002) derived (non-cancer effects) health-based limits (HBLs) from daily therapeutic doses using a composite uncertainty or application factor of 30 to account for the use of a lowest observed effect level (LOEL) to estimate a no observed effect level (NOEL) (factor 3) and inter-individual variability (factor 10). These HBLs were compared to worst-case aqueous media concentrations. In all cases, potential exposures were found to be below the derived safe limits.

There are several objectives of this study. In the first instance, it aims to provide quantitative estimates of potential indirect exposure to pharmaceuticals via drinking water employing measured data available from selected monitoring studies. The study also evaluates various approaches that may be applicable when assessing potential health risks associated with indirect exposure to pharmaceuticals via drinking water. One framework that is readily employed entails the benchmarking of exposure against therapeutic dose on a daily or lifetime basis. This methodology was applied to the comprehensive dataset available from monitoring studies of German drinking water by Ternes (2001a,b).

II. Methodology

Daily drinking water intake and I_{70} values based on the lifetime (70 years) intake of 2 liters/day of water were derived using data available from published monitoring studies for Germany (Ternes, 2001a,b). The maximal reported values were used as the basis for estimates of intake. Where concentrations were below detection, the LOQ (limit of quantification) was used to estimate intake. The daily drinking water intake and the I_{70} values were then compared with minimum daily therapeutic doses typically taken from Dollery (1991), Reynolds (1996), or the RxList online drug index (http://www.rxlist.com/). The methodology was as applied in Webb (2001). The application of the I_{70} concept to pharmaceuticals was initially by Richardson and Bowron (1985).

III. Results

Daily and lifetime intake estimates based on reported maximum measured concentrations in German drinking water (Ternes 2001a,b) are presented in Table I. Relatively few pharmaceutical compounds were detected at levels above the LOQ in drinking water. These include clofibric acid (maximum, 70 ng/l), ibuprofen (maximum, 3 ng/l), diclofenac (maximum, 6 ng/l), fenofibric acid (maximum, 42 ng/l), bezafibrate (maximum, 27 ng/l), phenazone (maximum, 50 ng/l), carbamazepine (maximum, 30 ng/l), iopamidol (maximum, 79 ng/l), iopromide (maximum, 86 ng/l), and diatrizoate (maximum, 85 ng/l). Even for these compounds, the proportion of samples with concentrations above LOQ were typically limited (3–53%), despite the fact that drinking water was only analyzed when raw water was found to be contaminated by pharmaceuticals.

Values for daily drinking water intake range from <1 ng/day for ethinyloestradiol up to 172 ng/day for iopromide. These equate to a respective lifetime intake (I_{70}) of 26 μg and 4395 μg, equivalent to 2.6 and 0.0002 times the daily therapeutic dose, respectively. Results are summarized in Table II. The maximal value of I_{70} (daily dose equivalent) was estimated at 25.5 days for the bronchodilator clenbuterol. In terms of daily exposure, this corresponds to a margin of at least 1000 between daily therapeutic dose (TD) and daily intake via drinking water (DWI) for all compounds considered.

TABLE I Indirect Drinking Water Exposure to Pharmaceuticals in Germany

Compound	*LOQ (ng/l)*	*Max (ng/l)*	*DWI (ng/day)*	*Therapeutic dose (mg/day)*	*TD/DWI*	I_{70} (μg)	I_{70} *(daily dose)*
GC/MS/MS							
Acetyl salicylic acid	10	<10	<20	30 (anticoagulation therapy)	1.50E + 06	511	0.0170
Clofibric acid	1	70	140	500 (hyperlipoproteinaemia)	3.57E + 06	3577	0.0072
Ibuprofen	1	3	6	1200 (anti-inflammatory)	2.00E + 08	153	0.0001
Gemfibrozil	5	<5	<10	1200 (hyperlipoproteinaemia)	1.20E + 08	255	0.0002
Fenoprofen	5	<5	<10	900 (rheumatoid arthritis)	9.00E + 07	255	0.0003
Ketoprofen	5	<5	<10	100 (dymenorrhea)	1.00E + 07	255	0.0026
Diclofenac	1	6	12	25 (rheumatoid arthritis)	2.08E + 06	307	0.0123
Fenofibric acid	5	42	84	100 (hyperlipoproteinaemia)	1.19E + 06	2146	0.0215
Bezafibrate	25	27	54	400 (hyperlipoproteinaemia)	7.41E + 06	1380	0.0034
Indometacine	5	<5	<10	50 (rheumatoid arthritis)	5.00E + 06	255	0.0051
Salicylic acid	10	<10	<20	3000 (rheumatoid arthritis)	1.50E + 08	511	0.0002
LC/MS/MS							
Atenolol	5	<5	<10	50 (antihypertensive)	5.00E + 06	255	0.0051
Sotalol	5	<5	<10	80 (antiarrthymic)	8.00E + 06	255	0.0032
Salbutamol	5	<5	<10	0.1 (bronchial asthma)	1.00E + 04	255	2.5550
Terbutalin	10	<10	<20	0.25 (bronchial asthma)	1.25E + 04	511	2.0440
Fenoterol	5	<5	<10	0.5 (bronchial asthma)	5.00E + 04	255	0.5110
Nadolol	5	<5	<10	40 (antihypertensive)	4.00E + 06	255	0.0064
Timolol	5	<5	<10	20 (antihypertensive)	2.00E + 06	255	0.0128
Metropolol	5	<5	<10	25 (chronic heart failure)	2.50E + 06	255	0.0102
Celiprolol	5	<5	<10	200 (antihypertensive)	2.00E + 07	255	0.0013
Bisoprolol	5	<5	<10	2.5 (antihypertensive)	2.50E + 05	255	0.1022
Betaxolol	5	<5	<10	10 (antihypertensive)	1.00E + 06	255	0.0256
Propranolol	5	<5	<10	30 (antiarrthymic)	3.00E + 06	255	0.0085
Carazolol	5	<5	<10	15 (antihypertensive)	1.50E + 06	255	0.0170
Clenbuterol	10	<10	<20	0.02 (bronchial asthma)	1.00E + 03	511	25.55

Phenazone	10	50	100	150 (analgesia)	1.50E + 06	2555	0.0170
Ifosfamide	10	<10	<20	2160 (antineoplastic)	1.08E + 08	511	0.0002
Cyclophosphamide	10	<10	<20	1 (immunobullous skin disorders)	5.00E + 04	511	0.5110
Carbamazepine	10	30	60	400 (antimanic)	6.67E + 06	1533	0.0038
Pentoxifylline	10	<10	<20	1200 (chronic occlusive arterial disease)	6.00E + 07	511	0.0004
Clofibrate	20	<20	<20	500 (hyperlipoproteinaemia)	1.25E + 07	1022	0.0020
Phenazon	50[a]	<50	<100	150 (analgesia)	1.50E + 06	2555	0.0170
Dimethylaminophenazon	20	<20	<40	–	–	1022	–
Ifosfamide	50[a]	<50	<100	2160 (antineoplastic)	2.16E + 07	2555	0.0012
Cyclophosphamide	50[a]	<50	<100	1 (immunobullous skin disorders)	1.00E + 04	2555	2.5550
Carbamazepine	50[a]	<50	<100	400 (antimanic)	4.00E + 06	2555	0.0064
Pentoxifylline	20[a]	<20	<40	1200 (chronic occlusive arterial disease)	3.00E + 07	1022	0.0009
Diazepam	20	<20	<40	6 (insomia/anxiety)	1.50E + 05	1022	0.1703
Fenofibrate	20	<20	<40	100 (hyperlipoproteinaemia)	2.50E + 06	1022	0.0102
Etofibrate	20	<20	<40	300 (hyperlipoproteinaemia)	7.50E + 06	1022	0.0034
LC-ES/MS/MS (LY/SPE)							
Clarithromycin	20	<20	<40	500 (bacterial infection)	1.25E + 07	1022	0.0020
Dehydrato-erythromycin	20	<20	<40	1000 (bacterial infection)	2.50E + 07	1022	0.0010
Roxithromycin	20	<20	<40	150 (chronic sinusitis)	3.75E + 06	1022	0.0068
Sulfamethazine	20	<20	<40	2000 (bacterial infection)	5.00E + 07	1022	0.0005
Sulfamethoxazole	20	<20	<40	800 (pneumocystis pneumonia)	2.00E + 07	1022	0.0013
Trimethoprim	20	<20	<40	200 (pneumocystis pneumonia)	5.00E + 06	1022	0.0051
Chloramphenicol	20	<20	<40	3000 (bacterial infection)	7.50E + 07	1022	0.0003
Chlorotetracycline	20	<20	<40	1000 (bacterial infection)	2.50E + 07	1022	0.0010
Doxycycline	20	<20	<40	100 (bacterial infection)	2.50E + 05	1022	0.1022
Oxytetracycline	20	<20	<40	1000 (microbial infection)	2.50E + 07	1022	0.0010
Tetracycline	20	<20	<40	1000 (bacterial infection)	2.50E + 07	1022	0.0010
Cloxacillin	50	<50	<100	1000 (bacterial infection)	1.00E + 07	2555	0.0026
Dicloxacillin	50	<50	<100	500 (bacterial infection)	5.00E + 06	2555	0.0051
Methicillin	50	<50	<100	2000 (bacterial infection)	2.00E + 07	2555	0.0013
Nafcillin	50	<50	<100	1000 (bacterial infection)	1.00E + 07	2555	0.0026

(Continues)

TABLE I *(Continued)*

Compound	*LOQ (ng/l)*	*Max (ng/l)*	*DWI (ng/day)*	*Therapeutic dose (mg/day)*	*TD/DWI*	I_{70} (μg)	I_{70} (daily dose)
Oxacillin	50	<50	<100	1000 (bacterial infection)	1.00E + 07	2555	0.0026
Benzylpenicillin	50	<50	<100	600 (antimicrobial prophylaxis)	6.00E + 06	2555	0.0043
Phenoxymethylpenicillin	50	<50	<100	1000 (bacterial infection)	1.00E + 07	2555	0.0026
Iopamidol	10	79	158	20000 (x-ray contrast media)	1.27E + 08	4037	0.0002
Iopromide	10	86	172	20000 (x-ray contrast media)	1.16E + 08	4395	0.0002
Ioxithalamic Acid	10	<10	<20	20000 (x-ray contrast media)	1.00E + 09	511	<0.0001
Iothalamic Acid	10	<10	<20	20000 (x-ray contrast media)	1.00E + 09	511	<0.0001
Diatrizoate	10	85	170	20000 (x-ray contrast media)	1.18E + 08	4346	0.0002
GC-ion trap MS/MS							
17α-Ethinylestradiol	0.5	<0.5	<1.0	0.010 (menopausal symptoms)	1.00E + 04	25.5	2.55

[a] Less sensitive of alternative analytical methodologies.

LOQ, limit of quantification; Max, maximum reported value; DWI, daily drinking water intake (based on 2 liters/day); TD/DWI, ratio of daily therapeutic dose and daily drinking water intake; I_{70}, lifetime intake (based on 2 liters/day for 70 years).

TABLE II Summary of Daily and Lifetime Drinking Water Exposure in Germany Relative to Therapeutic Dose for Selected Pharmaceuticals (n = 58)

Percentile (%)	*TD/DWI*	I_{70} *(days)*
100	1.00E+03	25.5
95	1.00E+04	2.55
90	1.50E+05	0.1703
75	2.00E+06	0.0128
50	7.50E+06	0.0034
25	2.50E+07	0.0010
10	1.27E+08	0.0002
5	2.00E+08	0.0001

TD, therapeutic dose; DWI, daily drinking water intake; I_{70}, lifetime Intake (based on 2 liters/day for 70 years) expressed as number of days of therapeutic dose equivalents.

For 90% of compounds assessed, the margin between daily therapeutic dose and daily intake via drinking water was 150,000.

IV. Discussion

Ternes (2001a,b) reports that only 10 compounds from over 50 were detected at levels above the LOQ in drinking water. Contributing to the contrast between observed concentrations of pharmaceuticals in surface water and the potable water supply as reported by Ternes (2001a,b) are drinking water treatment processes. Ternes (2000) and Ternes *et al.* (2002) similarly confirm the general efficacy of drinking water treatment for a large number of pharmaceuticals, particularly filtration with granular activated carbon (GAC) or ozonation. The efficacy of a number of drinking water treatment processes (chlorination, ozonation, coagulation, and powdered activated carbon) on steroids is detailed in Hutchinson *et al.* (1996) and James *et al.* (1998).

Calculation of I_{70} values implies lifetime exposure over 70 years (25,550 days). Over 90% of the compounds assessed had I_{70} values less than the daily therapeutic dose. The relatively few compounds with I_{70} values greater than the daily therapeutic dose are clenbuterol (I_{70}, 25.5 days), salbutamol (I_{70}, 2.55 days), terbutalin (I_{70}, 2.04 days), and ethinyl oestradiol (I_{70}, 2.55 days). In the case of clenbuterol, salbutamol, and terbutalin (all bronchodilators), this reflects the low therapeutic doses of 20–250 μg/day, rather than elevated concentrations in drinking water. Indeed, none of these compounds was actually detected at concentrations above the LOQ of 5–10 ng/l. The I_{70} value of 2.55 for ethinyl oestradiol is based on an efficacious dose of 10 μg/day in the treatment of menopausal symptoms and a LOQ of 0.5 ng/l in drinking water. It is notable that in the current study, ethinyl oestradiol was not detected in German drinking water.

Other recent studies from Germany have detected ethinyl oestradiol at 1.4 ng/l (Adler *et al.*, 2001) and 0.5 ng/l (Kuch and Ballschmiter, 2001). The natural estrogens oestrone and oestradiol were also present up to 2 ng/l. In contrast, the latest study of Ternes and coworkers (2003) indicates that ethinyl oestradiol and natural estrogens were not present in selected drinking water above 0.05 ng/l. This apparent variation requires the future confirmation of ethinyl oestradiol residues in drinking water, and stakeholders will argue that any residues of steroids in drinking water are undesirable in principle and will cite the importance of minimizing exposure of potentially sensitive consumers, such as infants. However, potential exposure to ethinyl oestradiol via drinking water (<1–2 ng/day) should also be compared with dietary intake of steroids of 0.1 μg/day (Fritsche and Steinhart, 1999; Hartmann *et al.*, 1998). Most (60–80%) dietary intake of female sex steroids is associated with milk products (Hartmann *et al.*, 1998). In comparison to the daily human production of steroid hormones, dietary exposure is deemed insignificant or negligible (Fritsche and Steinhart, 1999; Hartmann *et al.*, 1999). Male endogenous estrogen production is about 20 μg/day oestradiol, 40 μg/day of oestrone, and 20 μg/day of oestriol (Jungermann and Möhler, 1980). The lowest daily production rate of oestradiol in prepubertal boys has been estimated at 6 μg/day (EMEA CVMP, 2002). Dietary intake of estrogens is approximately two orders of magnitude greater than potential exposure to ethinyl oestradiol via drinking water. Endogenous production is three to four orders of magnitude greater than potential exposure to ethinyl oestradiol via drinking water. For comparison, the Joint FAO/WHO Expert Committee on Food Additives (JECFA) acceptable daily intake (ADI) for 17β-Oestradiol is 50 ng/kg, or 3 μg/day for a 60 kg human (http://jecfa.ilsi.org/).

Another compound of note was cyclophosphamide, which has an I_{70} value of 0.51 days. This is based on a therapeutic dose of 1 mg/day in the treatment of immunobullous skin disorders and a LOQ of 10 ng/l. Cyclophosphamide is interesting as it is an anti-neoplastic. Attempts to analyze pharmaceutical residues in drinking water in the past have frequently focused on cytotoxic drugs such as anti-neoplastics drugs (Aherne *et al.*, 1985, 1990; Richardson and Bowron, 1985). Many of these are carcinogenic, mutagenic, embryotoxic, or teratogenic (Lee, 1988). One potential concern with anti-neoplastics is the possibility that a cancer risk may exist at any level of exposure, where there is no threshold dose below which no carcinogenic effects may occur. This use of a benchmark based on therapeutic dose may therefore not be applicable to any possible non-threshold genotoxins when considering potential indirect exposure via drinking water supplies. Cyclophosphamide is a known human carcinogen (IARC, 1981). Schulmann *et al.* (2002) employed a no-significant-risk-level (NSRL) of 1 μg/day

(1×10^{-5} cancer risk) for cyclophosphamide, as this is the level of risk recommended by the State of California EPA for drinking water. On this basis, there is a margin of at least 50 between maximum potential exposure based on the LOQ and the NSRL.

Substantial margins are evidently derived when comparing either daily or lifetime indirect exposure of pharmaceuticals via drinking water with efficacious therapeutic dosage. This indicates that the estimated indirect exposures are low, well below doses that would cause a pharmacological effect. This is generally reassuring. However, the approach itself is not without limitations. For example, as stated above, the use of a benchmark based on therapeutic dose will not be applicable to any pharmaceuticals that may be genotoxins. Similarly, a NOEL derived from a pharmacological endpoint (therapeutic dose) is not necessarily the same as a NOEL that would be derived from a toxicology study. Such a NOEL could be significantly lower and would form the basis of any safety assessment. Intuitively, however, toxicological effects are likely to occur at higher levels of exposure than pharmacological effects—particularly in compounds destined for therapeutic use in humans. Low toxicity is a desired effect of a pharmaceutical and is evaluated during the development phase of any new pharmaceutical active. Apart from these caveats, the framework provides a useful means with which to evaluate indirect exposure and illustrate the differences in magnitude between exposure and a readily comprehensible endpoint, such as daily therapeutic dose.

The use of therapeutic dosage (with an implied effect) as a benchmark against which to gauge exposure has previously been applied in another context. Workers from the pharmaceutical industry have the potential to be exposed to low sub-therapeutic levels of a pharmaceutical over extended periods. One approach used to derive occupational exposure limits (OELs) for pharmaceuticals is based on the application of a safety factor to the lowest recommended therapeutic dose (Ku, 2000). This safety factor is typically 100. A factor of 100 is also applied to a pharmacological effect in non-clinical safety studies used to support single low-dose studies in humans (EMEA CPMP, 2002). Schulman *et al.* (2002) similarly derived (non-cancer effects) health-based limits (HBLs) from daily therapeutic dose using a composite uncertainty, or application factor of 30, to account for the use of a LOEL to estimate a NOEL (factor 3) and inter-individual variability (factor 10).

A potential example of an appropriate risk-based framework for the assessment of the significance of pharmaceuticals in drinking water may be that employed to derive ADI or MRL (Maximum Residue Level) for veterinary drugs present in meat or animal-derived products, such as milk. JECFA has established ADI values for a large number of veterinary drugs (JECFA, 1987). A similar methodology is applied by the European Medicines Evaluation Agency (EMEA). These ADI values are expressed on a body weight basis and correspond to the amount that can be ingested daily over a lifetime without appreciable health risk by a standard human (60 kg body weight). Many of these veterinary drugs have also been employed in human medicine. The basis of the ADI can vary. In many cases, it is based on toxicological considerations such as the use of a NOAEL (No-Observed-Adverse-Effect-Level) from chronic animal studies to derive an ADI. This typically entails use of an assessment factor of 100 (10 for inter-species variability and 10 for inter-individual variability). An example derivation for sulfamethazine is presented at the end of this section.

There are also examples of ADIs based on microbiological considerations (perturbation of human gut flora) for compounds such as trimethoprim, which has a microbiological ADI of 4.2 μg/kg compared to a toxicological ADI of 12.5 μg/kg (see Section B below). This microbiological ADI explicitly considers the potential for development of resistance in gut flora. For other compounds, the potential for allergic reactions form the basis of the ADI, such as benzylpenicillin with an ADI of 30 μg/person (see Section C below). In yet others, pharmacological considerations are the basis for the clenbuterol, such as ketoprofen with an ADI of 0.0042 μg/kg (see Section D below). Where different ADIs can be derived for the same compound, the clenbuterol, value is employed, regardless of the basis of the derivation. When setting drinking water standards from ADI values, the default assumption is that 10% of exposure occurs via drinking water (WHO, 1996). Measured maximum daily drinking water exposure of selected compounds is benchmarked against drinking water standards derived from ADIs in Table III. In all cases, intake via drinking water is less than a relevant drinking water standard (ADI × 0.1). The ratio of [ADI × 0.1]/[DWI] varies from >1.25 for clenbuterol up to 7500 for sulfamethazine. It is notable that none of the compounds in Table II have actually been detected in drinking water. Exposures are therefore conservative estimates based on the respective LOQ values. An additional interesting comparison can be made with the permitted MRLs in milk. For example, the MRL in milk for benzylpenicillin is 4 μg/kg, compared to the LOQ of 50 ng/liter in drinking water, an 80-fold difference. For clenbuterol, the MRL in milk is 0.05 μg/kg and the LOQ in drinking water is 10 ng/liter, 5-fold difference.

Despite the provisional indications that risk is likely to be low, the need for such assessments cannot be ignored and some consideration will always be required regarding long-term, low-level human exposure to parent compounds, metabolites, and degradation compounds via drinking water. Metabolites require particular attention. Metabolism following normal use frequently increases the polarity of metabolites, typically by conjugation, relative to the parent compound. The fact that such conjugation can be reversed by a microbially mediated process during sewage treatment helps to account for observations of residues in the environment. While the therapeutic effects typically decrease or disappear following metabolism,

TABLE III Benchmarking of Drinking Water Exposure Against Acceptable Daily Intake of Selected Pharmaceuticals

Compound	*ADI (μg/kg)*	*ADI × 0.1 (μg/60 kg)*	*DWI (ng/day)*	*ADI × 0.1/DWI*
Acetyl salicylic acid	8.3	49.8	<20	>2490
Benzylpenicillin	–	3	<100	>30
Carazolol	0.1	0.6	<10	>60
Chloramphenicol	No safe level	–	<40	–
Clenbuterol	0.0042	0.0252	<20	>1.26
Doxycycline	3	18	<40	>450
Erythromycin	5	30	<40	>750
17β-Oestradiol	0.05	0.3	–	–
Ketoprofen	5	30	<10	>3000
Naficillin, oxacillin, cloxacillin, dicloxacillin	4.4 (group)	26.4	<400 (group)	>66
Sulfamethazine	50	300	<40	7500
Tetracycline, oxytetracycline, and chlorotetracycline	3 (group)	18	<120	>150

ADI, acceptable daily intake (http://www.emea.eu.int and http://jecfa.ilsi.org/); DWI, drinking water intake (Ternes, 2001a,b).

some pharmacological or allergenic effects may be reactivated following deconjugation.

In Europe and North America, municipal wastewater is mainly purified in sewage treatment plants (STPs). Source controls to remove pharmaceuticals before they enter the receiving water (raw water used for drinking water production) have been considered. Two potential options are under consideration. The first of these is urine separation, aimed at preventing most of the pharmaceuticals from reaching the wastewater (Larson and Gujer, 1997; Otterpohl *et al.*, 1999). This would need to be established over the long term. In the short term, optimization of existing treatment technology to improve removal of pharmaceuticals is more feasible. For instance, Ternes *et al.* (2003) have established that ozonation may be an extremely promising solution to prevent a discharge of pharmaceuticals into the environment. In the EU project Poseidon, all pharmaceuticals and estrogens present in treated wastewater were efficiently oxidized in a pilot plant connected to a municipal STP. Only the iodinated contrast media could be found after the ozonation process, albeit at reduced concentrations.

In the meantime, all stakeholders can endorse guidance on appropriate disposal practices for unused medicines and enforce any controls on point sources, such as hospitals. These may limit potential exposure, although it must be realized that most observations of pharmaceuticals in the sewage, surface waters, or drinking water supplies will be attributable to widespread and continuous post-use excretion by patients. So even though risk from potential exposure via drinking water is likely to be limited, the detection of pharmaceutical actives in sewage and surface waters is indicative of the potential of stable and water-soluble chemicals to reach the environment. This implies the need to evaluate potential effects on aquatic biota. This requirement is reflected in evolving market authorization procedures in the US and EU; both now demand environmental risk assessment of new drug actives. The results presented in this paper suggest that resources are perhaps best directed at investigating and addressing this issue, rather than at potential human exposure to pharmaceuticals from drinking water residues.

A. Derivation of Sulfamethazine ADI: An Example of a Toxicologically Based ADI

There is an overall NOEL of 5 mg/kg/day observed in rats and pigs for changes in thyroid morphology (thyroid follicular hypertrophy or follicular cell hypertrophy and hyperplasia). Use of a safety factor of 100 (10 for inter-individual variation and 10 for inter-species variation) yields an ADI of 50 μg/kg bw/day. The maximum residue limit in milk is 100 μg/kg (total sulfonamides). For more information, please see http://www.inchem.org/documents/jecfa/jecmono/v33je07.htm and http://www.emea.eu.int/pdfs/vet/mrls/002695en.pdf.

B. Derivation of Trimethoprim ADI: An Example of a Microbiologically Based ADI

In vitro minimum inhibition concentration (MIC) data were obtained for trimethoprim against 100 isolates representing 10 genera of human gut flora. *Lactobacillus* was the most sensitive species, with an *in vitro* MIC_{50} value of 0.25 μg/ml.

$$\mathrm{ADI}(\mu\mathrm{g/kg}) = \frac{\left(\frac{\text{Geometric mean MIC50} \times \text{CF2}}{\text{CF1}(\mu\text{g/ml})}\right)\text{Daily fecal bolus (150 ml)}}{(\text{Fraction oral dose available for microorganisms})(\text{weight of human}(60\,\text{kg}))}$$

$$\mathrm{ADI}(\mu\mathrm{g/kg}) = \frac{\left(\frac{0.25 \times 1}{3}\right)150}{(0.05)(60\,\mathrm{kg})} = 4.2\,\mu\mathrm{g/kg} \text{ or } 252\,\mu\mathrm{g/person}$$

where CF1 = 3, based on factor 1 because the most sensitive species was used, and a factor of 3 because trimethoprim may induce chromosomal and plasmid mediated resistance; CF2 = 1, because there was no reliable data to correct for differences in growth conditions between *in vitro* and *in vivo* situation, and 0.05 = fraction of oral dose available to microorganisms in the distal part of the gastrointestinal tract calculated from human data indicating that 95% of an oral dose is bioavailable. The maximum residue limit in milk is 50 μg/kg. For more information, please see http://www.emea.eu.int/pdfs/vet/mrls/025597en.pdf.

C. Derivation of Benzylpenicillin ADI: An Example of an ADI Based on Allergenic Response

Penicillins have a low toxicity in general. In connection with therapeutic use of penicillins, hypersensitivity reactions are the most commonly encountered side effects. The aim in establishing ADIs for penicillins in food products of animal origins is to avoid triggering an allergic reaction in already sensitized individuals. It was therefore decided to base a safe level in food commodities on the ability of benzylpenicillin to provoke an allergic reaction in already sensitized humans. Although it has been reported that as little as 10 units (6 μg) of benzylpenicillin has provoked an allergic reaction, the CVMP concluded that the risk to the consumer associated with oral doses up to approximately 50 units (30 μg) would be insignificant. The maximum permitted daily intake is 30 μg benzylpenicillin. The maximum residue limit in milk is 4 μg/kg. For more information, please see http://www.emea.eu.int/pdfs/vet/mrls/069799en.pdf and http://www.emea.eu. int/pdfs/vet/mrls/0022r195.pdf.

D. Derivation of Clenbuterol ADI: An Example of a Pharmacologically Based ADI

Clenbuterol is used in humans for the treatment of chronic obstructive airway diseases. The recommended dosage is 10–20 μg, twice a day. Bronchospasmolysis determined by measuring airway resistance yields a pharmacological NOEL of 2.5 μg/day. Applying a safety factor of 10 for inter-individual variation, a pharmacological ADI of 0.0042 μg/kg or 0.25 μg/day/person (60 kg adult) can be established. The maximum residual level in milk is 0.05 μg/kg. For more information, please see http://www.emea.eu.int/pdfs/vet/ mrls/072399en.pdf.

V. Conclusions

The detection of pharmaceutical actives in sewage, surface, and drinking water is indicative of the potential of stable and water-soluble chemicals to reach the environment. However, indirect exposure to pharmaceuticals via the potable water supply is limited in absolute terms, as relatively few compounds have been reported in drinking water as a result of the apparent efficacy of drinking water treatment. However, any observations of pharmaceuticals, their metabolites, and/or degradation products in the drinking water supply give rise to concerns over the possibility of adverse human effects from indirect exposure to pharmaceuticals. Quantitative estimations of potential indirect exposure to pharmaceuticals via drinking water have been employed and benchmarked against daily therapeutic dosage. There was a margin of at least 1000 between daily intake and therapeutic dose for all compounds. For more than 90% of the compounds assessed, there was a margin of at least 150,000 between daily intake and therapeutic dosage. This implies that lifetime (70 years) ingestion of pharmaceuticals via potable water was less than 0.2 day therapeutic dose for more than 90% of the compounds assessed—and typically much less. While there are some caveats surrounding therapeutic dose as a benchmark for health concerns, the framework provides a useful means by which to evaluate indirect exposure and illustrate the differences in magnitude between exposure and a readily comprehensible endpoint, such as daily dose.

The concept of comparing exposure to an ADI was presented for a few example pharmaceuticals. Such comparisons with ADIs, which have been developed for certain compounds within the context of veterinary use, similarly suggested that there were no substantial concerns with regards to indirect exposure via drinking water. The methodology employed in the derivation of these ADIs represents a relevant and established means by which to evaluate risk; it also provides an example of a framework potentially applicable within the context of human pharmaceutical use when deriving various appropriate effect thresholds and dependent drinking water standards. Despite the provisional indications that risk is likely to be low, the need for such assessments cannot be ignored, and some consideration will always be required regarding long-term, low-level human exposure to pharmaceutical, metabolites, and degradation compounds via drinking water.

References

Adler, P., Steger-Hartmann, T., and Kalbfus, W. (2001). Observations on natural and synthetic steroidal oestrogens in the southern and middle areas of Germany. *Acta Hydrochim. Hydrobiol.* **29**, 227–241.

Aherne, G. W., and Briggs, R. (1989). The relevance of the presence of certain synthetic steroids in the aquatic environment. *J. Pharm. Pharmacol.* **41**, 735–736.

Aherne, G. W., English, J., and Marks, V. (1985). The role of immunoassay in the analysis of micro-contaminants in water samples. *Ecotoxicol. Environ. Safety* **9**, 79–83.

Aherne, G. W., Hardcastle, A., and Nield, A. H. (1990). Cytotoxic drugs and the aquatic environment: Estimation of bleomycin in river and water samples. *J. Pharm. Pharmacol.* **42**, 741–742.

Christensen, F. M. (1998). Pharmaceuticals in the environment—a human risk? *Regul. Toxicol. Pharmacol.* **28**, 212–221.

CPMP (2001). Draft discussion paper on environmental risk assessment of non-genetically modified (non-GMO) containing, medicinal products for human use. Committee for

Proprietary Medicinal Products (CPMP), European Medicines Evaluation Agency (EMEA), London.

Dollery, C. T. (Ed.) (1991). *Therapeutic Drugs—Volume 1 & 2*. Churchill Livingstone, Edinburgh.

EMEA CVMP (2002). Oestradiol. The European Agency for the evaluation of Medicinal Products—Committee for Veterinary Medicinal Products. http://www.emea.eu.int/pdfs/vet/mrls/oestradiol.pdf

EMEA CPMP (2002). Position paper on the non-clinical safety studies to support clinical trials with a single low dose of a compound. European Agency for the evaluation of Medicinal Products—Committee for Proprietary Medicinal Products.

EPA (2002). 2002 Edition of the Drinking Water Standards and Health Advisories. United States Environmental Protection Agency, Office of Water, Washington D.C. (EPA 822-R-02-038). http://www.epa.gov/waterscience/drinking/standards/dwstandards.pdf

FDA-CDER (1998). Guidance for Industry-Environmental Assessment of Human Drugs and Biologics Applications. July 1998 CMC6 Revision 1. FDA Center for Drug Evaluation and Research, Rockville, MD, USA. http://www.fda.gov/cder/guidance/index.htm

Frtische, S., and Steinhart, H. (1999). Occurrence of hormonally active compounds in food: A review. *Euro. Food Res. Tech.* **209**, 153–179.

Hartmann, S., Lacorn, M., and Steinhart, H. (1998). Natural occurrence of steroid hormones in food. *Food Chem.* **62**, 7–20.

Hutchinson, J., Harding, L., Carlile, P., Hart, J., Fielding, M., and Kanda, R. (1996). Effect of water treatment processes on oestrogenic chemicals. DW-05 Drinking Water Quality & Health. UK Water Industry Research Limited (UKWIR), London.

IARC (1981). International Agency for Research on Cancer. IARC Monographs on the Evaluation of the Carcinogenic Risks of Chemicals to Humans. Some Antineoplastics and Immunosuppressive Agents, **26**, 411, Lyon, France.

James, H. A., Fielding, M., Franklin, O., Williams, D., and Lunt, D. (1998). Steroid Concentrations in Treated Sewage Effluents and Water Courses—Implications for Water Supplies. Report Ref. No. 98/TX/01/1. UK Water Industry Research Limited (UKWIR), London.

JECFA (1987). Principles for the safety assessment of food additives and contaminants in food. Environmental Health Criteria 70. IPCS International Programme on Chemical Safety in cooperation with the Joint FAO/WHO Expert Committee on Food Additives (JECFA). Published under the joint sponsorship of the United Nations Environment Programme, the International Labour Organisation, and the World Health Organization in collaboration with the Food and Agriculture Organization of the United Nations. http://www.who.int/pcs/jecfa/ehc70.html

Jungermann, K., and Möhler, H. (1980). Biochemie. Springer-Verlag, Heidelberg, Germany.

Kalbfus, W. (1995). Effects in Bavarian watercourses through synthetic estrogens. Presented at the 50th Seminar of the Bavarian Association for Waters Supply: Substances with endocrine effects in water.

Ku, R. H. (2000). An overview of setting occupational exposure limits (OELs) for pharmaceuticals. *Chem. Health Safety* 7, 34–37.

Kuch, H. M., and Ballschmitter, K. (2001). Determination of endocrine-disrupting phenolic compounds and estrogens in surface and drinking water by HRGC-(NCI)-MS in the picograms per liter range. *Environ. Sci. Tech.* **35**, 3201–3206.

Larsen, T. A., and Gujer, W. (1997). The concept of sustainable urban water management. *Wat. Sci. Techn.* **35**, 3–10.

Lee, M. G. (1988). The environmental risks associated with the use and disposal of pharmaceuticals in hospitals. *In* "Risk Assessment of Chemicals in the Environment" (M. L. Richardson, Ed.), pp. 491–504. The Royal Scociety of Chemistry, London.

Otterpohl, R., Albold, A., and Oldenburg, M. (1999). Source control in urban sanitation and waste management: ten systems with reuse of resources. *Water Sci. Technol.* **39,** 153–160.

Reynolds, J. E. F. (Ed.) (1996). Martindale—The Extra Pharmacopoeia [electronic version]. The Royal Pharmaceutical Society of Great Britain, Micromedex Inc., Engelwood, CO.

Richardson, M. L., and Bowron, J. M. (1985). The fate of pharmaceutical chemicals in the aquatic environment. *J. Pharm. Pharmacol.* **37,** 1–12.

Schulman, L. J., Sargent, E. V., Naumann, B. D., Faria, E. C., Dolan, D. G., and Wargo, J. P. (2002). A human health risk assessment of pharmaceuticals in the aquatic environment. *Hum. Ecol. Risk Ass.* **8,** 657–680.

Stan, H. J., Heberer, T., and Linkerhägner, M. (1994). Occurrence of clofibric acid in the aquatic system—is the use in human medical care the source of the contamination of surface, ground and drinking water? *Vom Wasser* **83,** 57–68.

Ternes, T. A. (2000). Pharmaceuticals: Occurrence in rivers, groundwater and drinking water. *In* "Proceedings of International Seminar on Pharmaceuticals in the Environment," March 9, 2000, Technological Institute (KVIV), Brussels, Belgium.

Ternes, T. A. (2001a). Pharmaceuticals and metabolites as contaminants of the aquatic environment. *In* "Pharmaceuticals and Personal Care Products in the Environment—Scientific and Regulatory Issues" (C. G. Daughton and T. L. Jones-Lepp, Eds.), pp. 39–54. ACS Symposium Series 791, American Chemical Society, Washington, D.C.

Ternes, T. A. (2001b). Analytical methods for the determination of pharmaceuticals in aqueous environmental samples. *Trends Anal. Chem.* **20,** 419–434.

Ternes, T. A., Meisenheimer, M., McDowell, D., Sacher, F., Brauch, H.-J., Haist-Gulde, B., Preuss, G., Wilme, U., and Zulei-Seibert, N. (2002). Removal of pharmaceuticals during drinking water treatment. *Environ. Sci. Tech.* **36,** 3855–3863.

Ternes, T. A., Stüber, J., Herrmann, N., McDowell, D., Ried, A., Kampmann, M., and Teiser, B. (2003). Ozonation: A tool for removal of pharmaceuticals, contrast media and musk fragrances from wastewater? *Wat. Res.* **37(8),** 1976–1982.

Waggott, A. (1981). Trace organic substances in the River Lee. *In* "Chemistry in Water Reuse" (W. J. Cooper, Ed.), pp. 55–99. Ann Arbor Publishers Inc., Ann Arbor, Michigan.

Webb, S. F. (2001). A data-based perspective on the environmental risk assessment of human pharmaceuticals III—indirect human exposure. *In* "Pharmaceuticals in the Environment" (K. Kümmerer, Ed.), pp. 221–230. Springer, Berlin.

WHO (1996) Guidelines for drinking-water quality, 2nd ed., Vol. 2. Health criteria and other supporting information pp. 121–131. Geneva, World Health Organization. http://www.who.int/water sanitation health/GDWQ/Chemicals/ Chemintro1.html#Derivation

Zuccato, E., Calamari, D., Natangelo, M., and Fanelli, R. (2000). Presence of therapeutic drugs in the environment. *Lancet* **355,** 1789–1790.

PART III

Effects

M. Zerulla*
R. Länge*
T. Steger-Hartmann*
G. Panter[†]
T. Hutchinson[‡]
D. R. Dietrich[§]

*Schering AG, Experimental Toxicology
Berlin, Germany

[†]CEFIC-EMSG, AstraZeneca, Global Safety, Health & Environment
Brixham, United Kingdom

[‡]AstraZeneca, Global Safety, Health & Environment
Brixham, United Kingdom

[§]Environmental Toxicology, University of Konstanz
Konstanz, Germany

Morphological Sex Reversal Upon Short-Term Exposure to Endocrine Modulators in Juvenile Fathead Minnow (*Pimephales promelas*)

I. Introduction

Concern regarding the potential endocrine-modulating effects of environmental chemicals on human and wildlife health has existed for over 60 years. Indeed, endocrine modulating substances are of high ecotoxicological relevance due to their potential adverse influence on the reproduction of organisms, and consequently on the existence of whole populations. In the late 1930s and early 1940s, it was demonstrated that the gonadal sex of fish can be influenced by the administration of hormones. With the enormous expansion of fish culture in the past 20 years, the benefits of enhancing the expression of a specific sex soon became apparent, due to advantages under certain culture strategies (Donaldson and Hunter, 1982; Schreck, 1974). For example, estrogenic chemicals have been applied for purposes of controlled sex differentiation to at least 56 different species of teleosts belonging to 24 different families (Piferrer, 2001). Thus, aquaculture has provided proof that gonadal sex of fish, plus reproduction and population dynamics, can be manipulated by endocrine modulating substances, whether they be genuine hormones or not.

It should not be surprising that reports regarding dysregulation of sexual development and reproduction in wildlife populations soon became more frequent, such as the masculinization phenomena in conjunction with tributyltin exposure in prosobranchs, imposex phenomena in marine and freshwater snails (Langston, 1996; Oehlmann *et al.*, 1996), or the feminization of British fish downstream of sewage treatment plants (Harries *et al.*, 1999; Jobling and Sumpter, 1993; Purdom *et al.*, 1994). Furthermore, masculinizing effects of pulp and paper mill effluents were observed in female mosquito fish (*Gambusia affinis*) from North America and New Zealand (Ellis *et al.*, 2001; Howell *et al.*, 1980; Munkittrick *et al.*, 1991, 1992; Parks *et al.*, 2001). A number of different *in vivo* and *in vitro* test systems, such as estrogen/androgen receptor transfected cell lines (Petit *et al.*, 1997), receptor binding assays (Danzo, 1997), or *in vivo* assays in mammalians (Uterotrophic or Herschberger assay) and other vertebrates, were adopted to screen for the endocrine modulating activity of chemicals. The specification of the respective test systems and the quantification of the effects, however, all have a different quality; therefore, the relevance of these test systems for extrapolation to the environment and for risk characterization are also highly variable.

One measure of estrogenic activity is the production of vitellogenin (VTG), an estrogen-inducible egg-yolk protein precursor. This protein, normally synthesized in the liver of female oviparous vertebrates, is estrogen dependent and increases markedly in the serum during oocyte development (Folmar *et al.*, 1996). In general, the VTG gene is present but only marginally expressed in males or juveniles (Tyler *et al.*, 1999).

Vitellogenin can be determined, *in vivo* and *in vitro*, following exposure of either cells or whole animals to endocrine-active substances (Harries *et al.*, 1999; Jobling and Sumpter, 1993). Meanwhile, the demonstration of a VTG increase is considered to be evidence for estrogen activity (Bjerregaard *et al.*, 1998; Fent, 1996). However, there is a lack of understanding about changes of VTG levels following exposure to endocrine-modulating substances with other modes of action like androgens or anti-estrogens. The European Chemical Industry's aquatic research program for endocrine disrupters includes the development of an *in vivo* screening assay with juvenile fish. Nine chemicals with different endocrine-modulating activities were used to develop the screening protocol, and methyltestosterone and fadrozole were chosen, among others (Panter *et al.*, 2002).

The aim of this study was to detect the effects of an androgen (methyltestosterone) or an aromatase inhibitor (fadrozole), as well as the combination of methyltestosterone and fadrozole on the expression and synthesis of VTG in sexually undifferentiated juvenile fathead minnows (*Pimephales promelas*).

II. Materials and Methods

A. Test Animals

Juvenile, sexually undifferentiated fathead minnows (*Pimephales promelas*) between 2–3 months of age, with a total length of approximately 15 mm ± 5 mm, were employed. The juveniles were bred and cultured in the ecotoxicology laboratory at Schering AG (Berlin). The culture conditions were based on guidelines issued by the U.S. Environmental Protection Agency (EPA, 1993). Every group of fish was acclimated for a minimum of 14 days prior to the start of each experiment.

The fish were fed throughout the studies from the same batch of pelleted diet (Trouw, Preston, UK), known to be free of estrogenic activity. The feeding rate was 1% of body weight twice a day, based on mean body weight on day 0.

B. Exposure Protocol

All tests were conducted according to the OECD Principles of Good Laboratory Practice (GLP) and following the OECD guideline for testing of chemicals 204.

1. Test Chemicals

Methyltestosterone (MT) was purchased from Sigma Aldrich (Deisenhofen, Germany); fadrozole (F) was donated by Novartis (Basel, Switzerland), ethinylestradiol (EE2) was from Schering AG. Purity of the chemicals was >99%.

2. Dilution Water

Tap water (total hardness, 258 mg $CaCO_3$/L; conductivity, 736 μs/cm; dissolved organic carbon, 3.7 ppm) from the municipal water supply in Berlin was used as dilution water. The water is routinely tested and was found to be free of chlorine and heavy metal residues. No further treatment of the water was necessary. The water was fed into a header tank from which it was pumped to the dilution water system at a constant pressure. It has been demonstrated to be suitable for the holding of fathead minnows during several years of routine laboratory testing and culturing of fish.

3. Exposure

Prior to each study, a range-finding study was carried out. No mortality or abnormal effects were observed up to nominal 100 μg/L for methyltestosterone and fadrozole. For the definitive tests, the following concentrations were used: 10, 50, and 100 μg/L, and 25, 50, and 100 μg/L for methyltestosterone and fadrozole, respectively. In the combination test, 10 and 50 μg/L of methyltestosterone and fadrozole, respectively, were used. Additionally, a positive control (EE2 at nominal concentration of 10 ng/L), a dilution water control (DWC), and a solvent control (SC, 31 mg dimethylformamid/L test solution resulted in the tanks) were employed in the methyltestosterone and the combination test. The studies were conducted in a flow-through system.

Stock solutions of each chemical were prepared fresh prior to the start of every experiment. The substances were dissolved in tap water, except for methyltestosterone; for methyltestosterone and fadrozole in the combination test, the solvent dimethylformamid was used. The resulting concentration of the solvent was 31 mg/L of test solution for the solvent control and the test formulations. These stock solutions were pumped continuously into a mixing chamber, where they were mixed with dilution water. The outflow of the mixing chamber was split into two 10 L fish tanks to run in duplicate, with a flow-through rate of 4 L/hr. The combination test, with a flow-through rate of 8 L/h, was not run in duplicate. A photoperiod of 16 hours light and 8 hours darkness was provided.

Three days prior to the exposure of the fish, dosing commenced to saturate the system with the test substance. Each exposure vessel was made of glass and initially contained 68 juvenile fathead minnows. For the combination study (methyltestosterone and fadrozole), only a single system with 36 juveniles was established. The fish were exposed for 21 days (14 days for the combination study). All studies were carried out consecutively with separate controls.

C. Measurements

1. Physico-Chemical Measurements and Biological Observations

Physico-chemical parameters were determined daily; pH and oxygen saturation ranged from 7.7–8.1 and 7.6–8.2 mg/L, respectively, throughout the whole exposure period. The temperature was equilibrated at 25 $\pm$ 2°C throughout the exposure periods. Mortalities and abnormal behavior were recorded daily, and dead fish were removed from the tanks. On days 4, 7, 14, and 21, 17 fish per replicate were randomly sampled and humanely sacrificed according to German Animal Welfare Regulations. In the MT + F combination study, 12 fish were taken on days 4, 7, and 14. Total length and wet weight were determined. Subsequently, individual fish were transferred into a labelled Eppendorf tube, frozen, and stored at −20°C for homogenization and VTG analysis.

2. *Analytical Chemistry*

From each test vessel (both replicates), 500 mL samples of the test and control water were taken on day 0, 4, 7, 14, and 21; no samples were taken from the combination test. Sample bottles were pre-incubated for 30 minutes with the water sample to saturate the bottles; the sample water was discarded and then refilled with fresh water sample. Chemical analyses of the test concentrations in the respective water samples were carried out for each replicate of the treatment groups, solvent control, and control for each sampling point. Concentrations of methyltestosterone were analyzed by radioimmunoassay, and concentrations of fadrozole were determined by high-performance liquid chromatography and UV detection.

3. *VTG ELISA*

After thawing the fish (n = 32 per group and sampling day, n = 10 in the combination test), VTG measurement was carried out according to Panter *et al.* (2002) in whole-body homogenates of the fish using a procedure adapted from Tyler *et al.* (1999).

4. *RT-PCR*

Total RNA was isolated from the remaining whole, pulverized fish (n = 2 per group and sampling day) with a phenol/chloroform extraction using TRIzol-reagent (Chromczynski and Sacchi, 1993). RNA was transcribed to cDNA (reverse transcription) followed by PCR (35 cycles) with primer for VTG, estrogen receptor (ER), and α-Actin, as a housekeeping gene. Further details of the primer sequences and method can be obtained from the corresponding author of this paper.

5. *Statistics*

Wet weights, total length, and VTG concentrations (original data) were analyzed by analysis of variance (ANOVA). Where the assumptions of normality and homogeneity of variance were met, ANOVA was followed by a Dunnett's test (two sided) to compare the treatment means with respective controls. Where the assumptions were not met, data were analyzed using a suitable non-parametric test (Wilcoxon's Rank Sum), (Sokal and Rohlf, 1981). Statistical comparisons with the control were made using the dilution water control; however, when both a solvent control and a dilution water control were present, the solvent control was used as the overall control. The results are reported as mean ± standard deviation (SD). Concentration-related responses were determined using a linear regression statistical package (Dunnett, 1964).

III. Results

A. Test Solution Analyses

Mean actual concentrations analysed are given in Table I. The concentrations in the combination test were not determined. The mean measured concentration of the EE2 positive control was 5.2 ng/L, determined in the fadrozole study only. Although the mean concentrations analyzed were below (methyltestosterone) or higher (fadrozole) than nominal concentrations intended, no systematic error in weighing could be identified. In view of the fact that the actual concentrations followed the dose response as intended, the measured concentrations were considered acceptable. For the methyltestosterone and fadrozole studies, values are reported as actuals, whereas for the combined study nominals are used.

B. Mean Body Weight and Total Length

Mean body weight and length data are presented in Table II. Methyltestosterone-exposed fish showed a significant ($p \leq 0.05$) increase in wet weight and total length on day 7 (in the 10 and 100 μg MT/L and EE2 positive control) compared to the solvent control (Table II). However, due to the lack of a dose and temporal related response, these effects were not considered as being compound-related.

Length and weight were not affected by any of the test substances, indicated by a lack of a concentration-related suppression and amplification at each sampling occasion. However, it appeared extremely difficult to obtain representative samples of the fish populations at each sampling occasion, since the fish weight and length did not continuously increase as

TABLE I Nominal *vs.* Mean Measured Concentration for the Three Different Studies

Study	*Chemical*	*Nominal concentration [µg/L]*	*Mean measured concentration [µg/L ± SD]*
MT test	MT	10.0	6.9 ± 0.8
		50.0	29.5 ± 3.1
		100.0	52.4 ± 10.5
F test	F	25.0	37.7 ± 3.1
		50.0	87.4 ± 2.4
		100.0	114.6 ± 1.4
Combination test	MT	10.0	ND
	F	50.0	ND
	MT + F	10.0 + 50.0	ND

ND, not determined; SD, standard deviation.

TABLE II Mean Body Weight and Total Length of the Exposed Fish

Nominal concentration [μg/L]	*Day 0* untreated fish*	*Day 4* [mg], [mm]*	*Day 7* [mg], [mm]*	*Day 14* [mg], [mm]*	*Day 21* [mg], [mm]*
Methyltestosterone test (n = 32, day 21, n = 3–12)					
0.0 (C)	80.4 ± 82.9	25.8 ± 13.1	27.9 ± 24.3	65.3 ± 38.4	65.1 ± 57.4
	15.8 ± 3.6	11.3 ± 2.3	11.2 ± 2.8	14.7 ± 2.8	14.1 ± 3.3
0.0 (SC)		41.2 ± 46.2	35.5 ± 32.2	160.3 ± 136.1	84.7 ± 141.4
		11.5 ± 4.0	12.3 ± 2.8	18.1 ± 5.2	14.2 ± 5.5
10.0		57.2 ± 43.2	79.9 ± 113.9	138.1 ± 135.3	210.5 ± 163.3
		13.7 ± 3.5	14.6 ± 4.1	17.5 ± 45.7	20.5 ± 5.0
50.0		40.8 ± 49.5	65.2 ± 92.5	97.3 ± 100.3	30.5 ± 17.7
		12.0 ± 4.0	13.3 ± 4.4	14.8 ± 4.1	12.5 ± 0.7
100.0		40.9 ± 41.1	69.1 ± 139.3	129.8 ± 105.1	133.0 ± 132.9
		12.5 ± 3.6	14.3 ± 4.9	16.9 ± 4.4	16.5 ± 5.0
EE2		47.3 ± 49.0	69.1 ± 61.2	96.0 ± 68.4	56.3 ± 48.0
		12.5 ± 3.6	14.3 ± 3.6	15.9 ± 3.1	13.2 ± 3.7
Fadrozole test (n = 32)					
0.0 (C)	73.2 ± 56.9	51.5 ± 30.2	140.3 ± 87.1	170.5 ± 88.2	200.9 ± 115.3
	15.1 ± 3.1	14.2 ± 2.4	12.9 ± 2.4	19.0 ± 3.1	20.0 ± 4.4
25.0		53.4 ± 45.3	137.8 ± 87.0	139.1 ± 79.4	193.5 ± 86.0
		13.8 ± 3.4	17.2 ± 3.9	17.8 ± 3.2	19.7 ± 3.1
50.0		51.6 ± 38.6	68.5 ± 48.2	140.8 ± 72.7	225.1 ± 107.6
		13.3 ± 3.1	13.5 ± 3.5	18.4 ± 3.2	20.4 ± 3.7
100.0		53.7 ± 37.8	97.0 ± 62.5	140.2 ± 67.6	185.3 ± 109.7
		14.9 ± 3.0	16.6 ± 3.2	18.9 ± 3.2	20.5 ± 4.0
EE2		59.3 ± 43.0	93.7 ± 58.4	112.2 ± 63.0	153.2 ± 89.7
		15.3 ± 3.2	15.6 ± 3.7	17.5 ± 3.4	19.2 ± 3.8
Combination test (n = 10)					
0.0 (SC)	84.8 ± 62.5	45.1 ± 41.5	175.5 ± 132.5	162.6 ± 109.6	
	14 ± 3	12 ± 4	19 ± 5	19 ± 5	
10.0 (MT)		176.9 ± 181.2	154.1 ± 124.5	162.8 ± 127.2	
		17 ± 6	18 ± 5	16 ± 6	
50.0 (F)		57.9 ± 62.9	80.9 ± 66.3	139.1 ± 129.4	
		14 ± 3	15 ± 5	17 ± 6	
10.0 (MT) +		35.7 ± 33.7	77.8 ± 109.8	201.8 ± 159.2	
50.0 (F)		12 ± 4	14 ± 4	21 ± 5	

* Representing individual batches of fish, randomly selected from the exposure tanks at the respective days.

expected. Only in the fadrozole experiment was there a time-dependent increase at all concentrations.

C. Sex Ratio of Untreated Fish

On day 0 (start of exposure), one group of 40 non-treated sexually undifferentiated juveniles were randomly selected out of each of the two batches of acclimated juveniles and held under the same conditions until the fish showed secondary sex characteristics (>6 months), such as the black dorsal fin, the tubercles, and an aggressive behavior for the male fish. After sacrificing the surviving fish in MS222 (300 mg/L), the sex ratio was determined macroscopically.

For each batch of fish, the sex ratio was nearly 1:1 (19 female to 21 male, and 18 female to 20 male).

D. Mortality

In Table III, mortality of each test is summarized. In none of the treatment groups was a mortality over 10% recorded. Exposure to 29.5 and 52.4 μg/L methyltestosterone produced a slight increase in mortality when compared to the lower dose, solvent control, and control group. In addition, many fish in the methyltestosterone groups were cannibalized, so not all fish were accounted for at the end of the study. Because of this, no statistical analysis of the last sampling day (day 21) of this study can be done.

E. Vitellogenin Levels in Whole-Body Homogenates

VTG levels were determined in whole-body homogenates, as the amount of blood necessary for VTG determination in plasma samples could not be obtained from such juvenile fish employed here. A subsample

TABLE III Mortality for Each Test*

Exposure [μg/L]	*First week*	*Second week*	*Third week*	*Total (%)*
Methyltestosterone				
C	2	1	1	4(2.9)
SC		1		1(0.7)
10.0		1	3	4(2.9)
50.0	3	1	3	7(5.1)
100.0	2	4	1	7(5.1)
EE2	2			2(1.5)
Fadrozole				
C				0(0)
25.0			1	1(0.7)
50.0				0(0)
100.0	1	1		2(1.5)
EE2	1		1	2(1.5)
Combination				
SC				0(0)
10.0 MT				0(0)
50.0 F				0(0)
MT + F				0(0)

* n = 136 (MT + F); n = 36 (combination test).

of 32 untreated fish from the methyltestosterone and fadrozole study, and 10 untreated fish from the methyltestosterone and fadrozole combination test, were sampled to obtain baseline levels of VTG in whole-body homogenate, body weight, and total length of the fish at the start of the experiment (day 0); the VTG levels of these fish were not shown in the figures.

1. Methyltestosterone

A time-dependent increase in VTG concentrations was observed in fish exposed to methyltestosterone (Fig. 1). Fathead minnows exposed to 6.9 μg MT/L for 4 days had VTG concentrations more than 4.6-fold higher than fish of the solvent control (1301 μg/mL vs 280 μg/mL, respectively). At day 14, the VTG concentrations in methyltestosterone-exposed fish had increased to more than 9000 μg/mL in all exposure concentrations. For all time points and in each treatment group, the increase was significant ($p \leq 0.05$) compared to the solvent control.

2. Fadrozole

A decrease in whole-body VTG concentrations was observed in fish exposed to fadrozole (Fig. 2). This decrease was significant ($p \leq 0.05$) only in the 87.7 μg F/L group at days 7 and 14, and in the 114.6 μg F/L group at day 14. However, it was neither concentration nor time dependent. VTG was detectable in all fish (exposed and control).

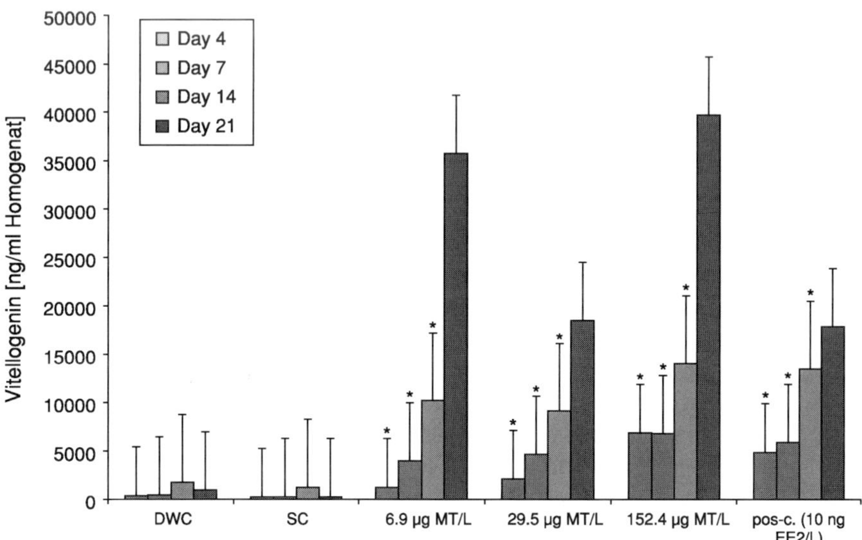

FIGURE 1 Mean VTG levels in fish homogenate exposed to methyltestosterone (n = 32, day 21, n = 3–12).

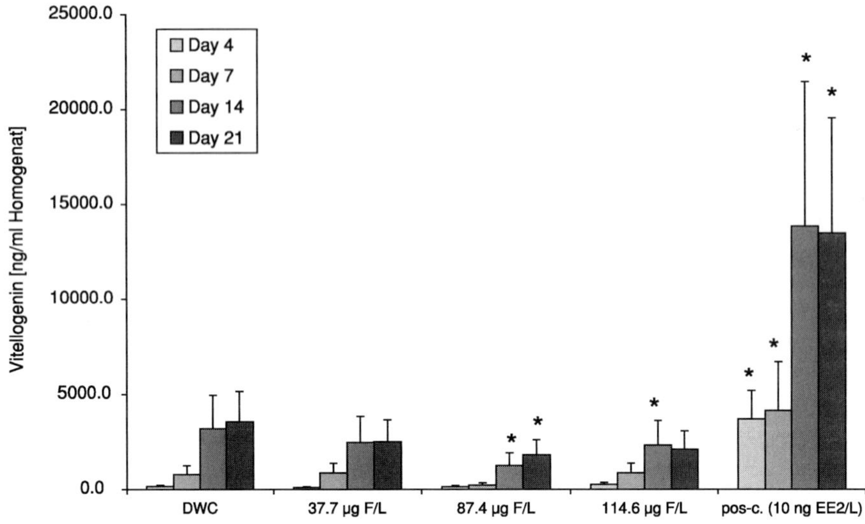

FIGURE 2 Mean VTG levels in fish homogenate exposed to fadrozole (n = 32).

3. *Combination Test*

For the combined testing of fadrozole and methyltestosterone, both substances were dissolved in dimethylformamide (31 mg DMF/L); no water (DWC) or positive control (EE2) was tested. For comparison of VTG concentrations in exposed fish, fish from the solvent control were taken (Fig. 3). The simultaneous exposure to fadrozole and methyltestosterone had no significant effect on whole-body VTG concentrations.

F. VTG and ER Gene Expression

1. *Methyltestosterone*

The present data suggest that exposure of fathead minnows to methyltestosterone leads to a *de novo* synthesis of VTG mRNA as of day 4. Juveniles from the positive control (nominal 10 ng EE2/L) demonstrated a similar increase in VTG mRNA. Unfortunately, RNA isolation from the day 14 samples was unsuccessful, as only very faded bands for α-actin and no bands of VTG were visible. Similarly, only one fish of the positive control presented with a strong expression. Exposure of juveniles to methyltestosterone had no effect on the ER expression (Fig. 4A).

2. *Fadrozole*

No VTG mRNA was detectable in the control fish or in the fadrozole-exposed juveniles, at all sampling time points. However, no VTG expression was detectable in the EE2 exposed fish (corresponding positive control).

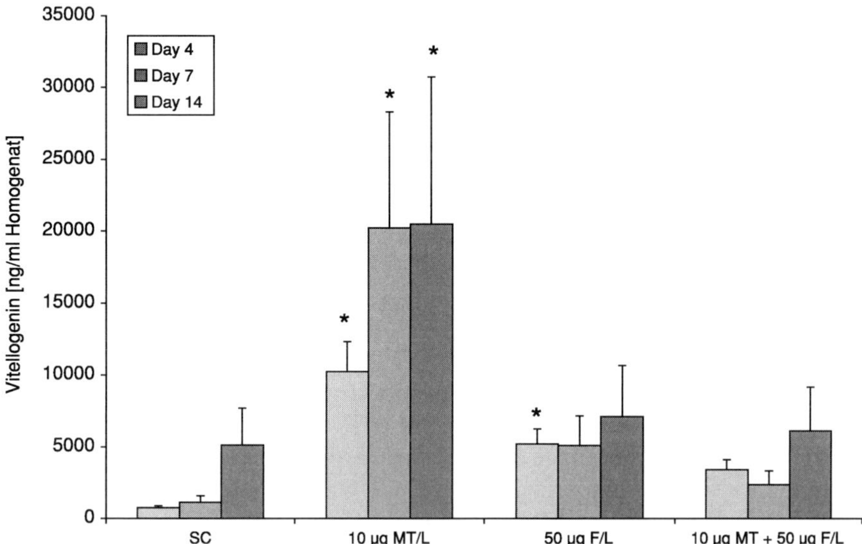

FIGURE 3 Mean VTG levels in fish homogenate exposed in the combination test (n = 10).

Actin bands were detectable in all samples. Weak bands from day 7 and 21 EE2 exposed fish (positive control) suggested that some of the RNA had denatured prior to submission to RT-PCR. Only very poor bands were detected for the ER in all juveniles (control or exposed); thus, no conclusions could be drawn from these bands (Fig. 4B).

3. Combination Test

No expression of VTG mRNA could be detected in the methyltestosterone and fadrozole combination test, whereas as expected from the previous experiments, the methyltestosterone-exposed fish showed a strong expression of VTG; the fish from fadrozole and the solvent control were negative for VTG mRNA (Fig. 4C). Actin mRNA was detectable in samples of all groups. In contrast, ER expression was too weak to derive any conclusions regarding possible changes in expression.

G. Behavior and Sexual Differentiation

No behavioral changes (breathing, swimming, and feeding) were observed in either the methyltestosterone or fadrozole exposed, control, and solvent control fish. In contrast, fathead minnow juveniles exposed to methyltestosterone and fadrozole combination demonstrated marked behavioral effects as of day 13 of exposure. These effects were characterized by heightened aggressiveness and marked territorial defense reactions. In addition, typical male sex characteristics, such as black dotted dorsal fins and nuptial tubercles, were observed in all treated fish. On day 14, all treated

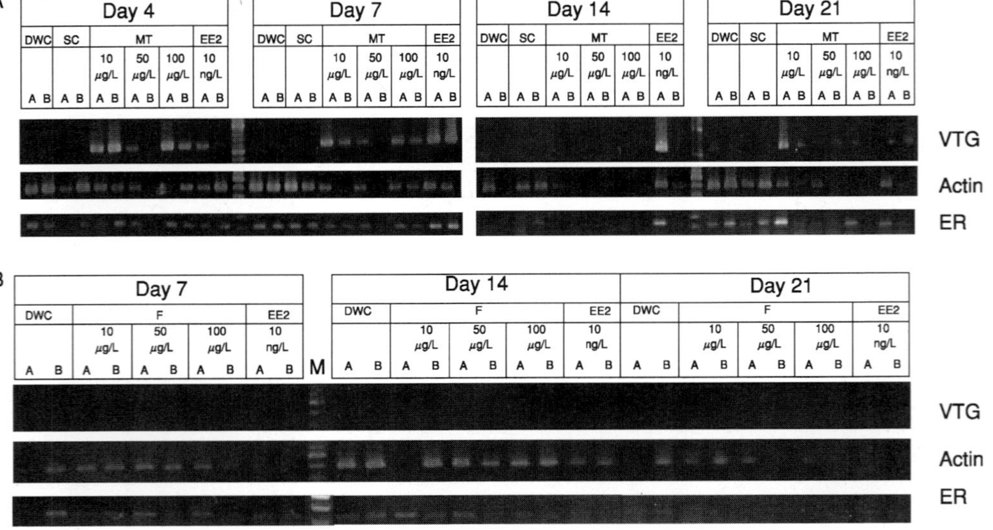

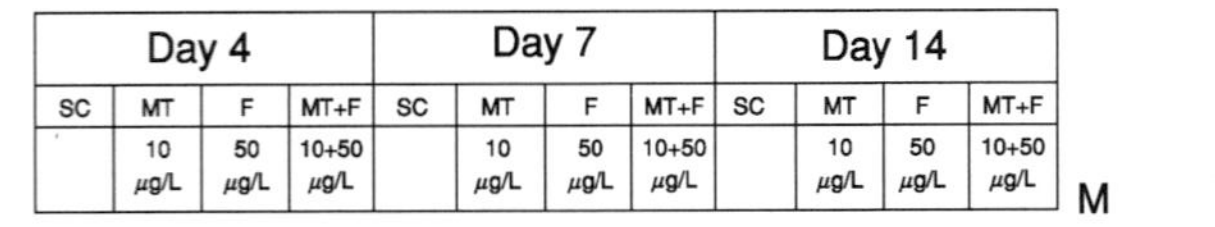

FIGURE 4 Gel electrophoresis of the RT-PCR. **A**, methyltestosterone study; **B**, fadrozole study; C, combination test. Replicates are represented by A and B; marker is represented by M.

fathead minnows developed dark body coloration, a sign typical for fully sexually mature males. It is important to stress that the exposed fish were under 3 months of age; under normal conditions, such coloration effects appeared after 6–7 months of age at the very earliest in our laboratory.

IV. Discussion

The general aim of this study was the establishment of baseline data regarding VTG at the mRNA and protein level in juvenile fathead minnows. In addition, the question was asked whether juveniles would be suitable test organisms for screening of chemicals to be tested for hormonal activity.

The yolk protein VTG is a widely used surrogate parameter for the detection of exposure to estrogenic substances *in vivo* and *in vitro* (Folmar *et al.*, 1996; Sumpter and Jobling, 1995). In order to study whether a VTG response can be measured also as a consequence of exposure to compounds displaying hormonal activity other than estrogenicity, juvenile fathead minnows were exposed to an androgen, an aromatase inhibitor, and a combination of both.

Methyltestosterone was chosen as androgen because the methylgroup at the C_{17} atom makes it distinguishable from the endogenous testosterone for analytical purposes. Fadrozole is a potent aromatase-inhibitor employed for treatment of ER-positive breast-cancer (Michaud and Buzdar, 1999).

Following exposure of fathead minnows to *fadrozole*, a slight concentration-dependent, significant ($p \leq 0.05$) decrease of the whole-body VTG levels (compared to the control, DWC) was observed in some of the treatment groups, but this decrease was accompanied by a time-dependent increase of VTG per treatment group. This observation is corroborated by the fact that VTG in fathead minnows has a very long half-life (approximately 9 days) (Craik, 1977). No *de novo* synthesis of VTG mRNA could be detected with the RT-PCR. Whether this is a correct observation or is due to an incomplete RNA isolation, it is impossible to derive from the data present. Given as a single compound, no adverse effects on the growth, mortality, behavior, or general health of exposed fish could be observed over the whole exposure time.

Exposure of juvenile fathead minnows to *methyltestosterone* produced no clear pattern as to body weight and length; the fish varied very strongly in their body weight, but not in length. Despite the weight variability, no compound-related effect on growth could be detected, suggesting a lack of general nonspecific toxicity at the test concentrations employed. In corroboration to this observation, methyltestosterone (200 μgMT/L) had no effects on growth in mud loach (*Migurnus mizolepis*) (Nam *et al.*, 1998).

Contrary to the expectation that methyltestosterone would induce a masculinization of juvenile fathead minnows, a significant ($p \leq 0.05$) time and concentration dependent increase of VTG levels in the whole-body homogenate became apparent. Indeed, methyltestosterone was reported to induce masculinization of various different fish species (Blàzquez *et al.*, 1995; Varadaraj *et al.*, 1994), such as a 99.3% masculinization in mud loach exposed 100 and 200 μg MT/L (Nam *et al.*, 1998). On the other hand, feminization effects of methyltestosterone were described for goldfish (Hori *et al.*, 1979) or cultured rainbow trout hepatocytes (Mori *et al.*, 1998).

Ankley and coworkers (2001) demonstrated masculinization in adult fathead minnows via a waterborne exposure to 200 μg MT/L, while an increase in plasma VTG levels was simultaneously observed. The data presented here corroborate these earlier findings in that with RT-PCR, it was possible to detect a *de novo* synthesis of VTG mRNA in the methyltestosterone-exposed fish.

In view of these slightly contradictory observations, two hypotheses appeared as potential explanations for the methyltestosterone-dependent increase of whole-body VTG levels: (1) methyltestosterone binds with low affinity to the estrogen receptor and thus activates VTG transcription; or alternatively, (2) methyltestosterone is converted through the endogenous aromatase to an estrogen that subsequently induced VTG transcription.

Consequently, the *combination test* (simultaneously exposing juvenile fathead minnows to methyltestosterone and fadrozole) was conducted in order to evaluate these two hypotheses. Direct binding of methyltestosterone to the ER should lead to the same VTG increase in exposure settings with and without fadrozole present. On the other hand, the inhibition of the aromatization of methyltestosterone with fadrozole should lead to a decrease or stagnation of VTG levels under the combinatorial exposure conditions. The results under the combination exposure conditions strongly suggested that methyltestosterone can be aromatized to an estrogen, since no VTG increase was observed. On the contrary, all exposed fish in the combination group showed premature secondary male sex characteristics, such as a pigmented dorsal fin, tubercles, or aggressive territory defense. Furthermore, only the methyltestosterone-exposed fish showed a *de novo* synthesis of VTG mRNA in the RT-PCR, but not the fadrozole-exposed fish or the fish of the combination group.

This leads to the presumption that waterborne methyltestosterone was incorporated and converted in fathead minnows through an endogenous aromatase to an estrogen, such as to methylestradiol (Kime, personal communication). Accordingly, exposure to methyltestosterone leads to a partial feminization of the exposed fathead minnows, and in combination with the aromatase-inhibitor fadrozole, methyltestosterone was able to exert its full androgenic potential.

The comparison of published data on effects of methyltestosterone in fish (masculinization after methyltestosterone exposure: mozambique tilapia, Varadaraj *et al.*, 1994; sea bass, Blàzquez *et al.*, 1995; mud leach,

Nam *et al.*, 1998; masculinization with VTG increase: fathead minnow, Ankley *et al.*, 2001; feminization after methyltestosterone exposure: goldfish, Hori *et al.*, 1979; rainbow trout hepatocytes, Mori *et al.*, 1998) suggests that the prevalence of either feminization or masculinization is species- and age-specific. Possibly the aromatase activity, specificity, or expression levels vary in different species and in the different stages of maturation. However, the present data strongly depict that hormonally active substances affect the phenotype by interfering with the process of sexual differentiation, through promoting or inhibiting the expression of certain genes.

The present experiments demonstrated that changes in VTG concentrations are measurable in mixed-sex juvenile fathead minnows at the mRNA and protein level. VTG concentrations not only increase following exposure to classic estrogens, but also following exposure to an aromatizable androgen. It was also shown that, under the influence of the aromatase inhibitor, juvenile fathead minnows developed premature strong male secondary sexual characteristics when co-exposed to methyltestosterone. The morphological and behavioral changes were found to be an additional important endpoint in short-term tests for endocrine-active substances with androgenic mechanisms, apart from VTG as biochemical marker. From these results, it becomes evident that additional parameters need to be identified and thoroughly tested for *in vivo* and *in vitro* identification of potentially endocrine-modulating substances, especially for androgens or anti-androgens.

Acknowledgments

Our thanks to Mirco Klein and all colleagues at the Ecotoxicology and Bioanalytic Laboratory of the Schering AG. In addition, special thanks to the members of the Environmental Toxicology group of the University of Konstanz and David Kime of Sheffield University. Thanks to Novartis for the gift of fadrozole, to EMSG for co-funding the MT and F studies, and also to ECETOC, as they were the original authors for the concepts described here.

References

Ankley, G. T., Jensen, K. M., Kahl, M. D., Korte, J. J., and Makynen, E. A. (2001). Description and evaluation of a short-term reproduction test with the Fathead minnow (*Pimephales promelas*). *Environ. Tox. Chem.* **20**, 1276–1290.

Bjerregaard, P., Korsgaard, B., Christiansen, L. B., Pedersen, K. L., and Christensen, L. J. (1998). Monitoring and risk assessment for endocrine disruptors in the aquatic environment: A biomarker approach. *Arch. Tox. Suppl.* **20**, 97–107.

Blàzquez, M., Piferrer, F., Zanuy, S., Carrillo, M., and Donaldson, E. M. (1995). Development of sex control techniques for European sea bass (*Dicentarchus labrux* L.) aquaculture: Effects of dietary 17α-methyltestosterone prior to sex differentiation. *Aquaculture* **135**, 329–342.

Chomczynski, P., and Sacchi, N. (1993). A reagent for the single-step simultaneous isolation of RNA and proteins from cell and tissue samples. *Biotechniques* **15**, 532–534.

Craik, J. C. A. (1977). Kinetic studies of vitellogenin metabolism in the Elasmobranch *Scyliorhinus canicula* L. *Comp. Biochem. Physiol.* **61A**, 355–361.

Danzo, B. J. (1997). Environmental xenobiotics may disrupt normal endocrine function by interfering with the binding of physiological ligands to steroid receptors and binding proteins. *Environ. Health Persp.* **105**, 294–301.

Donaldson, E. M., and Hunter, G. A. (1982). Sex control in fish with particular reference to salmonides. *Can. J. Fish. Aquat. Sci.* **39**, 99–110.

Dunnett, C. W. (1964). New tables for multiple comparisons with a control. *Biometrics* **26**, 482–491.

Ellis, R. J., van den Heuvel, M. R., Stuthridge, T. R., McCarthy, L. H., Ling, N., Hogg, I. D., and Dietrich, D. R. (2001). Androgenic responses in two fish species following exposure to a New Zealand pulp and paper mill wastewaters. Annual Meeting of the Society of Toxicology, San Francisco, March 25–29, 2001. *Toxicol. Sci.* **60**, 771.

EPA (1993). EPA Guideline, 4th edition. "Methods for measuring the acute toxicity of effluents and receiving waters to freshwater and marine organisms," pp. 1–293. EPA/600/4-90/027F, Washington, D.C.

Fent, K. (1996). Endocrinically active substances in the environment: State of the art. *Texte Umweltbundesamt Berlin* **3**, 69–80.

Folmar, L. C., Denslow, N. D., Rao, V., Chow, M., Crain, D. A., Enblom, J., Marcino, J., and Guillette, L. J., Jr. (1996). Vitellogenin induction and reduced serum testosterone concentrations in feral male Carp (*Cyprinus carpio*) captured near a major metropolitan sewage treatment plant. *Comp. Biochem. Physiol.* **104**, 1096–1101.

Harries, J. E., Janbakhsh, A., Jobling, S., Matthiessen, P., Sumpter, J. P., and Tyler, C. R. (1999). Estrogenic potency of effluent from two sewage treatment works in the United Kingdom. *Environ. Tox. Chem.* **18**, 932–937.

Hori, S. H., Kodama, T., and Tanahashi, K. (1979). Induction of vitellogenin synthesis in goldfish by massive doses of androgens. *Gen. Comp. Endocrinol.* **37**, 306–320.

Howell, W. M., Black, D. A., and Bortone, S. A. (1980). Abnormal expression of secondary sex characteristics in a population of Mosquitofish, *Gambusia affinis holbrooki*: Evidence for environmentally induced masculinization. *Copeia* **4**, 676–681.

Jobling, S., and Sumpter, J. P. (1993). Detergent components in sewage effluent are weakly oestrogenic to fish: An *in vitro* study using Rainbow trout (*Oncorhynchus mykiss*). *Aquatic Tox.* **27**, 361–372.

Langston, W. J. (1996). Recent developments in TBT ecotoxicology. *Toxico. Ecotoxico. News* **3**, 179–187.

Michaud, L. B., and Buzdar, A. U. (1999). Risks and benefits of aromatase inhibitors in postmenopausal breast cancer. *Drug Saf.* **21**, 297–309.

Mori, T., Matsumoto, H., and Yokota, H. (1998). Androgen-induced vitellogenin gene expression in primary cultures of Rainbow trout hepatocytes. *J. Steroid Biochem.* **67**, 133–141.

Munkittrick, K. R., Portt, C. B., van der Kraak, G., Smith, R. J., and Rokosh, D. A. (1991). Impact of bleached kraft mill effluent on population characteristics, liver MFO activity, and serum steroid levels of a Lake Superior White Sucker (*Catostomus commersoni*) Population. *Can. J. Fish. Aquat. Sci.* **48**, 1371–1380.

Munkittrick, K. R., van der Kraak, G., McMaster, M. E., and Portt, C. B. (1992). Response of hepatic MFO activity and plasma sex steroids to secondary treatment of bleached kraft pulp mill effluent and mill shutdown. *Environ. Tox. Chem.* **11**, 1439–1452.

Nam, Y. K., Noh, C. H., and Kim, D. S. (1998). Effects of 17 alpha-methyltestosterone immersion treatments on sex reversal of Mud loach, *Migurnus mizolepis*. *Fish. Sci.* **64**, 914–917.

Oehlmann, J., Stoben, E., Schulte-Oehlmann, U., Bauer, B., Fioroni, P., and Markert, B. (1996). Tributyltin biomonitoring using Prosobranchs as sentinel organism. *Fresenius J. Anal. Chem.* **354,** 540–545.

Panter, G. H., Hutchinson, T. H., Länge, R., Zerulla, M., Lye, C. M., Sumpter, J. P., and Tyler, C. R. (2002). Utility of a juvenile Fathead minnow screening assay for detecting (anti-) estrogenic substances. *Environ. Tox. Chem.* **21,** 319–326.

Parks, L. G., Lambright, C. S., Orlando, E. F., Guillette, L. J., Jr., Ankley, G. T., and Gray, L. E., Jr. (2001). Masculinization of female Mosquitofish in kraft mill effluent-contaminated Fenholloway river water is associated with androgen receptor agonist activity. *Tox. Science* **62,** 257–267.

Petit, F., Le Goff, P., Cravédi, J.-P., Valotaire, Y., and Pakdel, F. (1997). Two complentary bioassays for screening the estrogenic potency of xenobiotics: Recombinant yeast for trout estrogen receptor and trout hepatocyte cultures. *J. Mol. Endocrin.* **19,** 321–335.

Piferrer, F. (2001). Endocrine sex control strategies for the feminization of teleost fish. *Aquaculture* **197,** 229–281.

Purdom, C. E., Hardiman, P. A., Bye, V. J., Eno, N. C., Tyler, C. R., and Sumpter, J. P. (1994). Estrogenic effects of effluents from sewage treatment works. *Chem. Ecol.* **8,** 275–285.

Schreck, C. B. (1974). Hormonal treatment and sex manipulation in fishes. *In* "Control of Sex in Fishes" (C. B. Schreck, Ed.), pp. 84–106. Virginia Polytechnic Institute and State University Sea Grant Program.

Sokal, R. R., and Rohlf, F. J. (1981). Biometry: The principals and practice of statistics in biological research. Freeman & Co, New York.

Sumpter, J. P., and Jobling, S. (1995). Vitellogenesis as a biomarker for estrogenic contamination of the aquatic environment. *Environ. Health Persp.* **103,** 173–178.

Tyler, C. R., van Aerle, R., Hutchinson, T. H., Maddix, S., and Trip, H. (1999). An *in vivo* testing system for endocrine disruptors in fish early life stages using induction of vitellogenin. *Environ. Tox. Chem.* **18,** 337–347.

Varadaraj, K., Kumari, S. S., and Pandian, T. J. (1994). Comparison of conditions for hormonal sex reversal of Mozambique tilapias. *Prog. Fish Cultur.* **56,** 81–90.

T. Schmid*,†
J. Gonzalez-Valero†
H. Rufli†
D. R. Dietrich*

*Environmental Toxicology
University of Konstanz
78457 Konstanz, Germany

†Syngenta Crop Protection AG
Ecological Sciences
4002 Basel, Switzerland

Determination of Vitellogenin Kinetics in Male Fathead Minnows (*Pimephales promelas*)

I. Introduction

Currently, one of the most frequently used endpoints to detect effects mediated by estrogenic substances is the measurement of the induction of hormone-dependent protein synthesis, such as proteins synthesized under the strict control of specific hormones. Vitellogenin (VTG), a prominent member of this class of biomarkers (Sumpter and Jobling, 1995), is under estrogenic control. It is a glycoprotein that is synthesized in the liver of female oviparous vertebrates (Wahli *et al.*, 1981), released into the blood, sequestered from the blood by the growing oocyte, and cleaved there to yield the main egg yolk proteins—phosvitin and lipovitellin (Mommsen and Walsh, 1988). VTG levels in the plasma of female fish undergo seasonal variations, reaching concentrations of up to several milligrams per milliliter in some species during oocyte development (Tyler *et al.*, 1996).

The VTG gene is also present in male fish, but under normal conditions is not expressed, possibly due to low concentrations of estrogens in the blood. Nevertheless, upon stimulation with estrogenic substances, blood plasma VTG protein levels in male fish can attain the same range as in mature females (Korte *et al.*, 2000; Purdom *et al.*, 1994). As methods to detect VTG synthesis both at the protein (Folmar *et al.*, 2000; Monteverdi and Di Giulio, 1999; Mourot and Le Bail, 1995; Parks *et al.*, 1999; Tyler *et al.*, 1999) and at the mRNA level (Flouriot *et al.*, 1996; Folmar *et al.*, 2000; Korte *et al.*, 2000) have been developed in recent years, this model offers a sensitive and specific means to detect estrogenic influences in male fish.

No satisfactory characterization of the accumulation, and especially the clearance, kinetics in male fish has been carried out yet. Most current studies involve short-term exposure regimens of a maximum duration of 21 days without examining the depuration. This neither allows the accurate determination of the accumulation kinetics, including the possibility of a plateau being approached, or gives any information concerning possible elimination mechanisms. These parameters are, however, indispensable to the determination of the value of this biomarker, with particular reference to how long influences of estrogenic substances are sustained and detectable following the actual exposure. This might be even more important for male fish, because they do not have the ability to remove VTG protein from their blood via uptake into the oocytes, and until now, no specific metabolic pathways for VTG protein have been found. Thus, the elimination of this rather large protein (approximately 156,000 Da. in fathead minnows) (Parks *et al.*, 1999) could prove difficult for the male organism.

Another neglected aspect of VTG protein synthesis in male fish is the correlation of this "unphysiological" process with the harmful outcome of estrogenic exposure on the organism. The main focus so far has been the possible reproductive impairment, leaving possible adverse effects on the health of the animal largely unnoticed. Such findings would put the reproductive effects in perspective, because the most important effect would be, of course, the direct reduction of the fitness of the individual.

To elucidate these issues, it was decided to expose fathead minnows to 17α-ethinylestradiol (EE_2) at a concentration of 50 ng/l. Fathead minnow was chosen as test species because it is one of the recommended test species for long-term exposure experiments in the testing of pesticides (EPA, 1996a,b,c; OECD, 1992, 1996). The concentration used is rather high, though still within the range found in some German rivers (Stumpf *et al.*, 1996) and even lower than that used in recent studies, which claimed concentrations of 100 ng/l to be "on the high end of levels found in the environment" (Bowman *et al.*, 2000; Denslow *et al.*, 2001a,b). This concentration was chosen mainly for two purposes: (1) to achieve a rapid increase in VTG levels, and (2) if possible, to attain a plateau level of the protein in the plasma. In contrast to previous studies, the exposure phase was extended to 35 days with a subsequent depuration phase of 35 days, which again is in accordance with guidelines for bioaccumulation studies (EPA, 1996a; OECD, 1996). This time frame was chosen to allow elucidation of both the accumulation as well as the clearance pattern of the VTG protein. A plateau of VTG protein in the plasma was targeted, to allow for a proper analysis of the depuration kinetics, the main focus of this study. According to previous studies, VTG mRNA levels require a somewhat shorter time frame to attain plateau and reattain control levels (Bowman *et al.*, 2000).

The results obtained should allow for a further validation of the biomarker qualities of VTG in the assessment of estrogens in male fish. In addition, monitoring of the mortality during the study, as well as the fitness levels of sampled fish, should allow for further predictions concerning the correlation between VTG levels and adverse affects on the male organism induced by estrogenic influences.

II. Materials and Methods

A. Test Organisms

Sexually mature, male fathead minnows (*Pimephales promelas*) were obtained from Osage Beach Catfisheries (Missouri). Prior to and during the experiment, they were maintained under standardized conditions (16 h light, 8 h dark, 30 min transition periods; active-charcoal filtered, dechlorinated water). Controlled parameters included temperature ($24 \pm 1°C$), oxygen saturation ($95 \pm 15\%$), and pH (8.2 ± 0.2). Feeding consisted of frozen artemia larvae at a rate of 10–15% of body weight per day in two portions.

B. Test Substance

EE_2 was obtained from Sigma-Aldrich (Buchs, Switzerland). It was shown to be pure with respect to the presence of other estrogens (estrone (E_1) and 17β-estradiol (E_2)) by comparing the GC-MS analysis to that of a certified batch of EE_2 provided by Schering AG (Berlin, Germany) and batches of E_1 and E_2 obtained from Sigma-Aldrich.

C. Chemicals

N,N-dimethylformamide (DMF) obtained from Fluka (Buchs, Switzerland) was used as a vehicle. Chemicals for reverse transcription (RT) were obtained from Perkin-Elmer (Weiterstadt, Germany), and chemicals for polymerase chain reaction (PCR) were purchased from Roche Molecular Biochemicals (Rotkreuz, Switzerland). All other chemicals used, unless otherwise stated, were of the highest quality commercially available.

D. Experimental Design

Fish in a flow-through system were exposed to either 50 ng EE_2/l (0.0001% DMF used as vehicle), 0.0001% DMF, or filtered water only. At the beginning, 144 fish were kept in each of the three 92-liter tanks. The exposure lasted 35 days, followed by a depuration phase of another 35 days during which all fish received filtered water.

The concentration of EE_2 was determined via comparison of the pump rates of water and substance pumps, which were measured daily (Monday to Friday) and corrected if necessary. The mean concentration was determined to be 48.3 ($\pm$6.3) ng/l.

Fish samples ($n \geq 9$) were taken on days 3, 7, 14, 21, 28, 35, 36, 38, 42, 49, 56, 63, and 70 from each tank. Fish were terminally narcotized using a 100 μg/l MS222-solution (Fluka, Switzerland). Before blood samples were taken, the length and weight of each fish were measured. Fitness was calculated using the following equation:

$$\text{Fitness factor} = 100 \times \text{weight [g]} \times (\text{length [cm]})^{-3}$$

Mortality was monitored daily and added to a 7-day-accumulative mortality for each sampling day. Subsequently, blood samples were taken by cardiac puncture with a sterile, heparinized syringe. The plasma was gained by centrifugation at 3000 $\times$ g for 30 minutes and thereafter stored at -20°C. Liver samples for the determination of the VTG mRNA content were stored at $-20\,^{\circ}$C in RNAlater (Ambion, Austin, TX) upon sampling.

E. mRNA Analyses

Liver samples were homogenized in the frozen state in liquid nitrogen using a mortar and pestle. Total RNA was extracted using the Perfect RNA Eukaryotic extraction kit (miniscale) from Eppendorf (Schönenbuch, Switzerland). After extraction, the samples were diluted to 0.5 μg RNA/μl in DMPC-treated, sterile water. RT was carried out with 0.5 μg of RNA from each sample. PCR was performed using carp α-actin primers (Watabe *et al.*, 1995), which were responsive in fathead minnow as well, and fathead minnow VTG primers (Korte *et al.*, 2000) in a LightCycler from Boehringer (Mannheim, Germany). The LightCycler PCR was always performed until all samples reached a plateau level of the amplified DNA; thus, the number of cycles used differed among runs. Individual samples were evaluated for their VTG mRNA content with respect to a serial dilution of one sample of an exposed fish, sampled on day 35; the values given in the results are always relative to this individual sample, which was assigned a value of 1 (100%). This was done because no standard cDNA was commercially available; samples of this day were assumed to have VTG mRNA levels approximately equal to the maximum α-actin that was used to verify that the same amount of cDNA was used for each sample in the PCR.

The identity of the amplified fragment as VTG was verified by agarose gel separation of the PCR products and subsequent extraction of the desired fragment, followed by sequencing of this product performed by GATC GmbH (Konstanz, Germany). The sequence was identified using FASTA3 of the EMBL Outstation of the European Bioinformatics Institute (Pearson and Lipman, 1988). On the basis of the protein data, only mRNA from the water and solvent control fish of days 0, 35, and 70 were analyzed.

F. Protein Analyses

VTG protein was analyzed using the preliminary version of the carp-VTG enzyme-linked immunosorbent assay (ELISA) No. 103 from Biosense (Bergen, Norway), which was developed on the basis of the assay described by Tyler and co-workers (Tyler *et al.*, 1999). This assay was performed as a competitive ELISA. For this purpose, plasma samples were thawed on ice and diluted in blocking buffer. The dilution factor for the control samples (DMF- or water-exposed animals), which was dependent on the amount of plasma gained from the individual animal, was kept at a minimum between 1:20 and 1:100. Plasma from exposed fish was diluted between 1:100 and 1:1,000,000, depending on the stage of the study.

G. Statistics

Statistical analyses were performed with GraphPad Prism version 3.00 for Windows (GraphPad Software, San Diego, CA). Protein and mRNA levels were compared using a one-way ANOVA with Tukey-Kramer's post-test. Protein kinetics were calculated and statistically evaluated using nonlinear regression with a Run's test.

III. Results

A. Fitness Status and Mortality

Fish from control tanks yielded fitness values between 0.78 and 0.93 g/cm^3, with an average of 0.87 ($\pm$0.04) g/cm^3. In the EE_2-exposure tank, the values decreased to an average of 0.69 g/cm^3 at the end of the exposure phase (35 days), re-attaining control values of 0.88 g/cm^3 at day 70 of the study (Fig. 1). The decrease in the fitness of the exposed fish in comparison to the mean of the control fish proved to be statistically significant only for fish sampled on days 14, 35, 36, 38, 42, 49, and 56.

Mortality accumulated to a total of 3.5% and 2.8% in the control and the solvent control, respectively. In the EE_2 exposure tank, total mortality reached 12.5%, with a peak mortality of 13.3% in the 7 days preceding day 36 (Fig. 2); values dropped to 0% within 2 weeks following removal of the test substance. All mortalities in this tank occurred between days 20 and 36.

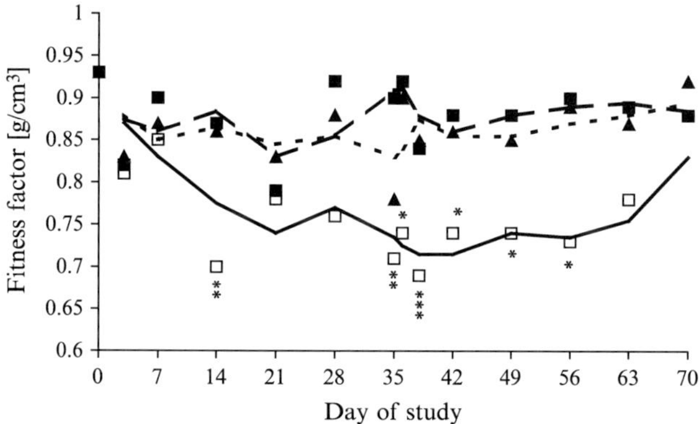

FIGURE 1 Fitness factors of the sampled fish from the control (■ and dashed line), the solvent control (▲ and dotted line), and the EE_2 exposure tanks (□ and solid line). The asterisks indicate that values are significantly different compared to the other values obtained on the same day, as determined using a one-way ANOVA with Tukey-Kramer's post test (*, $p < 0.05$; **, $p < 0.01$; ***, $p < 0.001$M; $n \geq 9$ for each data point). The lines represent the moving average value for each treatment.

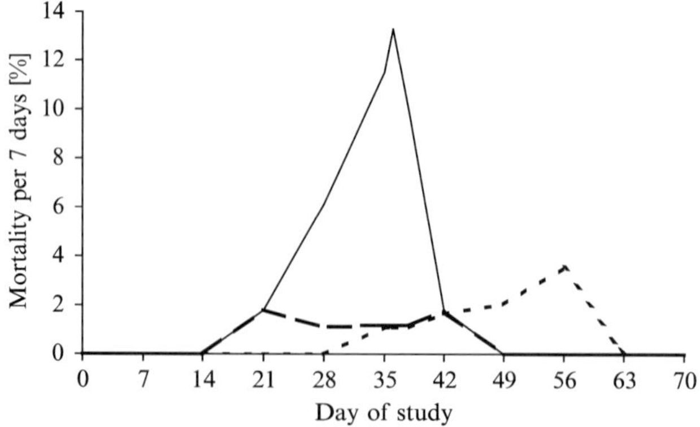

FIGURE 2 Mortality per 7 days in control (dashed line), solvent control (dotted line), and EE_2-exposure tanks (solid line), as a percentage of the remaining fish in the tanks.

B. VTG mRNA Development

α-actin mRNA levels of all samples showed similar results. Thus, the total mRNA used for the determination of VTG mRNA was deemed to be equal for all samples. All VTG mRNA values (from exposed and control animals) are given relative to the VTG mRNA level of a randomly selected standard sample from an EE_2-exposed fish, taken on day 35 of the exposure phase and assigned a value of 1.

Because the protein levels did not vary to any significant degree between control samples, VTG mRNA levels were just determined for water and solvent control fish from days 0, 35, and 70. No significant changes in VTG mRNA levels could be detected when comparing these control values, from either the solvent or the water control. Thus, all control values were pooled to yield a more representative control level, determined to be 2.05×10^{-3} (as compared to the defined standard sample).

Fish exposed to EE_2 showed a rapid increase in VTG mRNA within 3 days, from the control level to a level of 6.11×10^{-1}. The level of VTG mRNA did not vary to any significant degree until day 1 following cessation of EE_2 exposure (day 36), with maximal levels of 9.16×10^{-1} at day 28. The VTG mRNA values from exposed fish decreased significantly to 1.33×10^{-1} on day 38; on day 42, values reached levels similar to the control value of 2.7×10^{-3} (Fig. 3). From day 42 onward, no significant difference between control values and values from the EE_2-exposed animals could be detected.

Following the PCR reaction, the amplified fragments were separated on an agarose gel; the main fragment was subsequently sequenced and was confirmed to be 98.587%, identical to the gene used for the alignment of the primer. This was deemed to be the desired VTG mRNA.

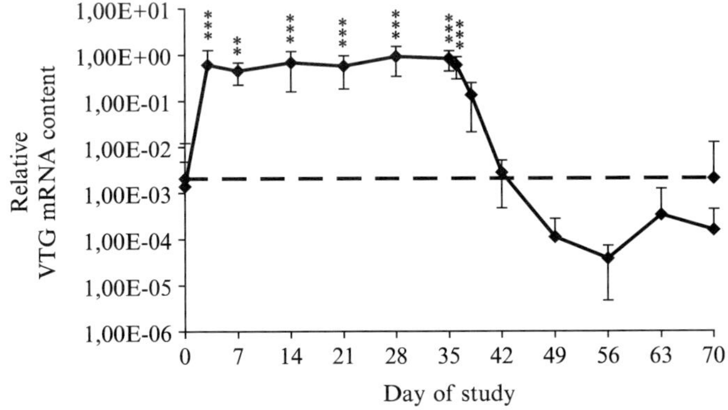

FIGURE 3 Relative VTG mRNA content in pooled liver samples from control and solvent control fish (dashed line), as compared to EE_2 exposed fish (solid line). Values are given relative to a self-defined "standard sample," which was set as 1. The asterisks indicate that values are significantly different to the pooled control as determined using a one-way ANOVA with Tukey-Kramer's post test (**, $p < 0.01$; ***, $p < 0.001$; $n \geq 9$ for each data point).

C. VTG Protein Development

Samples from control animals (water and solvent) from all timepoints consistently displayed VTG protein levels at or below the detection limit, which varied between 0.5 and 250 ng VTG/ml. The measurements could not be carried out to any more accurate degree, since the samples had to be diluted at least 20-fold due to their small volumes. The variation in the detection limit accounts for the strong differences within the control groups on individual sampling days. The VTG protein concentration in control fish remained at an average level (including all control fish of all time points) of 175 ± 156 ng/ml plasma for the entire duration of the experiment. No significant differences were observed between the water control and the vehicle control and, in contrast to previous reports (Ren *et al.*, 1996), no estrogenic influence of DMF could be demonstrated.

EE_2-exposed fish showed an increase in plasma VTG protein over control levels beginning on day 3 of exposure (first sampling). The most pronounced was observed between days 14, 21, and 28 of the study. Starting on day 28, a plateau appeared to be reached, with a level of approximately 47 mg VTG/ml plasma. This level persisted until day 38 (3 days following cessation of EE_2 exposure). Thereafter, VTG protein levels were observed to decrease, reaching a value of 4.6 ± 1.8 mg VTG/ml plasma on day 35 of the clearance phase (day 70 of the study) (Fig. 4). The increase in VTG protein concentrations in the exposed animals as compared to the control fish was significant ($p < 0.001$) for fish sampled between days 21 and 42, and for those sampled on day 49 of the study ($p < 0.01$).

The increase in plasma VTG protein concentration from days 0 to 38 could be described by a two-phased accumulation model (Fig. 4A). The first phase was calculated using the data from days 0 to 14, and consisted of an exponential increase of the protein in the blood following the equation:

$$C_T = C_0 \times e^{(KT)},$$

where C_T, concentration of VTG in the plasma [mg/ml] at time T; C_0, concentration of VTG in the plasma [mg/ml] at $T = 0$; K, rate constant of the exponential increase [days^{-1}]; and T, time [days]. Calculation of the initial accumulation phase yielded a concentration of 10.74 mg/ml on day 15. The starting concentration C_0 was calculated to be 4000 ng/ml and the rate constant to be 0.5256 day^{-1}, resulting in the doubling of VTG protein every 1.319 days.

The second accumulation phase was calculated for the data from days 14 to 38 using a one-compartment saturation model with the following equation:

$$C_T = C_{15} + C_{sat}\left(1 - e^{(-KT)}\right),$$

where C_T, concentration of VTG in the plasma [mg/ml] at time T; C_{15}, concentration of VTG in the plasma [mg/ml] at day 15 (start); C_{sat},

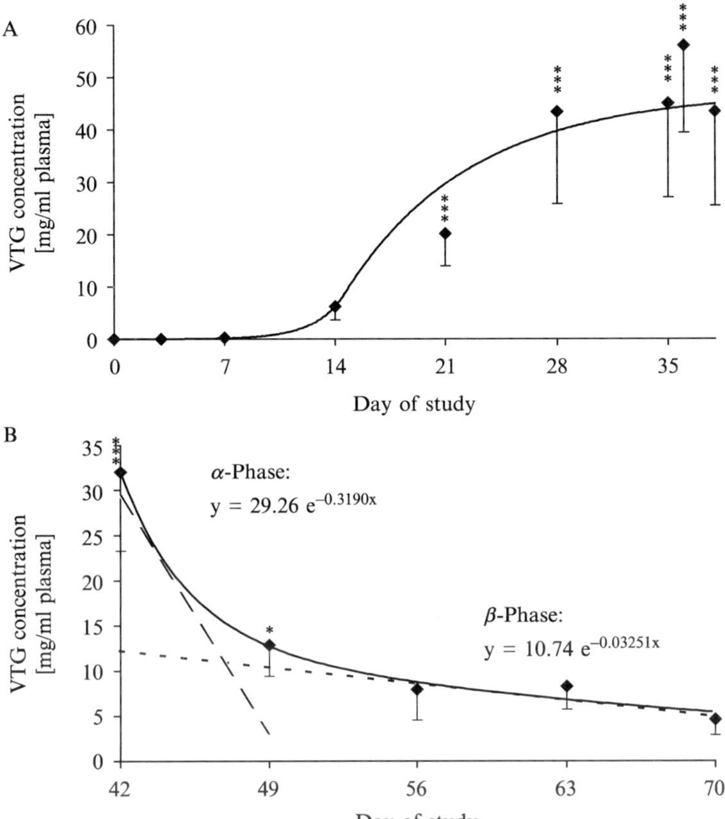

FIGURE 4 Development of VTG protein plasma levels (±SD) in fish sampled from EE_2-exposure tank (dots). (A) Accumulation phase and proposed kinetics model (solid line). (B) Depuration phase including a proposed kinetics model (solid line). In addition, linear regression analyses of the two elimination phases (α-phase, dashed line; β-phase, dotted line) obtained by assuming a two-compartment model are given. The numerical values calculated for the model parameters were: A $= 29.26$ mg VTG ml^{-1}; $\alpha = 0.3190$; B $= 10.74$ mg VTG ml^{-1}; $\beta = 0.03251$. Control values were not included in the graph, since they were too low (mean 175 (±156) ng VTG ml^{-1}) to illustrate any additional information. The asterisks indicate that values are significantly different to the pooled control as determined using a one-way ANOVA with Tukey-Kramer's post test (*, $p < 0.05$; **, $p < 0.01$; ***, $p < 0.001$; $n \geq 9$ for each data point).

concentration of VTG in the plasma [mg/ml] at saturation; K, rate constant of the increase [$days^{-1}$]; and T, time [days]. The saturation concentration C_{sat} was kept at a constant value of 47 mg/ml, which was the mean plateau concentration, and the starting concentration C_{15} was set to 10.74 mg/ml. This resulted in an increase rate-constant K of 0.1233 day^{-1}.

The clearance kinetics of the protein were calculated using a two-compartment open model. The concentrations of the protein in the plasma from days 42 to 70 were fitted to a biexponential expression (Klaassen *et al.*, 1995), adding a final plateau level approached by the decay (Fig. 4B):

$$C_p = Ae^{(-\alpha T)} + Be^{(-\beta T)} + \text{Plateau},$$

where C_p, concentration of VTG in the plasma [mg/ml] at time T; A and B, y-intercepts; α, elimination constant I ($\alpha = \ln 2/t_{1/2}$); β, elimination constant II ($\beta = \ln 2/t_{1/2}$); T, time [days]; and Plateau, concentration approached through the decay. The plateau was normalized to the control level of 175 ng VTG/ml plasma, resulting in y-intercepts of $A = 29.26$ and $B = 10.74$. The elimination constants were calculated to be $\alpha = 0.3190$ and $\beta = 0.03251$. This yielded half-lives of 2.17 days for the first phase and 21.32 days for the second phase of the elimination phase. The fixed plateau level yielded a curve, fitting the actual data with $R^2 > 0.99$ and a p-value of 1.00 for the Run's test.

IV. Discussion

Our results on the VTG mRNA levels suggest a rapid induction of gene expression and attainment of a plateau level in response to EE_2 exposure. The existence of a plateau level is in accordance with previous reports (Folmar *et al.*, 2000) and may be due to an initial boost of VTG gene expression, followed by a period where almost no additional mRNA synthesis occurs and the plateau level observed could be attributed to an extended half-life of VTG mRNA. Previous studies have reported an extension of the half-life of VTG mRNA under estrogenic influences (Blume and Shapiro, 1989; Brock and Shapiro, 1983) from 16–33 hours under normal conditions, to about 500 hours in the presence of E_2 for *Xenopus laevis* hepatocytes in culture. In our study, the mRNA levels decreased to control level within 7 days of EE_2 removal, yielding a half-life of 20–30 hours. Similar results have been reported for other endocrine disrupters (EDCs) such as nonylphenol (Lech *et al.*, 1996; Ren *et al.*, 1996).

Though the time course of the experiment was not repeated, a model that was developed might accurately represent the actual situation, due to the rather large number of samples. It was possible to describe the development of VTG protein concentrations in the plasma by a biphasic accumulation model. The first phase was characterized by an exponential increase in VTG concentrations, and the second phase followed a saturation kinetics model. Although days 36 and 38 were actually part of the depuration phase, they were included in the accumulation kinetics, because they still displayed strongly elevated VTG mRNA levels, resulting in relatively constant protein levels. Because of the elevated VTG mRNA levels on the first seven days after removal of the inducing substance from the water, these timepoints were not included in the determination of the depuration kinetics for the protein. Thereafter, the depuration phase yielded a two-phased decreasing curve of VTG plasma concentrations. The first phase of depuration yielded a half-life for VTG of 52.1 h; the second phase yielded a half-life of 511.7 h.

This two-phased clearance curve might be explained by a primary elimination phase, during which the VTG excretion is still under the influence of the metabolizing enzymes of the liver induced by EE_2. Steroids are known to induce the activity of certain members of the cytochrome P450-dependent monooxygenase system, such as the NADPH cytochrome P450 reductase (Solé *et al.*, 2000; Stegeman and Woodin, 1994). This highly unspecific elimination system might, even if not directly induced by VTG, also be responsible for the elimination of this protein. The second phase would then represent the pure elimination phase, via yet unknown mechanisms without an additional induction of the cytochrome P450-system by EE_2. This decrease in kinetics might also be caused by resorption mechanisms; for example, VTG protein might be filtered, excreted via the kidney, and resorbed again. These mechanisms might either be more effective at lower concentrations of VTG protein, or are already saturated at the high levels observed during the first elimination phase. Thus, the second phase would describe the VTG excretion under physiological conditions. In both cases, the second phase could be used to estimate the actual half-life of VTG in the male organism.

The rather long half-life of approximately 21 days in the second phase confirms the proposed lack of specific excretion mechanisms in male fish. An even longer half-life of VTG in male fathead minnow of more than 40 days has been proposed by Korte *et al.* (2000), which is similar to the half-life of 40 days described by Tata (1976) for male *Xenopus laevis*. On the other hand, Allen *et al.* (1999) reported a linear elimination of VTG from the blood with a calculated half-life of 13.5 days in male flounder (*Platichthys flesus*).

Maximum VTG protein levels reported for plasma of fish exposed to EDCs are relatively consistent for various studies, within a range of 1–120 mg/ml depending on species (Allen *et al.*, 1999; Folmar *et al.*, 2000; Harries *et al.*, 1999; Jobling *et al.*, 1996; Korte *et al.*, 2000; Parks *et al.*, 1999). All of the mentioned studies employed exposure scenarios of 16–21 days. There are, however, contrary opinions concerning the existence of a relationship between exposure concentration of the EDC and VTG protein "signal" (Folmar *et al.*, 2000; Panter *et al.*, 1998). The data actually suggest that a concentration dependence exists, as long as the VTG level does not attain maximum levels. After reaching a certain induction level this dependence ceases, resulting in the aforementioned relatively consistent maximum plasma VTG concentrations, seemingly independent of the strength of the stimulus. This threshold value might be due to the fact that male fish possibly die at higher VTG plasma levels. The fact that mortality in the EE_2 exposure tank occurred entirely between days 20 and 36 of the exposure phase suggests a connection with the exposure to the test substance, or even the high VTG levels observed. This is supported by the finding that 35 days were needed after exposure to completely recover the fitness status of the exposed fish.

The main difference between EE_2-exposed and control fish was the possible induction of the synthesis of estrogen-controlled proteins, mainly VTG. Because VTG is not synthesized under normal conditions in male fish, it could conceivably cause negative effects such as kidney dysfunction leading to death (Herman and Kincaid, 1988). Another factor that might decrease fitness status are metabolizing enzymes in the liver, such as the cytochrome P450-dependent monooxygenase system. These enzymes play a key role in the oxidative metabolism of steroids. Therefore, an increased activity of these enzymes would be expected following exposure to EE_2 (Pajor *et al.*, 1990; Snowberger *et al.*, 1991). A strong induction of these enzymes might cause an increased basic turnover and thereby lead to weight loss, a sign of decreased fitness status of the fish. A similar trend has been described by Korsgaard and Mommsen (1993), who observed a strong decline in gluconeogenesis starting after one week of exposure to EE_2, reaching a maximum after two weeks. In addition, high plasma concentrations of a large protein might cause a blocking of small capillaries, as present in the glomeruli of the kidney and also in the liver. This would explain the histopathological changes of the main excretion (kidney) and metabolizing (liver) organs following the exposure to EDCs described elsewhere (Hori *et al.*, 1979; Lewis *et al.*, 1976; Nicholls *et al.*, 1968; Schwaiger *et al.*, 2000; Schweinfurth *et al.*, 1997), and could also contribute to the observed mortality. The slightly increased mortality in the solvent control (between days 35 and 56) lies within the natural variability. The level remained well below 5%, which is a mortality considered to be still acceptable by other studies (Zerulla *et al.*, 2002).

The results of the present study suggest that there is a connection between increased VTG protein concentrations in the plasma (or an increased synthesis of VTG protein) and decreased fitness values with increased mortality in male fathead minnows. The kinetics of VTG protein versus mRNA suggest large differences between these parameters in respect to their biomarker qualities. The slow clearance of VTG protein offers the possibility to even detect influences that occurred long before the measurement. This long half-life does not allow for delimiting an estrogenic influence to a specific time point. VTG mRNA measurement, however, provides a method that allows for faster detection of estrogenic influences and moreover, to delimit the possible exposure to a relatively short period of time. This is due to the shorter time (<3 days) needed for a detectable increase to occur, and to the fast return to control levels after removal of the EDC (≤ 7 days). An observation of both mRNA and protein levels offers the possibility to monitor longer timeframes (protein) while at the same time determining if an exposure occurred more recently, or even if the exposure is continuing at the moment of sampling (mRNA).

VTG has been implied to be a sensitive and specific biomarker for estrogenic influences. This study gives more information on its applicability for future studies. The information might prove to be of particular interest for the design of field studies and studies used for registration purposes.

Acknowledgments

We would like to thank Biosense Laboratories (Bergen, Norway) for the rewarding cooperation in the testing and optimization of the VTG-ELISA-kit. This study was partially performed at and financially supported by Syngenta Crop Protection AG, Ecological Sciences (Basel, Switzerland) and by the EUREGIO Ecotoxicology Service Laboratory (Konstanz, Germany). In addition, we would like to thank Dr. E. O'Brien for critically reading and correcting the manuscript.

References

Allen, Y., Matthiessen, P., Scott, A. P., Haworth, S., Feist, S., and Thain, J. E. (1999). The extent of oestrogenic contamination in the UK estuarine and marine environments – further surveys of flounder. *Sci. Tot. Environ.* **233**, 5–20.

Blume, J. E., and Shapiro, D. J. (1989). Ribosome loading, but not protein synthesis, is required for estrogen stabilization of *Xenopus laevis* vitellogenin mRNA. *Nucl. Acids Res.* **17**, 9003–9014.

Bowman, C. J., Kroll, K. J., Hemmer, M. J., Folmar, L. C., and Denslow, N. D. (2000). Estrogen-induced vitellogenin mRNA and protein in sheepshead minnow (*Cyprinodon variegatus*). *Gen. Comp. Endocrinol.* **120**, 300–313.

Brock, M. L., and Shapiro, D. J. (1983). Estrogen stabilizes vitellogenin mRNA against cytoplasmic degradation. *Cell* **34**, 207–214.

Denslow, N. D., Bowman, C. J., Ferguson, R. J., Lee, H. S., Hemmer, M. J., and Folmar, L. C. (2001). Induction of gene expression in sheepshead minnows (*Cyprinodon variegatus*) treated with 17β-estradiol, diethylstilbestrol, or ethinylestreadiol: The use of mRNA fingerprints as an indicator of gene regulation. *Gen. Comp. Endocrinol.* **121**, 250–260.

Denslow, N. D., Lee, H. S., Bowman, C. J., Hemmer, M. J., and Folmar, L. C. (2001). Multiple responses in gene expression in fish treated with estrogen. *Comp. Biochem. Physiol. B* **129**, 277–282.

EPA (1996a). Fish BCF. United States Environmental Protection Agency (U.S. EPA), Washington, D.C.

EPA (1996b). Fish early-life stage toxicity test. United States Environmental Protection Agency (U.S. EPA), Washington, D.C.

EPA (1996c). Fish life cycle toxicity. United States Environmental Protection Agency (U.S. EPA), Washington, D.C.

Flouriot, G., Pakdel, F., and Valotaire, Y. (1996). Transcriptional and post-transcriptional regulation of rainbow trout estrogen receptor and vitellogenin gene expression. *Mol. Cell. Endocrinol.* **124**, 173–183.

Folmar, L. C., Hemmer, M., Hemmer, R., Bowman, C., Kroll, K., and Denslow, N. D. (2000). Comparative estrogenicity of estradiol, ethynyl estradiol and diethylstilbestrol in an *in vivo*, male sheepshead minnow (*Cyprinodon variegatus*), vitellogenin bioassay. *Aquat. Toxicol.* **49**, 77–88.

Harries, J. E., Janbakhsh, A., Jobling, S., Matthiessen, P., Sumpter, J. P., and Tyler, C. R. (1999). Estrogenic potency of effluent from two sewage treatment works in the United Kingdom. *Environ. Toxicol. Chem.* **18**, 932–937.

Herman, R. L., and Kincaid, H. L. (1988). Pathological effects of orally administered estradiol to rainbow trout. *Aquaculture* **72**, 165–172.

Hori, S. H., Kodama, T., and Tanahashi, K. (1979). Induction of vitellogenin synthesis in goldfish by massive doses of androgens. *Gen. Comp. Endocrinol.* **37**, 306–320.

Jobling, S., Shaehan, D., Osborne, J. A., Matthiessen, P., and Sumpter, J. P. (1996). Inhibition of testicular growth in rainbow trout (*Oncorhynchus mykiss*) exposed to estrogenic alkylphenolic chemicals. *Environ. Toxicol. Chem.* **15**, 194–202.

Klaassen, C. D., Amdur, M. O., and Doull, J. (1995). Casarett's & Doull's Toxicology; The Basic Science of Poisons. McGraw-Hill, New York.

Korsgaard, B., and Mommsen, T. P. (1993). Gluconeogenesis in hepatocytes of immature rainbow trout (*Oncorhynchus mykiss*): Control by estradiol. *Gen. Comp. Endocrinol.* **89**, 17–27.

Korte, J. J., Kahl, M. D., Jensen, K. M., Pasha, M. S., Parks, L. G., LeBlanc, G. A., and Ankley, G. T. (2000). Fathead minnow vitellogenin: Complementary DNA sequence and messenger RNA and protein expression after 17β-estradiol treatment. *Environ. Toxicol. Chem.* **19**, 972–981.

Lech, J. J., Lewis, S. K., and Ren, L. (1996). *In vivo* estrogenic activity of nonylphenol in rainbow trout. *Fund. Appl. Toxicol.* **30**, 229–232.

Lewis, J. A., Clemens, M. J., and Tata, J. R. (1976). Morphological and biochemical changes in the hepatic endoplasmic reticulum and golgi apparatus of male *Xenopus laevis* after induction of egg-yolk protein synthesis by oestradiol-17β. *Mol. Cell. Endocrinol.* **4**, 311–329.

Mommsen, T. P., and Walsh, P. J. (1988). Vitellogenesis and oocyte assembly. *In* "Fish Physiology" (W. S. Hoar and D. J. Randell, Eds.), pp. 347–406. Academic Press, London.

Monteverdi, G. H., and Di Giulio, R. T. (1999). An enzyme-linked immunosorbent assay for estrogenicity using primary hepatocyte cultures from the channel catfish (*Ictalurus punctatus*). *Arch. Environ. Contamin. Toxicol.* **37**, 62–69.

Mourot, B., and Le Bail, P.-Y. (1995). Enzyme-linked immunosorbent assay (ELISA) for rainbow trout (*Oncorhynchus mykiss*) vitellogenin. *J. Immunoassay* **16**, 365–377.

Nicholls, T. J., Follett, B. K., and Evennett, P. J. (1968). The effects of oestrogens and other steroid hormones on the ultrastructure of the liver of *Xenopus laevis Daudin. Zeitschrift für Zellforschung und Mikroskopische Anatomie* **90**, 19–27.

OECD (1992). Guideline 210: Fish, early-life stage toxicity test. Organization for Economic Cooperation and Development, Paris, France.

OECD (1996). Guideline 305: Bioconcentration: Flow-through fish test. Organization for Economic Cooperation and Development, Paris, France.

Pajor, A., Stegeman, J., Thomas, P., and Woodin, B. (1990). Feminization of the hepatic microsomal cytochrome P-450 system in brook trout by estradiol, testosterone, and pituitary factors. *J. Exper. Zool.* **253**, 51–60.

Panter, G. H., Thompson, R. S., and Sumpter, J. P. (1998). Adverse reproductive effects in male fathead minnows (*Pimephales promelas*) exposed to environmentally relevant concentrations of the natural oestrogens, oestradiol and oestrone. *Aquat. Toxicol.* **42**, 243–253.

Parks, L. G., Cheek, A. O., Denslow, N. D., Heppell, S. A., McLachlan, J. A., LeBlanc, G. A., and Sullivan, C. V. (1999). Fathead minnow (*Pimephales promelas*) vitellogenin:

Purification, characterization and quantitative immunoassay for the detection of estrogenic compounds. *Comp. Biochem. Physiol. C* **123,** 113–125.

Pearson, W. R., and Lipman, D. J. (1988). Improved tools for biological sequence comparison. *Proc. Nat. Acad. Sci.* **85,** 2444–2448.

Purdom, C. E., Hardiman, P. A., Bye, V. J., Eno, N. C., Tyler, C. R., and Sumpter, J. P. (1994). Estrogenic effects of effluents from sewage treatment works. *Chem. Ecol.* **8,** 275–285.

Ren, L., Lewis, S. K., and Lech, J. J. (1996). Effects of estrogen and nonylphenol on the post-transcriptional regulation of vitellogenin gene expression. *Chem. Biol. Inter.* **100,** 67–76.

Schwaiger, J., Spieser, O. H., Bauer, C., Ferling, H., Mallow, U., Kalbfus, W., and Negele, R. D. (2000). Chronic toxicity of nonylphenol and ethinylestradiol: Haematological and histopathological effects in juvenile common carp (*Cyprinus carpio*). *Aquat. Toxicol.* **51,** 69–78.

Schweinfurth, H., Länge, R., Miklautz, H., and Schauer, G. (1997). Umweltverhalten und aquatische Toxizität von Ethinylestradiol. *In* "Stoffe mit endokriner Wirkung im Wasser." R. Oldenbourg Verlag, München, Germany.

Snowberger, E. A., Woodin, B. R., and Stegeman, J. J. (1991). Sex differences in hepatic monooxygenases in winter flounder (*Pseudopleuronectes americanus*) and scup (*Stenotomus chrysops*) and regulation of P450 forms by estradiol. *J. Exper. Zool.* **259,** 330–342.

Solé, M., Porte, C., and Barceló, D. (2000). Vitellogenin induction and other biochemical responses in carp, *Cyprinus carpio*, after experimental injection with 17α-ethynylestradiol. *Arch. Environ. Contamin. Toxicol.* **38,** 494–500.

Stegeman, J. J., and Woodin, B. R. (1994). Biochemistry and molecular biology of monooxygenases: Current perspectives on forms, functions and regulation of cytochrome P450 in aquatic species. *In* "Aquatic Toxicology: Molecular, Biochemical, and Cellular Perspectives" (D. C. Malins and G. K. Ostrander, Eds.), pp. 87–204. Lewis Publishers, Boca Raton, FL.

Stumpf, M., Ternes, T. A., Haberer, K., and Baumann, W. (1996). Nachweis von natürlichen und synthetischen Östrogenen in Kläranlagen und Fliessgewässern. *Vom Wasser* 87, 251–261.

Sumpter, J. P., and Jobling, S. (1995). Vitellogenesis as a biomarker for estrogenic contamination of the aquatic environment. *Environ. Health Persp.* **103,** 173–178.

Tata, J. R. (1976). The expression of the vitellogenin gene. *Cell* **9,** 1–14.

Tyler, C. R., Aerle, R. V., Hutchinson, T. H., Maddix, S., and Trip, H. (1999). An *in vivo* testing system for endocrine disruptors in fish early life stages using induction of vitellogenin. *Environ. Toxicol. Chem.* **18,** 337–347.

Tyler, C. R., van der Erden, B., Jobling, S., Panter, G., and Sumpter, J. P. (1996). Measurement of vitellogenin, a biomarker for exposure to oestrogenic chemicals, in a wide variety of cyprinid fish. *J. Comp. Physiol. B* **166,** 418–426.

Wahli, W., Dawid, I. B., Ryffel, G. U., and Weber, R. (1981). Vitellogenesis and the vitellogenin gene family. *Science* **212,** 298–304.

Watabe, S., Hirayama, Y., Imai, J., Kikuchi, K., and Yamashita, M. (1995). Sequences of cDNA clones encoding α-actin of carp and goldfish skeletal muscles. *Fish. Sci.* **61,** 998–1003.

Zerulla, M., Länge, R., Steger-Hartmann, T., Panter, G., Hutchinson, T., and Dietrich, D. R. (2002). Sex reversal upon short-term exposure to endocrine modulators in juvenile fathead minnow (*Pimephales promelas*). *Toxicol. Lett.* **131,** 39–50.

Thomas H. Hutchinson
AstraZeneca Brixham Environmental Laboratory Freshwater Quarry
Brixham, Devon
TQ5 8BA United Kingdom

Integrated *In Vivo* and *In Vitro* Assessment of Reproductive and Developmental Effects of Endocrine Disrupters in Invertebrates

I. Introduction

Invertebrates comprise approximately 95% of all terrestrial and aquatic animal species (Wilson, 1999). Therefore, it is clearly necessary to take this biodiversity into account when addressing the potential developmental and reproductive hazards posed by environmental endocrine disrupters (EDCs). While knowledge into potential endocrine-mediated impacts in invertebrates is growing steadily (de Fur *et al.*, 1999), practical and scientific limitations in ecotoxicology are such that it remains necessary to continue to extrapolate from laboratory data based on a few invertebrate species to the wider range of invertebrate species in nature. In many cases, the aquatic organisms used for regulatory ecotoxicity testing purposes have been selected, at least in part, from knowledge of aquaculture (rotifers, daphnids, and bivalve mollusks) and therefore may not be ideal representatives of the major invertebrate groups (Ingersoll *et al.*, 1999).

The concern over endocrine disruption in aquatic invertebrates, in addition to the international debate over EDCs and human health, has been highlighted by the example of tributyltin (TBT)–induced reproductive damage and population declines in mollusks (Alzieu, 2000; Matthiessen and Gibbs, 1998). From the initial observations of dramatic impacts in French oyster populations as early as 1975, the biochemical and physiological mechanisms of TBT action remains the subject of ongoing research. Based on evidence that exposure of mollusks to TBT is associated with an increase in testosterone levels in females and induction of imposex (penis growth in females), it is considered *most* likely that TBT may be inhibiting cytochrome P450 aromatase (Alzieu, 2000; Matthiessen and Gibbs, 1998; Vos *et al.*, 2000). Other factors thought to contribute to TBT-induced imposex in gastropods include the inhibition of testosterone excretion and suppression of release of a neuroendocrine factor from the pleural ganglia (de Fur *et al.*, 1999; Matthiessen and Gibbs, 1998). This example underscores the fact that the developmental and reproductive health of invertebrates is dependent upon a highly integrated neuroendocrine system, as is the case for all animals (Laessig *et al.*, 1999).

Interestingly, attention has been raised regarding antidepressants and their potential reproductive toxicity in mollusks due to neuroendocrine disruption (Fong, 2001). Nice *et al.* (2003) have reported putative evidence of endocrine disruption in Pacific oysters (*Crassostrea gigas*); exposure to 1 and 100 μg 1–1 nonylphenol at days 7–8 post-fertilization results in a change of sex ratio toward females and an increase in the incidence of hermaphroditism. Ten months later, up to 30% of the resulting adults were fully functional hermaphrodites. Gamete viability in these oysters is also affected, resulting in poor embryonic and larval development (up to 100% mortality) of the subsequent generation.

II. Invertebrate Diversity

In large part due to the TBT example outlined above, there is a significant international research effort into the impacts of mammalian EDCs on aquatic invertebrates and the underlying modes of action (MOA) in non-target organisms. Of the major evolutionary invertebrate phyla (Fig. 1), those most widely used for chronic ecotoxicity testing of potential EDCs include arthropods (namely crustaceans and insects), annelids (including earthworms and polychaetes), and mollusks (including gastropods and bivalves) (de Fur *et al.*, 1999). In part prompted by available genomic information, there is also growing interest in using the Pseudocoelomate nematode, *Caenorhabdites elegans*, in ecotoxicity testing (Anderson *et al.*, 2001; de Fur *et al.*, 1999).

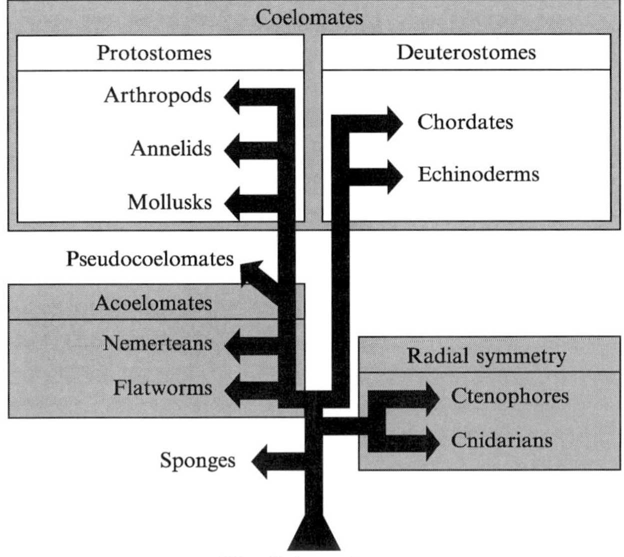

FIGURE 1 Phylogenic relationships within the animal kingdom (adapted from Solomon *et al.*, 1996).

Although of the same evolutionary lineage as the mammals and other Chordates, echinoderms are not widely used in ecotoxicity testing. Nonetheless, basic research shows that mammalian steroids are important in echinoderm reproductive endocrinology (de Fur *et al.*, 1999; Wasson *et al.*, 2000) and limb development (Carnevali *et al.*, 2001), while echinoderm embryos have been used in pharmaceutical research (Pagano *et al.*, 2001; Vogel *et al.*, 1995). de Fur *et al.* (1999) provide a comprehensive review of the basic endocrinology of all the major and minor invertebrate phyla.

Overall, there are examples of environmental chemicals affecting development and reproduction in many invertebrate groups; however, in many cases the only mechanistic information available is that inferred from mammals and fish. Progress in understanding EDCs and appropriate extrapolation of invertebrate chronic toxicity data to the wider world critically depends upon understanding MOA in a taxonomic context. This need for scientifically based extrapolation underscores the importance of the Weybridge workshop's definition of an endocrine-disrupting chemical as being "an exogenous substance that causes adverse effects in an intact organism, or its progeny, consequent to changes in endocrine function" (European Commission, 1996). Given the extensive use of arthropods (amphipods, chironomids, and daphnids) in the environmental hazard assessment of active pharmaceutical ingrediants (APIs) and other synthetic chemicals, the subsequent discussion uses relevant examples from the available literature to highlight the need for a MOA approach to ecotoxicity testing for potential invertebrate EDCs.

III. Ecotoxicity Testing with Crustaceans and Insects

Both laboratory ecotoxicity data and field observations relating to abnormal sexual development in crustaceans and insects are reviewed by de Fur *et al.* (1999). Moreover, many environmental testing schemes for APIs or effluents from pharmaceutical manufacturing plants use test protocols with daphnids and other aquatic organisms. The endocrinology of crustaceans and insects is reviewed by de Fur *et al.* (1999) with a clear emphasis on the functional importance of ecdysteroids and juvenile hormone.

A. Amphipods

A number of standardized laboratory protocols exist for chronic toxicity testing using marine (*Corophium volutator*) and freshwater species (the amorous *Gammarus* sp., *Hyalella* sp., etc.) (Ingersoll *et al.*, 1999). Recently, Gross *et al.* (2001) reported abnormalities in sexual development of the amphipod *Gammarus pulex* found below sewage treatment works in England. They observed a highly significant number of abnormal females (abnormal structure of oocytes in vitellogenesis) collected from a site known to elicit high estrogenic responses in fish. Amphipod body size was significantly shorter, and male:female size differential was significantly reduced below one of the discharges. Watts *et al.* (2002) exposed laboratory populations of *G. pulex* to 17α-ethinylestradiol (EE2) (0.104–7.5 μg/L) for 100 days, observing significant increases in population size and in the proportion of female amphipods at higher EE2 exposures.

B. Chironomids

Freshwater chironomids are widely used in Europe and North America for assessing the toxicity of contaminated sediments. Watts *et al.* (2002) described chronic exposure of *Chironomus riparius* to EE2 and Bisphenol A (BPA) over two generations in chronic sediment exposure assays. Results showed that emergence time and the percentage of adult emergence were affected by EE2 and BPA exposure. These effects were primarily associated with the second generation of test animals, and most notably in the BPA study, where the emergence of male and female adults was significantly ($P < 0.05$) delayed at concentrations ranging from 78 ng/L to 0.75 mg/L. At very low concentrations (1 ng/L) of EE2, both the first and second generation of adults emerged significantly earlier than control animals. The authors noted, however, that although certain responses were significantly affected, their results in general do not suggest that the apical criteria examined could be used to detect the estrogenic effects of EE2 and BPA on chironomids. Also, Meregalli and Ollevier (2001) exposed *Chironomus riparius* larvae to EE2 (1, 10, and 100 μg/L as nominal concentrations) but saw no significant impacts in survival and mouthpart deformities.

C. Cladocerans

For several decades, species such as *Daphnia magna* have been widely used for regulatory environmental assessments of APIs and other chemicals, including some compounds that are established as EDCs in fish and mammals (Baldwin and LeBlanc, 1994; LeBlanc, 1999; LeBlanc and McLachlan, 1999). For example, laboratory studies by Baldwin *et al.* (1995) demonstrate reduced molting frequency in daphnids exposed to diethylstilbestrol (DES) (500 μg/L) but no reduction in survival or fecundity. Exposure to nonylphenol and its related polyethoxylate has also been shown to affect daphnid fecundity and sex determination, while studies on many other environmental contaminants and daphnids are reviewed by Ingersoll *et al.* (1999).

More recently, Olmstead and LeBlanc (2001) reported methoprene (a juvenile hormone agonist) toxicity to *Daphnia magna*, observing multiple mechanisms of toxicity and low-exposure concentration effects. Peterson *et al.* (2001) also exposed *Daphnia pulex* to the methoprene (10 and 100 μg/L) and saw effects on the incidence of all-male broods compared with controls. Since methoprene has been found to bind to the mammalian retinoid X receptor, Peterson *et al.* also tested the effects of retinoic acid on daphnid reproduction; however, neither 9-cis-retinoic acid nor all-trans-retinoic acid had any observable effect. Exposure of daphnids to 20-hydroxyecdysone (1, 10, or 100 μg/L) suggest that juvenile hormone and ecdysteroids might play a role in daphnid sex determination. Mu and LeBlanc (2002a) show that *D. magna* molt frequency was delayed by exposure to testosterone (nominally 8 μM, equivalent to 2307 μg/L); at higher nominal testosterone concentrations (nominally 25 μM, equivalent to 7210 μg/L), this effect was mitigated by co-exposure to the ecdysteroid 20-hydroxyecdysone.

Testosterone exposure concentrations that interfered with molting also elicited developmental abnormalities among neonatal organisms produced by maternal organisms continuously exposed to testosterone, or among embryos that were removed from unexposed mothers and exposed directly to the hormone. Mu and LeBlanc (2002a) concluded that the embryo toxicity of such high nominal exposures of testosterone to daphnids relates to disturbed ecdysteroid control of development. Moreover, testosterone significantly antagonized the action of 20-hydroxyecdysone in an ecdysone-responsive cell line, but had no discernable impact on endogenous ecdysone levels in daphnids.

Mu and LeBlanc (2002b) also reported that the fungicide fenarimol exhibits antiecdysteroidal activity to *D. magna* by lowering endogenous ecdysone levels and delaying molting (a response that is mitigated by co-exposure to exogenous 20-hydroxyecdysone). In these studies, developmental abnormalities were associated with suppressed ecdysone levels in the daphnid embryos; the abnormalities could be prevented by co-exposure to 20-hydroxyecdysone. Developmental abnormalities were also associated with reduced parental fecundity. These comprehensive studies demonstrate the need to consider both metabolic modulators and classical receptor-mediated endocrine-disrupter mechanisms when seeking to understand the potential impact of xenobiotics on invertebrates and other taxa.

D. Copepods

Prompted by reports of abnormal intersex copepods found near a UK sewage effluent discharge, Hutchinson *et al.* (1999a,b) used a marine copepod (*Tisbe battagliai*) 21-day life-cycle test to evaluate ecdysteroid and estrogen agonists and a pharmaceutical anti-estrogen (ZM189,154). While 20-hydroxyecdysone and DES were highly toxic (21 d LC50 values of 53.4 and 31.6 μg/L, respectively), compounds such as 17β-oestradiol and oestrone (10–100 μg/L) produced no adverse effects in copepod populations. Bechman (1999) also demonstrated the toxicity of nonylphenol (NP) to *Tisbe battagliai*; however, no mechanism-specific data were provided. More recently, Andersen *et al.* (2001) studied the copepod *Acartia tonsa* larval development during exposure to known vertebrate steroid hormone agonists (including Bisphenol A, oestradiol, oestrone, 17α-ethinylestradiol, p-octylphenol, progesterone, and testosterone) and antagonists (including flutamide and tamoxifen). Of the natural chemicals and pharmaceuticals studied, potent inhibitors of naupliar development (based on 5 d EC10 nominal values) were EE2 (46 μg/L), octylphenol (5.2 μg/L), and tamoxifen (8.7 per L). Testosterone and progesterone did not inhibit development; however, based on 5 d EC10 nominal values, inhibitory responses were seen for flutamide (135 μg/L) and hydroxyflutamide (5.3 μg/L). The ecdysteroid agonist 20-hydroxyecdysone did not produce any effect up to a nominal concentration of 1.5 mg/L. Finally, Brietholz and Bengtsson (2001) reported an absence of significant chronic effects in *Nitocra spinipes* exposed for 18 days to either DES (0.3–30 μg/L), EE2 (0.5–50 μg/L), or E2 (0.5–50 μg/L).

E. Decapods

Prompted by interest in biofouling, a number of research groups have addressed endocrine mechanisms of the development in barnacles. Interest has extended to environmental estrogens, especially the effects of 4-n-nonylphenol

(NP) and 17β-oestradiol (E2) on vitellin-like protein induction and settlement of *Balanus amphitrite* and early development of *Elminius modestus* (Billinghurst *et al.*, 1998, 2000). Barnacle early life-stages (nauplii and cyprids) were exposed in the laboratory to NP or E2; both compounds caused disruption of the timing of larval development with evidence that the timing of exposure is critical. Exposures for a period of 12 months caused long-term effects. Based on these and related studies, Billingshurst *et al.* (1998, 2000) suggest that vitellin-like proteins may act as biomarkers for estrogen exposure in oviparous crustaceans.

F. Mysids

In view of their importance in marine food chains, regulatory test protocols have been successfully developed in North America for a range of mysid species, most notably *Americamysis bahia* (previously *Mysidopsis bahia*). For example, mysids have been shown to be especially sensitive to pesticides, such the juvenile hormone agonist methoprene (McKenney *et al.*, 1996), while *Neomysis integer* has been used to study hormonal and metabolic responses following exposure to tributyltin (Verslycke *et al.*, 2003). To date, there appears to be no data available from mysid tests on human pharmaceuticals, perhaps due to the practical limitations of conducting high volume flow-through tests with these often expensive compounds. For example, in the author's experience, a six-week chronic flow-through test using *Americamysis bahia* would generate approximately 8,000 litres of effluent from the experimental system whereas a three week lifecycle test using marine copepods (Hutchinson *et al.*, 1999a) would produce less than two litres of test system effluent (also reflecting the approximate 4000-fold saving in the quantity of pharmaceutical compound required for copepod versus mysid lifecycle testing). Finally, for an excellent review of the ecology, ecotoxicology, endocrinology, and metabolic capability of mysids, see Verslycke *et al.* (2004).

G. *In Vitro* Studies and Mechanistic Data

While the crustacean and insect database for the apical developmental and reproductive effects of mammalian EDCs is greater than for other invertebrate taxa, there is no known functional role for mammalian-type steroids in arthropods (de Fur *et al.*, 1999). This analysis has led to recent multidisciplinary links between insect biochemists and ecotoxicologists, with the aim of using *in vitro* methods to identify true arthropod EDCs. Recently, Dinan *et al.* (2001) have used an *in vitro* ecdysteroid receptor-based (EcR) screening assay to evaluate a wide range of compounds, including mammalian steroids and some related APIs. All of the estrogens tested (including DES and alkylphenols) showed no response *in vitro*, suggesting

TABLE I Suggested Reference Chemicals for Evaluating Relative Endpoint or Invertebrate Species Sensitivity to Potential EDCs

Possible MOA	*Chemical*
Juvenile hormone (JH) agonist	Methoprene
JH antagonist	Precocene
Ecdysone agonist	20-hydroxyecdysone
Ecdysone antagonist	Homobrassinolide, luteolin
Aromatase inhibitor	Fadrozole
Androgen agonist	Methyltestosterone
Androgen antagonist	Flutamide
Estrogen agonist (weak)	4-*tert*-pentylphenol
Estrogen antagonist (strong)	17α-Ethinylestradiol
Estrogen antagonist	ZM189, 154

that the toxicity of these compounds to crustaceans (see above) is not mediated via the EcR. Compared with the APIs tested in crustaceans *in vivo*, Dinan *et al.* (2001) observed very weak antagonistic activity for EE2 only (equivalent to mg/L levels). In terms of estrogen antagonists, tamoxifen has not been tested using the EcR screen. However, Pagano *et al.* (2001) suggest that the embryo toxic effects seen in marine echinoderms may be mediated via oxygen radical overproduction that occurs in tamoxifen metabolic activation. Whether this mechanism is relevant to copepods (Andersen *et al.*, 2001) or other invertebrate taxa remains to be studied. Further work is needed using a combination of such *in vitro* and *in vivo* methods, ideally including use of the reference EDCs recommended by Ingersoll *et al.* (1999) (Table I).

IV. Conclusions and Recommendations

As in other aspects of ecotoxicology, attempts to help protect natural invertebrate populations from the potential adverse effects of synthetic chemicals involves testing a limited suite of species, followed by extrapolation to the wider world via the predicted no-effect concentration (PNEC). For potential EDCs, including a minority of APIs, such extrapolation will be most effective if there are robust chronic test protocols available, using a variety of species that encompass the reproductive endocrine systems present in the invertebrate fauna (Hutchinson *et al.*, 2000). At present, there is a limited but useful range of test methods available for measuring the developmental and reproductive effects of APIs and related chemicals to a given invertebrate group (copepods, daphnids, or oyster embryos) (de Fur *et al.*, 1999). However, as exemplified by the historical failure of tests to predict the chronic toxicity of TBT to marine mollusks, the greater challenge lies in having a sufficiently robust understanding of the MOA that will ensure accurate prediction of the PNEC for the relevant ecosystem. While this is a significant task, it is argued that the question of EDCs and invertebrate health has arguably helped to establish a new chapter of mechanistic ecotoxicology, to better avoid potential TBT scenarios of the future.

In the long term, an MOA approach to species selection for API testing is likely to be more cost-effective and may also have ethical benefits in reducing the need for vertebrate tests (Vogel *et al.*, 1995). More specifically, for the environmental assessment of potential invertebrate EDCs, including neuroendocrine disrupters, it is recommended that attention be given to: (1) using reference compounds to validate new invertebrate test methods and effect endpoints (Table I); (2) evaluating *in vitro* data from both mammalian-type and arthropod hormone receptor binding assays and assays of metabolic function; (3) utilizing small-scale test methods that do not generate effluent handling problems or use large quantities of APIs; (4) using environmentally relevant test concentrations and exposure routes, including supporting chemical analyses; and (5) considering improved test methods using species for which molecular tools are becoming available (the nematode *C. elegans*).

Finally, for the future environmental assessment of APIs, whatever their MOA, it is also highly desirable that emerging regulatory schemes are cost-effective and actively encourage an ethical approach to ecotoxicity testing. To achieve these aims, stakeholders in the environmental safety assessment of human and veterinary APIs need to allow the future use of validated novel ecotoxicity (especially invertebrate) testing methods that will improve upon today's limited range of standard protocols.

References

Alzieu, C. (2000). Impact of tributyltin on marine invertebrates. *Ecotoxicol.* **9,** 71–76.

Andersen, H. R., Wollenberger, L., Halling-Sørensen, B., and Kusk, K. O. (2001). Development of copepod nauplii to copepodites – a parameter for chronic toxicity including endocrine disruption. *Environ. Toxicol. Chem.* **20,** 2821–2829.

Anderson, G. L., Boyd, W. A., and Williams, P. L. (2001). Assessment of sublethal endpoints for toxicity testing with the nematode *Caenorhabditis elegans*. *Environ. Toxicol. Chem.* **20,** 833–838.

Baldwin, W. S., and LeBlanc, G. A. (1994). *In vivo* biotransformation of testosterone by phase I and II detoxification enzymes and their modulation by 20-hydroxyecdysone in *Daphnia magna*. *Aquat. Toxicol.* **29,** 103–117.

Baldwin, W. S., Milam, D. L., and LeBlanc, G. A. (1995). Physiological and biochemical perturbation in *Daphnia magna* following exposure to the model environmental estrogen diethylstilbestrol. *Environ. Toxicol. Chem.* **14,** 945–952.

Bechmann, R. K. (1999). Effect of the endocrine disrupter nonylphenol on the marine copepod *Tisbe battagliai. Sci. Tot. Environ.* **233,** 167–179.

Billingshurst, Z., Clare, A. S., Fileman, T., McEvoy, J., Readman, J., and Depledge, M. H. (1998). Inhibition of barnacle settlement by the environmental oestrogen 4-nonylphenol and the natural oestrogen 17β-oestradiol. *Mar. Poll. Bull.* **36,** 833–839.

Billingshurst, Z., Clare, A. S., Matsumura, K., and Depledge, M. H. (2000). Induction of cypris major protein in barnacle larvae by exposure to 4-*n*-nonylphenol and 17β-oestradiol. *Aquat. Toxicol.* **47,** 203–212.

Brietholz, M., and Bengtsson, B.-E. (2001). Oestrogens have no hormonal effect on the development and reproduction of the harpacticoid copepod Nitocra spinepes. *Mar. Poll. Bull.* **42,** 879–886.

Carnevali, M. D. C., Galassi, S., Bonasoro, F., Patruno, M., and Thorndyke, M. C. (2001). Regenerative response and endocrine disrupters in crinoid echinoderms: Arm regeneration in *Antedon mediterranea* after experimental exposure to polychlorinated biphenyls. *J. Exp. Biol.* **204,** 835–842.

de Fur, P., Crane, M., Ingersoll, C., and Tattersfield, L. (Eds.) (1999). *Endocrine disruption in invertebrates: Endocrinology, testing, and assessment.* SETAC Technical Publications Series, SETAC, Pensacola, Florida.

Dinan, L., Bourne, P., Whiting, P., Dhadialla, T. S., and Hutchinson, T. H. (2001). Screening of environmental contaminants for ecdysteroid agonist and antagonist activity using the *Drosophila melanogaster* B_{II} cell *in vitro* assay. *Environ. Toxicol. Chem.* **20,** 2038–2046.

European Commission (1996). European workshop on the impact of endocrine disrupters on human health and wildlife. *In* Report of proceedings from a workshop held in Weybridge, UK, December 2–4, 1996. Report reference EUR 17549, European Commission, DGXII, Brussels, Belgium.

Fong, P. P. (2001). Antidepressants in aquatic organisms: A wide range of effects. *In* "Pharmaceuticals and Personal Care Products in the Environment – Scientific and Regulatory Issues" (C. G. Daughton and T. L. Jones-Lepp, Eds.), p. 396. ACS Symposium Series 791, American Chemistry Society, Washington D.C.

Gross, M. Y., Maycock, D. S., Thorndyke, M. C., Morritt, D., and Crane, M. (2001). Abnormalities in sexual development of the amphipod *Gammarus pulex* (L.) found below sewage treatment works. *Environ. Toxicol. Chem.* **20,** 1792–1797.

Hutchinson, T. H., Pounds, N. A., Hampel, M., and Williams, T. D. (1999a). Life-cycle effects of 20-hydroxyecdysone and diethylstilbestrol on the marine copepod *Tisbe battagliai. Environ. Toxicol. Chem.* **18,** 2914–2920.

Hutchinson, T. H., Pounds, N. A., Hampel, M., and Williams, T. D. (1999b). Impact of ecdysteroids and oestrogens on developmental and reproductive parameters in the marine copepod *Tisbe battagliai. Sci. Tot. Environ.* **233,** 167–179.

Hutchinson, T. H., Brown, R., Brugger, K. E., Campbell, P. M., Holt, M., Länge, R., McCahon, P., Tattersfield, L. J., and van Egmond, R. (2000). Ecological risk assessment of endocrine disruptors. *Environ. Health Perspect.* **108,** 1007–1014.

Ingersoll, C. G., Hutchinson, T. H., Crane, M., Dodson, S., DeWitt, T., Gies, A., Huet, M-C., McKenney, C. L., Oberdörster, E., Pascoe, D., Versteeg, D. J., and Warwick, O. (1999). Laboratory toxicity tests for evaluating potential effects of endocrine disrupting compounds. *In* "Endocrine Disruption in Invertebrates: Endocrinology, Testing and Assessment. Proceedings of the EDIETA Workshop" (P. DeFur, M. Crane, C. Ingersoll, and L. Tattersfield, Eds.), pp. 107–197. Pensucola, Florida. SETAC Technical Publications Series ISBN 1-880611-27-9.

Laessig, S. A., McCarthy, M. M., and Silbergeld, E. K. (1999). Neurotoxic effects of endocrine disruptors. *Curr. Opin. Neurol.* **12,** 745–751.

LeBlanc, G. A. (1999). Steroid hormone-regulated processes in invertebrates and their susceptibility to environmental endocrine disruption. *In* "Environmental Endocrine Disruption: An

Evolutionary Perspective" (L. Guillette and D. A. Crain, Eds.), pp. 126–154. Taylor & Francis, New York.

LeBlanc, G. A., and McLachlan, J. B. (1999). Molt-independent growth inhibition of *Daphnia magna* by a vertebrate anti-androgen. *Environ. Toxicol. Chem.* **18,** 1450–1455.

Matthiessen, P., and Gibbs, P. E. (1998). Critical appraisal of the evidence for tributyltin-mediated endocrine disruption in mollusks. *Environ. Toxicol. Chem.* **17,** 37–43.

McKenney, C. L., Jr., and Celestial, D. M. (1996). Modified survival, growth and reproduction in an estuarine mysid (*Mysidopsis bahia*) exposed to a juvenile hormone analogue through a complete lifecycle. *Aquat. Toxicol.* **35,** 11–20.

Meregalli, G., and Ollevier, F. (2001). Exposure of *Chironomus riparius* larvae to 17alpha-ethynylestradiol: Effects on survival and mouthpart deformities. *Sci. Total Environ.* **269,** 157–161.

Mu, X., and LeBlanc, G. A. (2002a). Developmental toxicity of testosterone in the crustacean Daphnia magna involves anti-ecdysteroidal activity. *Gen. Comp. Endocrinol.* **129,** 127–133.

Mu, X., and LeBlanc, G. A. (2002b). Environmental antiecdysteroids alter embryo development in the Crustacean *Daphnia magna. J. Exp. Zool.* **292,** 287–292.

Nice, H. E., Morritt, D., Crane, M., and Thorndyke, M. (2003). Long-term and transgenerational effects of nonylphenol exposure at a key stage in the development of *Crassostrea gigas* – possible endocrine disruption? *Mar. Ecol. Prog. Ser.* **256,** 293–300.

Olmstead, A. W., and LeBlanc, G. L. (2001). Low exposure concentration effects of methoprene on endocrine-regulated processes in the crustacean *Daphnia magna. Toxicol. Sci.* **62,** 268–273.

Pagano, G., de Baise, A., Deeva, I. B., Degan, P., Doronin, Y. K., Iaccarino, M., Oral, R., Trieff, N. M., Warnau, M., and Korkina, L. G. (2001). The role of oxidative stress in developmental and reproductive toxicity of tamoxifen. *Life Sci.* **68,** 1735–1749.

Peterson, J. K., Kashian, D. R., and Dodson, S. I. (2001). Methoprene and 20-OH-ecdysone affect male production in *Daphnia pulex. Environ. Toxicol. Chem.* **20,** 582–588.

Solomon, E. P., Berg, L. R., Martin, D. W., and Villee, C. (1996). Biology. 4th ed., Saunders College Publishing, New York.

Verslycke, T. A., Fockedey, N., McKenney, C. L., Jr., Roast, S. D., Jones, M. B., Mees, J., and Janssen, C. R. (2004). Mysid crustaceans as potential test organisms for the evaluation of environmental endocrine disruption: A review. *Environ. Toxicol. Chem.* **23,** 1219–1234.

Verslycke, T., Poelmans, S., De Wasch, K., Vercauteren, J., Devos, C., Moens, L., Sandra, P., De Brabander, H. F., and Janssen, C. R. (2003). Testosterone metabolism in the estuarine mysid *Neomysis integer* (Crustacea: Mysidacae) following tributyltin exposure. *Environ. Toxicol. Chem.* **22,** 2030–2035.

Vogel, S. V., Beushausen, S., and Lester, D. S. (1995). Application of a membrane fusion assay for rapid drug screening. *Pharm. Res.* **12,** 1417–1422.

Vos, J. G., Dybing, E., Greim, H., Ladefoged, O., Lambré, C., Tarazona, J. V., Brandt, I., and Vetkaak, A. D. (2000). Health effects of endocrine disrupting chemicals on wildlife, with special reference to the European situation. *Crit. Rev. Toxicol.* **30,** 71–133.

Wasson, K. M., Gower, B. A., and Watts, S. A. (2000). Responses of ovaries and tests of *Lytechinus variegatus* (Echinodermata: Echinoidea) to dietary administration of estradiol, progesterone and testosterone. *Mar. Biol.* **137,** 245–255.

Watts, M. W., Pascoe, D., and Carroll, K. (2002). Population responses of the freshwater amphipod *Gammarus pulex* (L.) to an environmental oestrogen, 17α-ethinylestradiol. *Environ. Toxicol. Chem.* **21,** 445–450.

Wilson, E. O. (1999). *The Diversity of Life.* Penguin, London.

B. Köllner*[*]
B. Wasserrab[†]
U. Fischer[‡]
G. Kotterba[‡]
M. van den Heuvel[§]

*Federal Research Centre for Virus Diseases of Animals
Institute of Diagnostic Virology, Germany

[†]Environmental Toxicology, University of Konstanz, Konstanz, Germany

[‡]Federal Research Centre for Virus Diseases of Animals
Institute of Infectology Insel Riems, Germany

[§]Forest Research, Rotorua, New Zealand

How Can Toxic Effects of Pollution of the Aquatic Environment on the Immunocompetence of Fishes Be Detected? A Discussion on the Relevance of the Biomarkers Philosophy

I. Introduction

Pollution of the aquatic environment throughout the world with industrial, agricultural, or municipal sewage not only causes adverse effects in all aquatic organisms, but is also a problem for human health. Furthermore, it has negative influences on the efficient production of high-quality food in aquaculture for the increasing human population (Bols *et al.*, 2001; Bucher and Hofer, 1993; Burckhardt-Holm *et al.*, 1997; Christensen, 1998; Dunier and Siwicki, 1993; Zelikoff, 1998).

Although the concentration of pollutants in sewage treatment water (STW) and surface waters is usually low, the presence of a broad array of substances, combined with the high potential of some to disturb physiological functions, have raised concerns about possible toxic effects of pollutant mixtures. However, classical risk assessment cannot always predict the effects of such a mixture contamination on aquatic organisms (Daughton and Ternes, 1999; Halling-Sorensen *et al.*, 1998; Jones, 2001; Ternes, 1998). Impacts on fish populations have been observed due to alterations of the reproductive system (Jobling *et al.*, 1996; Larrson *et al.*, 1999; Matthiessen and Sumpter, 1998; Robinson *et al.*, 2003; Spies and Rice, 1988). Little is known about adverse effects of a mixture contamination on the immune system in fish (Fig. 1), possibly causing a decreased resistance against pathogens that results in mortality (Luebke *et al.*, 1997).

The use and acceptance of biological endpoints for environmental assessment by scientists and regulators has increased dramatically since the 1970s. This change has partially been driven by the observations that chemical criteria alone are not sufficient to predict or protect the ecological integrity of the environment. There are numerous examples demonstrating a high rate of ecological change or impairment when chemical criteria are met (Munkittrick *et al.*, 2002; Yoder and Rankin, 1998).

The development of biological endpoints in ecotoxicological science has led to the concept of biomarkers. Early definitions of biomarkers vary, but the concept of "economical, rapid, early warning indicators" seems to have persisted in the literature since some of the first definitions were developed (NRCC, 1985). McCarty and Munkittrick (1996) have termed this phenomenon "the search for the ultimate shortcut." The search is for a limited number of biological endpoints that can be measured in a subpopulation of organisms, in order to allow regulators or scientists to diagnose the health of a population of organisms. This approach, using a small number of selected immunological bioassays, has become increasingly common in the toxicological literature in both laboratory and field situations.

Chemically induced disease, associated with an increased probability of mortality, is the ecologically relevant endpoint subsequent to immune suppression. There are numerous examples of disease in fish observed after exposure to stressors such as pulp and paper mill effluent (Lindesjöö and Thulin, 1994; Sharples *et al.*, 1994; Wilklund and Bylund, 1996), petrochemical contamination (Burton *et al.*, 1984, van den Heuvel *et al.*, 2000), and municipal wastewater (Cross, 1984). In most cases, infections by opportunistic pathogens are observed; the pathology appears as nonlethal clinical signs, such as fin erosion or virally induced tumors. Little research is carried out in the field to demonstrate higher mortality rates in response to chemical stress, possibly due to the difficulties inherent in establishing such an event. Most field studies of overt diseases in fish exposed to chemicals are based on empirical observations, and there rarely is mechanistic evidence to establish direct causal inferences or to identify specific chemicals (if indeed they are the cause).

The mechanistic evaluation of aquatic environmental pollution by chemicals or drugs regarding potential adverse effects on the immunocompetence of fish depends on standardized test systems. Most

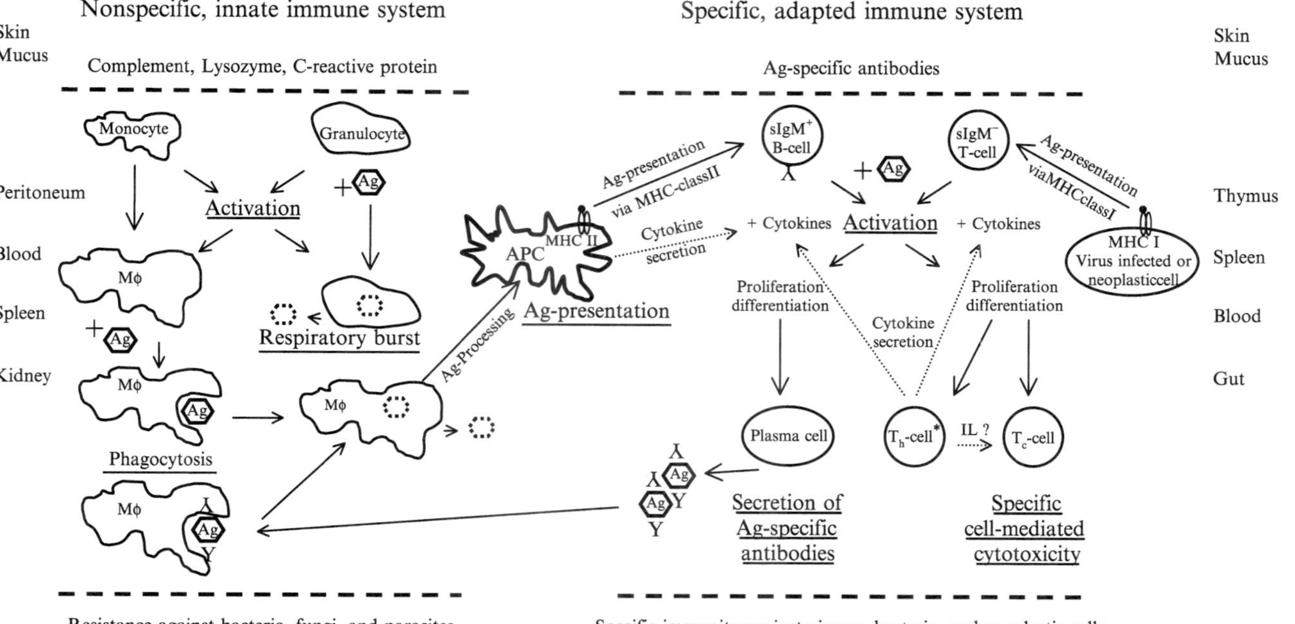

FIGURE 1 Schematic overview of known immune functions of bony fish. Dotted arrows indicate functions that are predicted from proliferation assays only. Ag, antigen; APC, antigen presenting cell; MHC, major histocompatibility complex; Mϕ, macrophage; sIgM, surface immunoglobulin M; T_h-cell, helper T lymphocytes; T_c-cell, cytotoxic T-lymphocytes, *, predicted from functional assays only and not shown as specific cell population.

immunotoxicological studies use only one or a limited number of parameters or functions as so-called biomarkers (Walker, 1998). Moreover, the studies only determine the overall ability of the resting immune system to react, but fail to investigate the influence on a specifically stimulated immune response (Twerdok *et al.*, 1996; Zelikoff *et al.*, 2000). Commonly used tests determine the amount of humoral factors of the innate (lysozyme, C-reactive protein, complement) and the acquired (immunoglobulin M, IgM) immune system, and some functional assays measure the phagocytotic activity or the proliferative response of immune cells. Complex, stimulated immune functions are generally not examined (Langezaal *et al.*, 2001; Luebke *et al.*, 1997; Luster *et al.*, 1992, 1993; Smith *et al.*, 1999).

Given the interconnectedness of physiological systems and the very complex regulation of immune responses, there is no effective way to predict the impact of changing individual immunological parameters on the host resistance. An increase or decrease of one or a limited number of parameters does not reflect the influence of a pollutant on the complex immune response involved in resistance to infections or cancer. The interactions between different immune functions, a possible overreaction of the immune system, or the compensation of one suppressed immune function by another one, cannot be revealed this way. Furthermore, pollutants seldom affect a single parameter or function (Keil *et al.*, 1999, 2001; Zelikoff *et al.*, 1995). The immune response of fish directly depends on both internal (age, sexual cycle) and external environmental (temperature, season) factors (Bly *et al.*, 1997; Köllner and Kotterba, 2002; Magnadottir *et al.*, 1999). Physiological stress mediated by cortisol also has well-known impacts on components of the immune system. Thus, it is often difficult to distinguish the direct effects of chemicals on the immune responses from indirect effects mediated by stress, or by toxicant impacts on sites of action rather than on specific immune components. (Fevolden *et al.*, 1994; Harris *et al.*, 2000; Pickering and Pottinger, 1989; Slater and Schreck, 1993; Kemenade *et al.*, 2000). These factors greatly influence immune functions, ranging from an almost non-reactive to a highly reactive state of all immune functions (Zapata and Amemiya, 2000, Zapata *et al.*, 1992). Stress factors, such as high population density and suboptimal environmental conditions, are one of the major problems in aquaculture, resulting in increased susceptibility of farmed fish to infectious diseases. Some of the influences are comparable to the situation of a polluted natural ecological system. In order to prevent dramatic loss of fish production due to infectious diseases, an extensive vaccination program has been introduced in aquaculture (Arkoosh *et al.*, 1998; Ellis, 1997; Gudding *et al.*, 1999). Unfortunately, the potency of these vaccines is not sufficient; one reason for their limited usefulness or benefit seems to be the above-mentioned stress and the resulting immunosuppression (Anderson, 1997).

Despite the limited knowledge of immune functions in fish, numerous functional assays have been developed to study antigen recognition, the interaction between immune cells (directly or via cytokines), and the regulation of effector mechanisms, (namely specific antibody secretion or cell-mediated cytotoxicity), in order to characterize natural immune responses to infectious agents (Arkoosh *et al.*, 1998; Fischer *et al.*, 1998; Köllner and Kotterba, 2002; Secombes *et al.*, 1990; Somamoto *et al.*, 2002; Zhou *et al.*, 2001).The methods discussed in this review can also be used to determine effects of pollutants affecting complex immune functions of fish by: (1) induction of a defined immune response to defined stimulants; (2) evaluation of the induced stimulation on different parts of the immune system from the first activation of immune cells to resistance against infectious microorganisms, as a result of a complex immune response; and (3) sensitivity of these methods to allow the detection of effects on distinct immune functions before a total loss (death of the effector cell), or a strong decrease or increase, of the function has occurred.

The *in vitro* methods described in this chapter are based on a specific *in vivo* triggering of the immune system before the measurement of cellular immune functions of different leukocyte populations. The influence of aquatic pollutants on immune functions can be determined by exposing the fish to a certain test substance, followed by a standardized *in vivo* stimulation of the immune response (pre-exposure) or by exposing the fish to the test substance simultaneously with the *in vivo* stimulation (co-exposure) (Fig. 2). The advantage of a standardized stimulation is the induction of a defined immune response, in which the difference between the real exposure to a certain pollutant and controls can be determined. In addition, several pollutants can be compared. In contrast to the *in vitro* determination of individual cellular based functions, such as phagocytosis and respiratory burst, the *in vivo* stimulation of the entire immune system reflects the capacity of complex immune functions. On the other hand, these complex immune reactions can be characterized in more detail using stimulants of known reactivity (bacteria, mitogens, allogeneic cells, virus derived peptides) (Fig. 2) yielding a stimulation of defined leukocyte subpopulations.

Parameters to be evaluated are activation and proliferation of leukocyte populations, phagocytosis and respiratory burst, secretion of antigen-specific antibodies, and specific cell-mediated cytotoxicity. Furthermore, *in vivo* challenge models with bacterial (*Aeromonas salmonicida*) and viral pathogens (viral hemorrhagic septicemia virus, VHSV) are presented (Fig. 3). Finally, the evaluation of a modified expression of genes involved in immune response by means of recently developed gene arrays will be discussed.

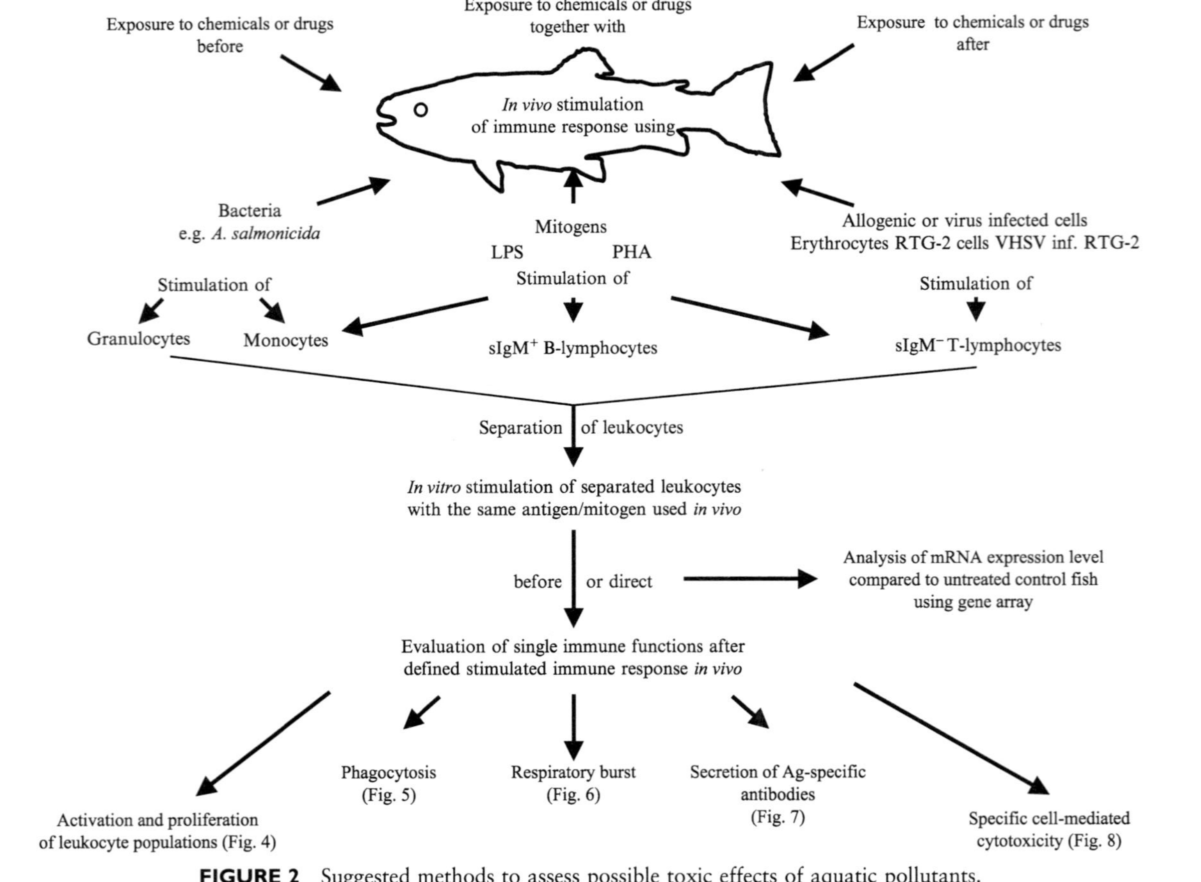

FIGURE 2 Suggested methods to assess possible toxic effects of aquatic pollutants.

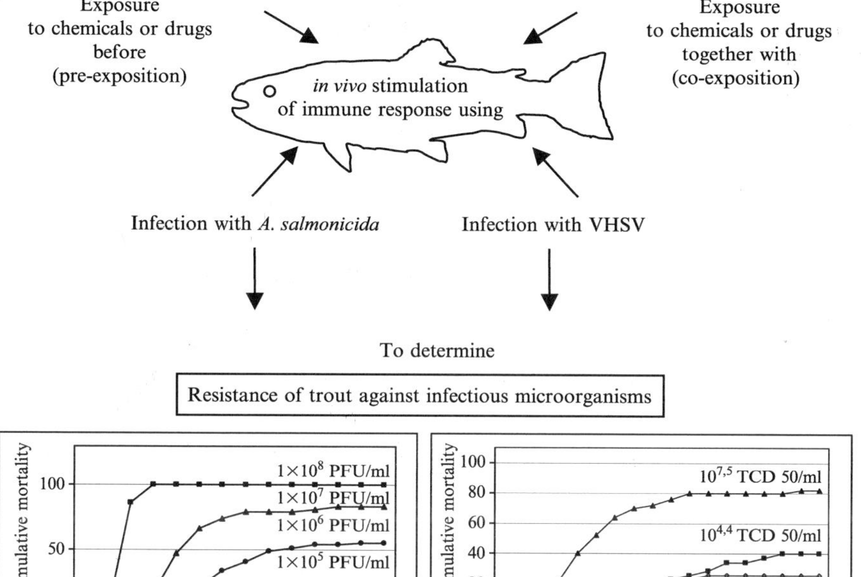

Cumulative mortality of healthy, unaffected trout after infection with different doses of *A. salmonicida*

Cumulative mortality of healthy, unaffected trout after infection with different doses of VHSV

FIGURE 3 Suggested evaluation model for the immunotoxic impact of pollutants by testing the altered resistance to infection with pathogenic microorganisms. TCD 50, tissue culture infectious dose; CFU, colony forming units; VHSV, viral hemorrhagic septicemia virus.

II. Evaluation of Natural Resistance or Immunity Against Bacterial and Viral Infection

Natural resistance against viral, bacterial, or fungal infection is mainly caused by a fully reactive innate immune system; immunity against such infections is based on an effective adaptive immune response (Fig. 1). Sewages affecting the innate and/or adaptive immune system by disturbing the function of single immune cell populations, the regulative compensation of affected immune cells, or disturbed immune functions lead to an increased mortality after microbial infection. Consequently, the correct determination of the possible immunotoxicity of a substance is based on the measurement of the inhibitory influence on the survival rate, after infection with virulent strains of bacteria or viruses (Zelikoff *et al.*, 1998, 2000).

A useful test to determine the resistance against a bacterial or viral infection is shown in Fig. 3. For these tests, highly virulent bacterial or virus strains, which cause mortality in untreated fish of a defined age under defined environmental conditions, should be used. The cumulative mortality after infection is easy to determine and can be adjusted to a value recommended for the testing of possible immunotoxic substances (Fig. 3). As stated before, the immunotoxic effects of aquatic pollutants can be determined using either a pre- or co-exposure method. To minimize inhibitory effects of the pollutant on bacteria or viruses, and to ensure an exact dose of microorganisms, an intraperitoneal rather than a water-born infection should be used.

The mortality test reflects a generalized toxic effect on the complex immune system. However, no data can be obtained on the level at which the immune system is affected. To answer this question, more detailed tests (described below) have to be performed.

III. Activation of Leukocytes and Leukocyte Subpopulations

The immune response against foreign antigens always starts with the activation and subsequent proliferation of leukocyte populations involved in the immune response (Rycyzyn *et al.*, 1998). The first cells to be activated are monocytes/macrophages and neutrophilic granulocytes, followed by IgM^+ Blymphocytes and IgM^- T-lymphocytes (Fig. 1). The activation depends on several internal (age, sexual cycle, antigenicity of invading antigen, health status) and external (water temperature, season) factors (Bly *et al.*, 1997). Inhibition or suppression of the ability to activate leukocyte populations leads to a "no-response-status" of the immune system to the pathogen, finally resulting in fatal outcome. Methods for evaluation of the activation/proliferation of leukocytes are based on different principles (Fig. 4): (1) measurement of the activity of mitochondrial enzymes using specific substrates (WST-1) (Blohm *et al.*, 2002); (2) measurement of the influx of Ca^{2+} ions into the cytoplasm using a fluorescent marker (Verburg-Van Kemenade *et al.*, 1998); (3) detection of the incorporation of ^3H thymidine into cellular DNA (Marsden *et al.*, 1995); and (4) detection of the percentage of leukocyte populations in lymphatic compartments (Köllner and Kotterba, 2002).

These methods are sensitive enough to distinguish the reactivity of leukocytes and leukocyte populations to different stimulants or different environmental temperatures, and should therefore also be useful for detecting stimulating or inhibiting effects of aquatic pollutants. One disadvantage of the use of leukocyte endpoints is their response to general stress.

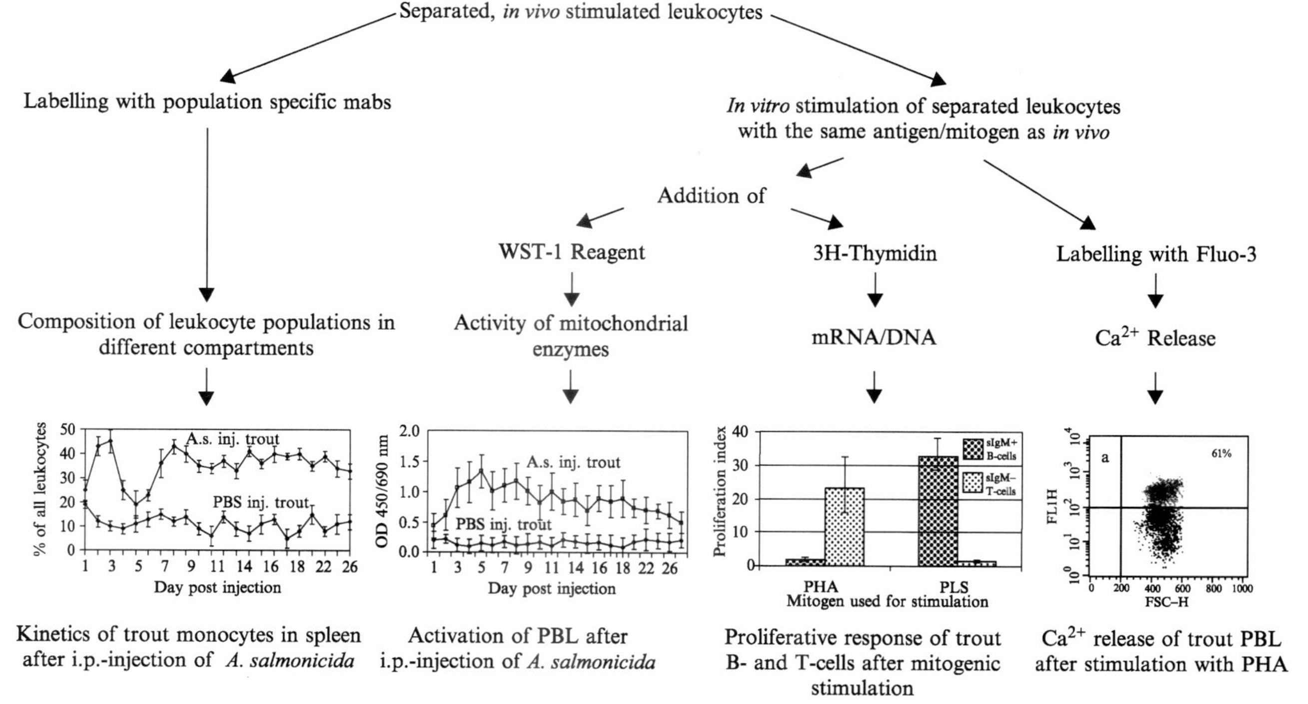

FIGURE 4 Methods to determine activation and proliferation of leukocyte populations.

Glucocorticoids, elevated during prolonged stress, are known to reduce the number of lymphocytes and monocytes while increasing the number of neutrophils (Bonga, 1997). Lymphocytes from cortisol-exposed fish have also been shown to be less responsive to mitogen-induced proliferation *in vitro* (Ellsaesser and Clem, 1987). Thus, it is recommended to monitor cortisol levels in order to distinguish a toxicant-induced leukocyte response gradient from a general stress-induced response, termed specific and nonspecific immune effects (Rice and Arkoosh, 2002).

IV. Phagocytosis

Phagocytosis is an innate immune function where phagocytes internalize, kill, and digest invading microorganisms. It has been shown in different fish species that circulating monocytes/macrophages and granulocytes form an integral immune defense network, capable of neutralizing a variety of invading pathogens and their secreted soluble factors by phagocytosis without prior activation (Fig. 1) (Ainsworth, 1992; Dannevig *et al.*, 1994; Secombes and Fletcher, 1992). It has also been shown that the activation of monocytes by bacterial lipopolysaccharide (LPS) or opsonization of microorganisms with complement components or antibodies leads to increased phagocytosis (Gudmundsdottir *et al.*, 1995; Solem *et al.*, 1995).

Furthermore, phagocytosis is the first step in the accessory function of monocytes and macrophages to stimulate lymphocyte response. Phagocytozed particles are processed and presented as antigenic peptides in association with class II MHC molecules on the surface of phagocytes. Subsequently, soluble mediators involved in lymphocyte activation, such as IL-1α, are secreted (Fig. 1) (Hong *et al.*, 2001). Inhibition of phagocytosis not only disturbs the clearance of bacteria, the processing, and presentation of antigens, but also cytokine secretion and subsequently the activation of a lymphocyte-based specific immune response.

The methods to measure phagocytosis are shown in Fig. 5. The sensitivity of both methods has been demonstrated by Solem *et al.* (1995), who detected a dose-dependent stimulation of phagocytosis using different concentrations of LPS.

V. Respiratory Burst

Free oxygen and nitrogen radicals kill invading microorganisms, either extracellularly without prior phagocytosis, or intracellularly after phagocytosis (Novoa *et al.*, 1996a,b; Secombes and Fletcher, 1992). This antigen-nonspecific immune function is executed by macrophages and granulocytes (Fig. 1). During respiratory burst, toxic oxygen products, such as superoxide and hydrogen peroxide, are produced and released into the surrounding

Phagocytosis

A process in which invading bacteria are internalized and killed by phagocytes. Phagocytosis initiates the processing and presentation of foreign antigens and the induction of cytokine secretion leading to lymphocyte activation

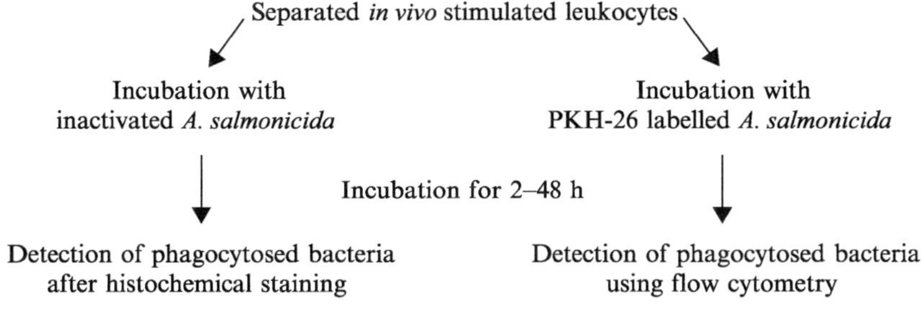

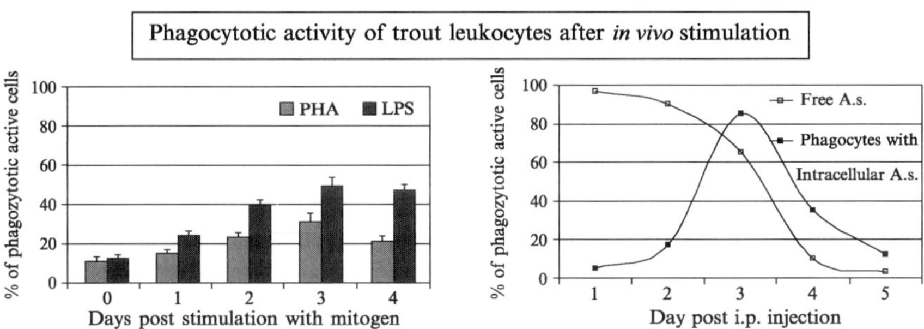

Reactivity of trout head kidney leukocytes in a phagocytosis assay after *in vivo* stimulation with different mitogens

Phagocytosis of *A. salmonicida* in trout peritoneal cavity by peritoneal leukocytes. At day 6 neither free *A. salmonicida* nor phagoctes with intracellular bacteria were detected (not shown).

FIGURE 5 Suggested methods to determine phagocytotic activity of fish leukocytes.

tissue (Jang *et al.*, 1995; Secombes, 1990). Inhibition of this process by immunotoxic substances leads to a prolonged survival of invading bacteria or fungi, followed by increased growth and a subsequent increase of pathogenic influence on normal physiology with a possible fatal outcome (Fig. 1). Most methods of measuring respiratory burst are based on the detection of intracellular superoxide production using tetrazolium salts (Fig. 6).

VI. Antibody Secretion

Antibodies are produced by B-lymphocytes directly after stimulation with antigen, or by antigenic peptides presented in association with MHC class II molecules on accessory cells. Cytokines, secreted by macrophages

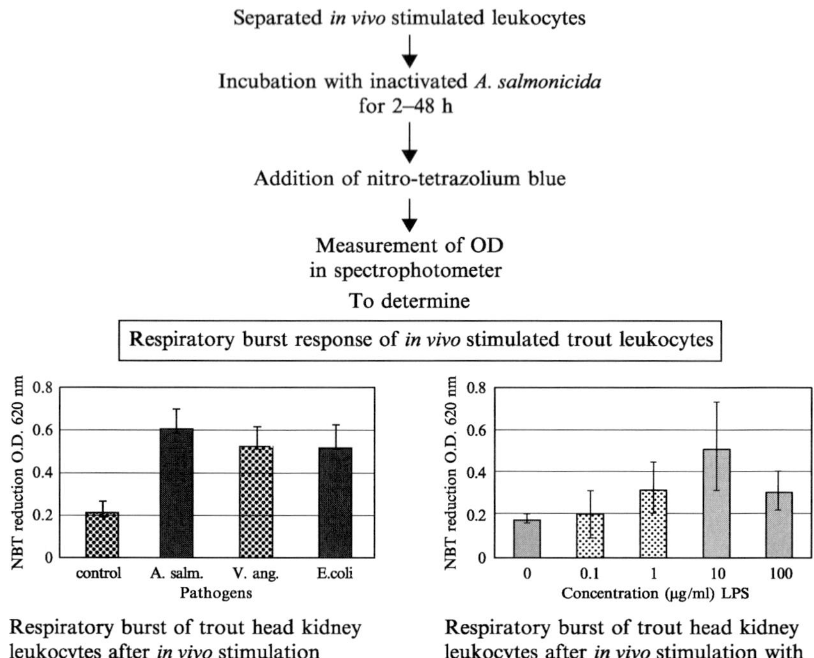

FIGURE 6 A method to determine respiratory burst.

and possibly $T_{(helper?)}$ cells that have not been clearly identified in fish yet, act as co-stimulatory molecules (Fig. 1) (Cain *et al.*, 2002; Christie, 1997). Strong antigens, such as bacterial LPS, induce an antibody response by direct activation of B-lymphocytes (Ellis, 1997). In fish, secreted antibodies in serum and mucus are of IgM only (Andersson *et al.*, 1995; Castillo *et al.*, 1993). After antigen binding, different effector functions are initiated: enhanced phagocytosis of the opsonized antigens by macrophages; complement-dependent cytolysis; antibody-dependent cell-mediated cytotoxicity (ADCC) via Fc-receptors on natural cytotoxic cells (NCC); and virus neutralization by blocking the entry into susceptible cells (Fig. 1) (Anderson and Jeney, 1992; Boudinot *et al.*, 1998; Emmenegger *et al.*, 1995; Hattenberger-Baudouy *et al.*, 1995; Kaattari *et al.*, 2002; Sharp *et al.*, 1992). Inhibition of the various functions of antibodies therefore results in an increased susceptibility to infectious diseases. One of the available tests used for the detection of antigen-specific antibody secretion

Secretion of antigen-specific antibodies

Specific immune response of sIgM+ B-lymphocytes, secreting antigen specific antibodies after presentation of foreign antigenic peptides via MHC class II by macrophages

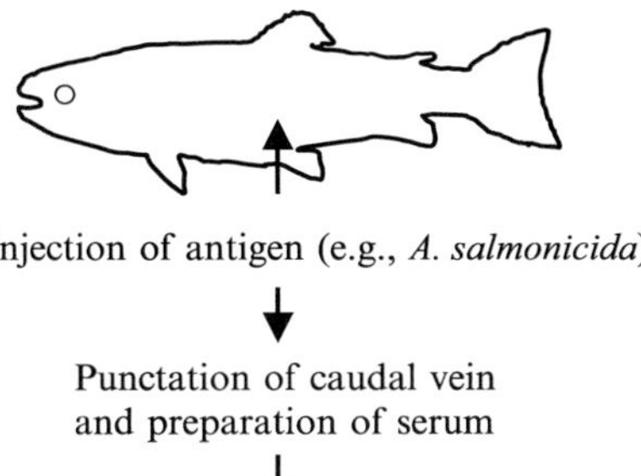

Titration of sera using ELISA

To determine

Antibody response of trout after single i.p.-injection of *A. salmonicida*

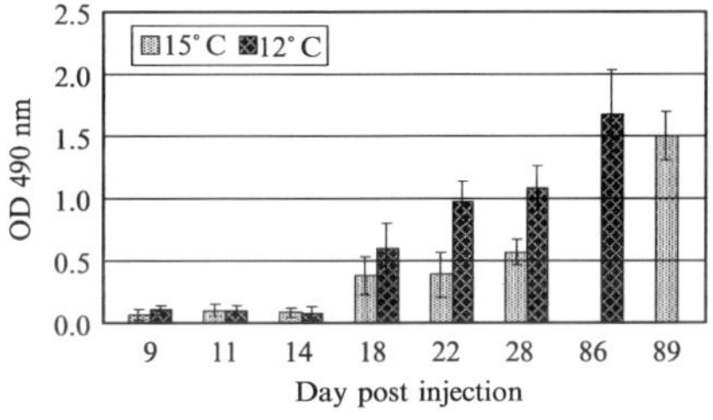

Temperature-dependent antibody response after i.p.-injection of inactivated *A. salmonicida*

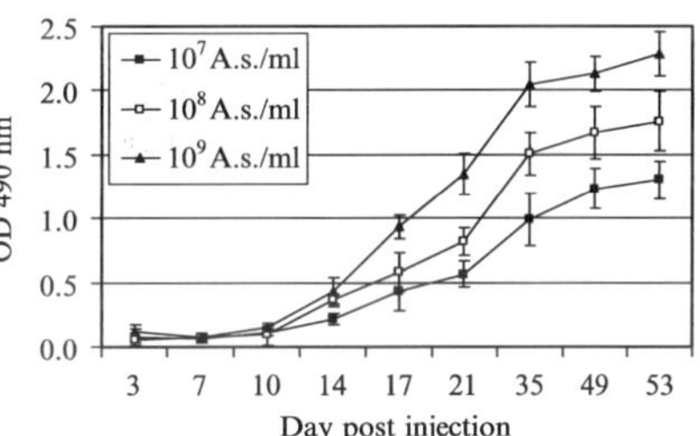

Dose-dependent antibody response after i.p.-injection of different amounts of inactivated *A. salmonicida*

FIGURE 7 Detection of antigen-specific antibodies in fish sera.

is based on an ELISA after a single immunization with the bacteria *A. salmonicida* (Köllner and Kotterba, 2002) (Fig. 7). The antibody response was strongly temperature-dependent (Fig. 7). Application of different amounts of inactivated bacteria resulted in direct dose-dependent antibody secretion (Gudmundsdottir *et al.*, 1995).

VII. Specific Cell-Mediated Cytotoxicity

Specific cell-mediated cytotoxicity, only recently shown in fish, is involved in the immune response against viral infection and cancer cells (Dijkstra *et al.*, 2001; Nakanishi *et al.*, 2002). Specifically sensitized sIgM$^-$ lymphocytes kill allogeneic cells or virus-infected fish cells (Fischer *et al.*, 1998, 2003; Somamoto *et al.*, 2000; Stuge *et al.*, 1997, 1995). *In vitro* assays of cell-mediated cytotoxicity are based on the release of hemoglobin from allogeneic erythrocytes (Fischer *et al.*, 1998), on the release of lactate dehydrogenase (LDH) from allogeneic or virus-infected cells (Fischer *et al.*, 1998), or on the release of ^{51}Cr from labelled target cells (Somamoto *et al.*, 2000).

The assay system for cell-mediated cytotoxicity is described in Fig. 8. For the detection of specific cell-mediated cytotoxicity against virus-infected cells, a system of MHC class I matching effector and target cells is required (MHC class I restriction of cytotoxicity). Preferably, clonal fish and target cell lines with a known MHC class I genotype are used. Furthermore, the use of clonal or inbred fish is required to obtain comparable data from differently treated groups. As shown in Fig. 8, the cytotoxic reaction is strongly specific and depends on prior sensitization.

VIII. Gene Array

Gene array analysis is a method to analyze the differential expression of thousands of genes simultaneously. For the purposes of immunological/toxicological assessment, this would typically be performed in infected vs noninfected, exposed vs unexposed, or at various developmental stages of an organism or tissue. It would give insight into: (1) patterns of genes expressed in a cell at its current state, where virtually all differences in cell state correlate with changes in mRNA levels of many genes (snapshot of the genome), or (2) matching experimental treatments with key genes or clusters of genes, based upon expression patterns (expression profiling).

The method involves reverse transcription of the complete sample of mRNA into labeled cDNA using chemiluminescent/fluorescent dyes, and hybridization of the resulting cDNA onto identical gene arrays. When the hybridized slide is fed into a scanner, the different light patterns created by the chemiluminescent dyes on the labeled cDNAs indicate whether certain genes are over- or under-expressed in comparison to other, unaffected genes. Exposure of certain genes to toxicants at a critical stage of development can affect their expression, or the process by which their encoded information is transferred to the organism's cells. Therefore, differences in expression patterns may indicate that a chemical has affected the organism. Genomic studies thus can provide evidence of the potential deleterious effects of a chemical on an organism at the molecular or functional level.

In the field of environmental toxicology, gene array slides could be used to examine the effects of toxic compounds in agricultural runoff, industrial effluent, and municipal wastewater on aquatic organisms. In the area of fish

Specific cell-mediated cytotoxicity

Specific immune response of sIgM⁻ T-Lymphocytes that kill allogeneic or infected isogenic cells after presentation of foreign antigenic peptides via MHC class I molecules

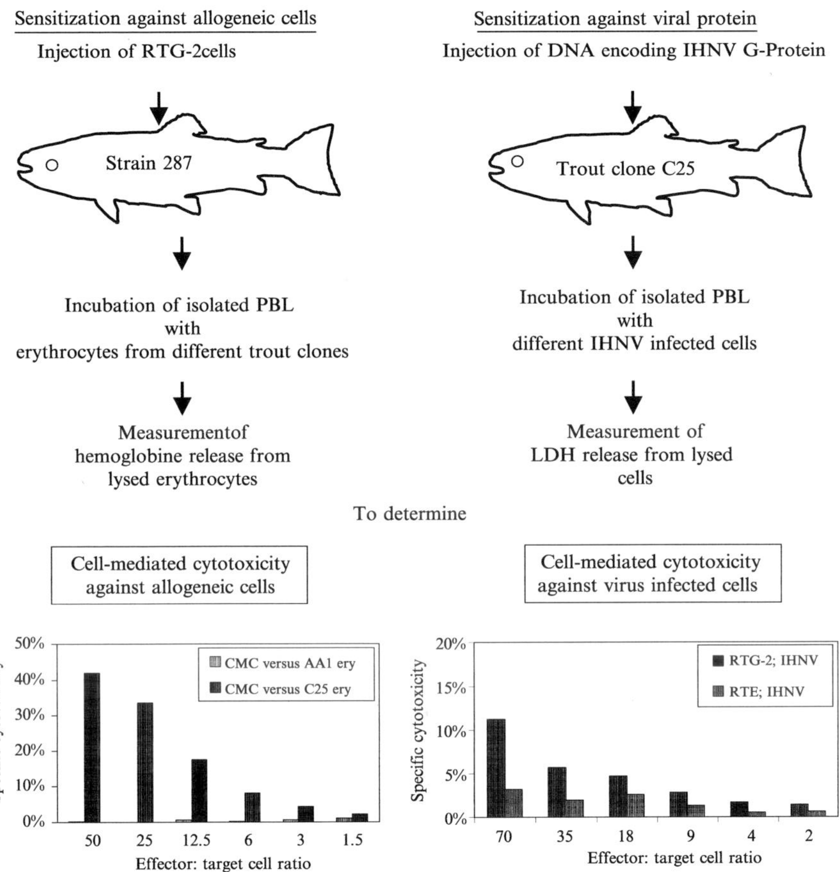

Cell-mediated cytotoxicity of trout leukocytes from strain 287 previously sensitized with allogeneic RTG-2 cells (MHC I = C25, ≠ 287 ≠ AA1) against MHC I matching (strain C25) and mismatching (strain AA1) erythrocytes

Cell-mediated cytotoxicity of trout leukocytes from strain C25 previously immunized with IHNV-DNA against IHNV infected MHC I matching (RTG-2, MHC I = C25) and mismatching (RTE MHC I ≠ C25) cells

FIGURE 8 Determination of specific cell-mediated cytotoxicity after *in vivo* sensitization with allogeneic cells or virus antigens.

toxicology, a gene array of about 150 genes involved in the response to chemical pollution for rainbow trout has been developed recently. Application of gene arrays might be specifically advantageous in the field of fish immunology because important molecules, like T-cell-receptors and CD8, have only been identified on the gene and mRNA level, and can therefore not be detected on the protein level (Hansen *et al.*, 1999; Partula *et al.*, 1996). Gene arrays thus have the potential to indicate changes in marker molecules, which could otherwise not be identified at all. Moreover, gene arrays allow a fast screening of a broad range of molecules, which can be regarded as first indicators of changes and potentially adverse effects, directing further functional studies into the right directions.

IX. Discussion

The knowledge of fish immunology and the tools available to investigate changes in immunocompetence, including changes caused by toxicants, continue to increase at a rapid pace. The increasing availability of these tools has led to their use as biomarkers in toxicological laboratory experimentation and field studies. The nature and complexity of the immune system pose several difficulties for this approach. The development of biomarkers was based on their ability to predict impacts at higher levels of biological organization. Ecologists would argue that as the level of biological complexity increases, emergent properties, or properties that could not be predicted based on the understanding of lower levels of organization, will occur (Kerr, 1976). Thus, predictions of more ecologically relevant impacts, particularly mortality in the case of immunological endpoints, may not always be possible. The term bioindicator has been proposed to indicate changes in order to estimate some ecotoxicological or environmental change of significance (Adams, 2002; McCarty and Munkittrick, 1996). A prerequisite for the use of bioindicators is that they be validated; links to effects at higher levels of biological organization must be demonstrated. Given that in most cases, the use of immunological endpoints does not include a measure of their capacity to respond to an immune challenge, the relevance of these endpoints is unclear. Immunocompetence can only be addressed through response to challenge. Thus, measurements in resting immune systems are not necessarily good bioindicators. This challenge can be accomplished in three ways: (1) *in vivo* stimulation *prior to* treatment of fish with a pollutant (pre-exposure stimulation) in order to evaluate *in vitro* the ability of a given pollutant to affect immune cells that were activated and proliferate in response to immune stimulants (antigens, mitogens, infectious microorganism); (2) *in vivo* stimulation *together with* treatment of fish with a pollutant (co-exposure stimulation) in order to determine *in vitro* the ability of a given pollutant to affect the response of immune cells to immune stimulants (antigens, mitogens, infectious microorganism after activation/stimulation); and (3) *in vivo* stimulation *after* treatment of fish with a pollutant (post-exposure stimulation) in order to evaluate *in vitro* the ability of probably affected immune cells to be activated and to proliferate in response to immune stimulants (antigens, mitogens, infectious microorganism).

These approaches provide an evaluation of immunotoxic effects of pollutants on a defined immune response to defined pathogens or stimulants.

In addition to the lack of validation, selected immunological endpoints seldom have strong causal links to specific toxicants. For this reason, they are not particularly suitable for defining cause and effect. The use of immunological endpoints as bioindicators is complicated, due to a number of characteristics that are inherent in the immune system and in the bioassays used: (1) the immune system is highly complex and multifaceted, and failure to reject the null hypothesis of no effect cannot be assumed unless components of all parts of the immune system are examined; (2) different components of the immune system have responses that vary significantly on a temporal scale, and measurement of endpoints at arbitrarily selected time points will not capture the full nature of an immune response; examining immune responses fully over a long time period is essential to understand the changes caused by an immune stressor; (3) nonspecific modifying factors such as stress and/or temperature (in poikilotherms) have dramatic effects on components of the immune system, and more detailed mechanistic research is required before true immunological impacts can be distinguished from indirect effects on the immunocompetence; and (4) use of immunological endpoints in the field still poses logistical difficulties due to the requirements of fresh tissue and specialized equipment for the many functional bioassays described here.

In summary, the use of bioassays, based on a defined stimulated immune response as bioindicators, has the potential to allow a better understanding of the risks of anthropogenic stress on the functionality of the immune system, but the basic understanding of the science has not yet evolved to this point. At present, the utility of this powerful tool should be restricted to laboratory studies directed at better understanding the mechanism of effects of different chemical classes on the immune system and their relationship to modifying factors of immunocompetence, or on establishing cause and effect where chemically induced disease is observed in the field.

References

Adams, S. M. (2002). Biological indicators of aquatic ecosystem stress: Introduction and overview. *In* "Biological Indicators of Aquatic Ecosystem Stress" (S. M. Adams, Ed.), pp. 1–11. American Fisheries Society, Bethesda, Maryland.

Ainsworth, A. J. (1992). Fish granulocytes: Morphology, distribution and function. *Ann. Rev. Fish Dis.* **2**, 123–148.

Anderson, D. P. (1997). Adjuvants and immunostimulants for enhancing vaccine potency in fish. *Dev. Biol. Stand.* **90**, 257–265.

Anderson, D. P., and Jeney, G. (1992). Immunostimulants added to injected *Aeromonas salmonicida* bacterin enhance the defense mechanisms and protection in rainbow trout (*Oncorhynchus mykiss*). *Vet. Immunol. Immunopathol.* **34**, 379–389.

Andersson, E., Peixoto, B., Tormanen, V., and Matsunaga, T. (1995). Evolution of the immunoglobulin M constant region genes of salmonid fish, rainbow trout (*Oncorhynchus mykiss*) and Arctic charr (*Salvelinus alpinus*): Implications concerning divergence time of species. *Immunogenetics* **41**, 312–315.

Arkoosh, M. R., Casillas, E., Clemons, E., Kagley, A. N., Olson, R., Reno, P., and Stein, J. E. (1998). Effect of pollution on fish diseases: Potential impact on salmonid populations. *J. Aquatic Animal Health.* **10**, 182–190.

Beaman, J. R., Finch, R., Gardner, H., Hoffmann, F., Rosencrance, A., and Zelikoff, J. T. (1999). Mammalian immunoassays for predicting the toxicity of malathion in a laboratory fish model. *J. Toxicol. Environ. Health. A* **56**, 523–542.

Blohm, U., Siegl, E., and Köllner, B. (2002). Rainbow trout (*Oncorhynchus mykiss*) slgM$^-$ leucocytes secrete an interleukin-2 like growth factor after mitogenic stimulation *in vitro*. *Fish Shellfish Immunol.*, submitted.

Bly, J. E., Quiniou, S. M., and Clem, L. W. (1997). Environmental effects on fish immune mechanisms. *Dev. Biol. Stand.* **90**, 33–43.

Bols, N. C., Brubacher, J. L., Ganassin, R. C., and Lee, L. E. (2001). Ecotoxicology and innate immunity in fish. *Dev. Comp. Immunol.* **25**, 853–873.

Bonga, S. E. W. (1997). The stress response in fish. *Physiol. Rev.* **77**, 591–625.

Boudinot, P., Blanco, M., de Kinkelin, P., and Benmansour, A. (1998). Combined DNA immunization with the glycoprotein gene of viral hemorrhagic septicemia virus and infectious hematopoietic necrosis virus induces double-specific protective immunity and nonspecific response in rainbow trout. *Virology* **249**, 297–306.

Bucher, F., and Hofer, R. (1993). The effects of treated domestic sewage on three organs (gills, kidney, liver) of brown trout (*Salmo trutta*). *Water Res.* **27**, 255–261.

Burckhardt-Holm, P., Escher, M., and Meier, W. (1997). Waste-water management plant effluents cause cellular alterations in the skin of brown trout. *J. Fish Biol.* **50**, 744–758.

Burton, D., Burton, M. P., and Idler, D. R. (1984). Epidermal conditions in post spawned winter flounder, Pseudopleuronectes americanus (Walbaum), maintained in the laboratory and after exposure to crude petroleum. *J. Fish. Biol.* **25**, 593–606.

Cain, K. D., Jones, D. R., and Raison, R. L. (2002). Antibody-antigen kinetics following immunization of rainbow trout (*Oncorhynchus mykiss*) with a T-cell dependent antigen. *Dev. Comp. Immunol.* **26**, 181–190.

Castillo, A., Sanchez, C., Dominguez, J., Kaattari, S. L., and Villena, A. J. (1993). Ontogeny of IgM and IgM-bearing cells in rainbow trout. *Dev. Comp. Immunol.* **17**, 419–424.

Christensen, F. M. (1998). Pharmaceuticals in the environment—a human risk? *Reg. Toxicol. Pharmacol.* **28**, 212–221.

Christie, K. E. (1997). Immunization with viral antigens: Infectious pancreatic necrosis. *Dev. Biol. Stand.* **90**, 191–199.

Cross, N. J. (1984). Fin erosion among fishes collected near a southern California municipal wastewater outfall (1971–1982). *Fish. Bull U.S.* **83**, 195–206.

Dannevig, B. H., Lauve, A., Press, C., and Landsverk, T. (1994). Receptor-mediated endocytosis and phagocytosis by rainbow trout head kidney sinusoidal cells. *Fish Shellfish Immunol.* **4**, 3–18.

Daughton, C. G., and Ternes, T. A. (1999). Cradle-to-cradle stewardship of drugs for minimizing their environmental disposition while promoting human health. I. Rationale for and avenues toward a green pharmacy. *Envir. Health Perspect.* **111**, 757–774.

Dijkstra, J. M., Fischer, U., Sawamoto, Y., Ototake, M., and Nakanishi, T. (2001). Exogenous antigens and the stimulation of MHC class I restricted cell-mediated cytotoxicity: Possible strategies for fish vaccines. *Fish Shellfish Immunol.* **11**, 437–458.

Dunier, M., and Siwicki, A. K. (1993). Effects of pesticides and other organic pollutants in the aquatic environment on immunity of fish: A review. *Fish Shellfish Immunol.* **3**, 423–438.

Ellis, A. E. (2001). Innate host defense mechanisms of fish against viruses and bacteria. *Dev. Comp. Immunol.* **25**, 827–839.

Ellis, A. E. (1997). Immunization with bacterial antigens: Furunculosis. *Dev. Biol. Stand.* **90**, 107–116.

Ellsaesser, C. F., and Clem, L. W. (1987). Cortisol-induced hematologic and immunologic changes in channel catfish, Ictalurus punctatus. *Comp. Biochem. Physiol. A.* **87**, 405–408.

Fevolden, S. E., Røed, K. H., and Gjerde, B. (1994). Genetic components of post-stress cortisol and lysozyme activity in Atlantic salmon; correlations to disease resistance. *Fish Shellfish Immunol.* **4**, 507–519.

Fischer, U., Ototake, M., and Nakanishi, T. (1998). *In-vitro* cell mediated cytotoxicity against allogeneic erythrocytes in gimbuna crucian carp and goldfish using a non-radioactive assay. *Dev. Comp. Immunol.* **22**, 195–206.

Fischer, U., Utke, K., Ototake, M., Dijkstra, J. M., and Köllnern, B. (2003). Adaptive cell-mediated cytotoxicity against allogeneic targets by CD8-positive lymphocytes in rainbow trout (*Oncorhynchus mykiss*). *Dev. Comp. Immunol.* **27**, 323–337.

Gudding, R., Lillehaug, A., and Evensen, O. (1999). Recent developments in fish vaccinology. *Vet. Immunol. Immunopathol.* **72**, 203–212.

Gudmundsdottir, S., Magnadottir, B., and Gudmundsdottir, B. K. (1995). Effects of antigens from *Aeromonas salmonicida* ssp *achromogenes* on leukocytes from primed and unprimed Atlantic salmon (*Salmo salar* L.). *Fish Shellfish Immunol.* **5**, 493–504.

Halling-Sorensen, B., Nors Nielsen, S., Lanzky, P. F., Ingerslev, F., Holten Lutzhoft, H. C., and Jorgensen, S. E. (1998). Occurrence, fate and effects of pharmaceutical substances in the environment—a review. *Chemosphere* **36**, 357–393.

Hansen, J. D., Strassburger, P., Thorgaard, G. H., Young, W. P., and Du Pasquier, L. (1999). Expression, linkage, and polymorphism of MHC-related genes in rainbow trout, *Oncorhynchus mykiss*. *J. Immunol.* **163**, 774–786.

Harris, J., and Bird, D. J. (2000). Modulation of the fish immune system by hormones. *Vet. Immunol. Immunopathol.* **77**, 163–176.

Harris, P. D., Soleng, A., and Bakke, T. A. (2000). Increased susceptibility of salmonids to the monogenean *Gyrodactylus salaris* following administration of hydrocortisone acetate. *Parasitology* **120**, 57–64.

Hattenberger-Baudouy, A. M., Danton, M., Merle, G., and de Kinkelin, P. (1995). Serum neutralization test for epidemiological studies of salmonid rhabdoviruses in France. *Vet. Res.* **5–6**, 512–520.

Hong, S., Zou, J., Crampe, M., Peddie, S., Scapigliati, G., Bols, N., Cunningham, C., and Secombes, C. J. (2001). The production and bioactivity of rainbow trout (*Oncorhynchus mykiss*) recombinant IL-1 beta. *Vet. Immunol. Immunopathol.* **81**, 1–14.

Jang, S. I., Hardie, L. J., and Secombes, C. J. (1995). Elevation of rainbow trout *Oncorhynchus mykiss* macrophage respiratory burst activity with macrophage-derived supernatants. *J. Leukoc. Biol.* **57**, 943–947.

Jobling, S., Sheahan, D., Osborne, J. A., Matthiessen, P., and Sumpter, J. P. (1996). Inhibition of testicular growth in rainbow trout (*Oncorhynchus mykiss*) exposed to estrogenic alkylphenolic chemicals. *Environ. Toxicol. Chem.* **15**, 194–202.

Jones, S. R. (2001). The occurrence and mechanisms of innate immunity against parasites in fish. *Dev. Comp. Immunol.* **25**, 841–852.

Kaattari, S. L., Zhang, H. L., Khor, I. W., Kaattari, I. M., and Shapiro, D. A. (2002). Affinity maturation in trout: Clonal dominance of high affinity antibodies late in the immune response. *Dev. Comp. Immunol.* **26**, 191–200.

Keil, D., Luebke, R. W., Ensley, M., Gerard, P. D., and Pruett, S. B. (1999). Evaluation of multivariate statistical methods for analysis and modeling of immunotoxicology data. *Toxicol. Sci.* **51**, 245–258.

Keil, D., Luebke, R. W., and Pruett, S. B. (2001). Quantifying the relationship between multiple immunological parameters and host resistance: Probing the limits of reductionism. *J. Immunol.* **167**, 4543–4552.

Kerr, S. R. (1976). Ecological analysis and the Fry paradigm. *J. Fish. Res. Bd. Can.* **33**, 329–339.

Köllner, B., and Kotterba, G. (2002). Temperature dependent activation of leukocyte populations of rainbow trout, *Oncorhynchus mykiss*, after intraperitoneal immunisation with *Aeromonas salmonicida*. *Fish Shellfish Immunol.* **12**, 35–48.

Langezaal, I., Coecke, S., and Hartung, T. (2001). Whole blood cytokine response as a measure of immunotoxicity. *Toxicol. In Vitro.* **15**, 313–318.

Larrson, D., Adolfsson-Ericiu, M., Parkonen, J., Petterson, M., Berg, A. H., Olsson, P.-E., and Förlin, L. (1999). Ethinyloestradiol—an undesired fish contraceptive? *Aquatic Toxicol.* **45**, 91–97.

Lindesjöö, E., and Thulin, J. (1994). Histopathology of skin and gills of fish in pulp mill effluents. *Dis Aquat. Org.* **18**, 81–93.

Luebke, R. W., Hodson, P. V., Faisal, M., Ross, P. S., Grasman, K. A., and Zelikoff, J. (1997). Aquatic pollution-induced immunotoxicity in wildlife species. *Fundam. Appl. Toxicol.* **37**, 1–15.

Luster, M. I., Portier, C., Pait, D. G., White, K. L., Jr., Gennings, C., Munson, A. E., and Rosenthal, G. J. (1992). Risk assessment in immunotoxicology. I. Sensitivity and predictability of immune tests. *Fundam. Appl. Toxicol.* **18**, 200–210.

Luster, M. I., Portier, C., Pait, D. G., Rosenthal, G. J., Germolec, D. R., Corsini, E., Blaylock, B. L., Pollock, P., Kouchi, Y., and Craig, W. (1993). Risk assessment in immunotoxicology. II. Relationships between immune and host resistance tests. *Fundam. Appl. Toxicol.* **21**, 71–82.

Magnadottir, B., Jonsdottir, H., Helgason, S., Bjornsson, B., Jorgensen, T. O., and Pilström, L. (1999). Humoral immune parameters in Atlantic cod (*Gadus morhua* L.) I. The effects of environmental temperature. *Comp. Biochem. Physiol. B Biochem. Mol. Biol.* **122**, 173–180.

Mathiessen, P., and Sumpter, J. P. (1998). Effects of estrogenic substances in the aquatic environment. *Fish Ecotoxicol.* **86**, 319–335.

Marsden, J. M., Vaughan, L. M., Foster, T. J., and Secombes, C. J. (1995). Proliferative response of rainbow trout, *Oncorhynchus mykiss*, T and B cells to antigens of *Aeromonas salmonicida*. *Fish Shellfish Immunol.* **5**, 199–210.

McCarty, L. S., and Munkittrick, K. R. (1996). Environmental biomarkers in aquatic toxicology: Fiction, fantasy, or functional? *Hum. Ecol. Risk. Assess.* **2**, 268–274.

Munkittrick, K. R., McGeachy, S. A., McMaster, M. E., and Courtney, S. C. (2002). Overview of freshwater fish studies from the pulp and paper environmental effects monitoring programme. *Wat. Qual. Res. J. Can.* **37**, 39–47.

Nakanishi, T., Fischer, U., Dijkstra, J. M., Hasegawa, S., Somamoto, T., Okamoto, N., and Ototake, M. (2002). Cytotoxic T cell function in fish. *Dev. Comp. Immunol.* **26**, 131–139.

National Research Council of Canada (NRCC), Associate Committee on Scientific Criteria for Environmental Quality (1985). The role of biochemical indicators in the assessment of ecosystem health: Their development and validation. National Research Council of Canada, Ottawa, Canada. Publication No. NRCC 24371.

Novoa, B., Figueras, A., and Secombes, C. J. (1996a). Effects of *in vitro* addition of infectious pancreatic necrosis virus (IPNV) on rainbow trout *Oncorhynchus mykiss* leucocyte responses. *Vet. Immunol. Immunopathol.* **51**, 365–376.

Novoa, B., Figueras, A., Ashton, I., and Secombes, C. J. (1996b). *In vitro* studies on the regulation of rainbow trout (*Oncorhynchus mykiss*) macrophage respiratory burst activity. *Dev. Comp. Immunol.* **20,** 207–216.

Partula, S., de Guerra, A., Fellah, J. S., and Charlemagne, J. (1996). Structure and diversity of the TCR alpha-chain in a teleost fish. *J. Immunol.* **157,** 207–212.

Pickering, A. D., and Pottinger, T. G. (1989). Stress responses and disease resistance in salmonid fish: Effects of chronic elevation of plasma cortisol. *Fish Physiol. Biochem.* **7,** 253–258.

Rice, C. D., and Arkoosh, M. R. (2002). Immunological indicators of environmental stress and disease susceptibility in fishes. *In* "Biological Indicators of Aquatic Ecosystem Stress" (S. M. Adams, Ed.), pp. 187–220. American Fisheries Society, Bethesda, MD.

Robinson, C. D., Brown, E., Craft, J. A., Davies, I. M., Moffat, C. F., Pirie, D., Robertson, F., Stagg, R. M., and Struthers, S. (2003). Effects of sewage effluent and ethynyl oestradiol upon molecular markers of oestrogenic exposure, maturation and reproductive success in the sand goby (*Pomatoschistus minutes*, Pallas). *Aquatic Toxicol.* **62,** 119–134.

Rycyzyn, M. A., Wilson, M. R., Bengten, E., Warr, G. W., Clem, L. W., and Miller, N. W. (1998). Mitogen and growth factor-induced activation of a STAT-like molecule in channel catfish lymphoid cells. *Mol. Immunol.* **35,** 127–136.

Secombes, C. J. (1990). Isolation of salmonid macrophages and analysis of their killing activity. *In* "Techniques in Fish Immunology" (J. S. Stolen, T. C. Fletcher, A. F. Rowley, J. T. Zelikoff, S. L. Kaattari, and S. A. Smith, Eds.), pp. 139–154. SOS Publications, Fair Haven, New Jersey.

Secombes, C. J., and Fletcher, T. C. (1992). The role of phagocytes in the protective mechanisms of fish. *Annual Rev. Fish Dis.* **2,** 53–71.

Sharp, G. J., Pike, A. W., and Secombes, C. J. (1992). Sequential development of the immune response in rainbow trout (*Oncorhynchus mykiss*, Walbaum, 1792) to experimental plerocercoid infections of *Diphyllobothrium dendriticum* (Nitzsch, 1824). *Parasitology* **104,** 169–178.

Sharples, A. D., Campin, D. N., and Evans, C. W. (1994). Fin erosion in a feral population of goldfish (*Carassius auratus L.*) exposed to bleached kraft mill effluent. *J. Fish Disease* **17,** 483–493.

Slater, C. H., and Schreck, C. B. (1993). Testosterone alters the immune response of chinook salmon, Oncorhynchus tshawytscha. *Gen. Comp. Endocrinol.* **89,** 291–298.

Smith, D. A., Schurig, G. G., Smith, S. A., and Holladay, S. D. (1999). Tilapia (*Oreochromis niloticus*) and rodents exhibit similar patterns of inhibited antibody production following exposure to immunotoxic chemicals. *Vet. Hum. Toxicol.* **41,** 368–373.

Solem, S. T., Jorgensen, J. B., and Robertson, B. (1995). Stimulation of respiratory burst in Atlantic salmon (*Salmo salar* L.) macrophages by lipopolysaccharide. *Fish Shellfish Immunol.* **5,** 475–491.

Somamoto, T., Nakanishi, T., and Okamoto, N. (2000). Specific cell-mediated cytotoxicity against a virus-infected syngeneic cell line in isogeneic ginbuna crucian carp. *Dev. Comp. Immunol.* **24,** 633–640.

Somamoto, T., Nakanishi, T., and Okamoto, N. (2002). Role of specific cell-mediated cytotoxicity in protecting fish from viral infections. *Virology* **297,** 120–127.

Spies, R. B., and Rice, D. W. (1988). Effects of organic contaminants on reproduction of the starry flounder, *Platichtys stellatus*, in San Francisco Bay. II. Reproductive success of fish captured in San Francisco Bay and spawned in the laboratory. *Marine Biol.* **98,** 191–200.

Stuge, T. B., Miller, N. W., and Clem, L. W. (1995). Channel catfish cytotoxic effector cells from peripheral blood and pronephros are different. *Fish Shellfish Immunol.* **5,** 469–471.

Stuge, T. B., Yoshida, S. H., Chinchar, V. G., Miller, N. W., and Clem, L. W. (1997). Cytotoxic activity generated from channel catfish peripheral blood leukocytes in mixed leukocyte cultures. *Cell. Immunol.* **177,** 154–161.

Ternes, T. A. (1998). Occurrence of drugs in German sewage treatment plants and rivers. *Water Res.* **32,** 3245–3260.

Twerdok, L. E., Beaman, J. R., Curry, M. W., and Zelikoff, J. T. (1996). Health status and monitoring in an aquatic model *Oryzias latipes* using immunotoxicological testing. *In* "Modulators of Immune Response: The Evolutionary Trail" (J. S. Stolen, T. C. Fletcher, C. J. Bayne, C. J. Secombes, J. T. Zelikoff, L. E. Twerdok, and D. P. Anderson, Eds.), pp. 417–424. SOS Publications, Fair Haven, NJ.

van den Heuvel, M. R., Power, M., Richards, J., MacKinnon, M., and Dixon, D. G. (2000). Disease and gill lesions in yellow perch (*Perca flavescens*) exposed to oil-sands mining associated waters. *Ecotoxicol. Environ. Safety.* **46,** 334–341.

Verburg-Van Kemenade, B. M. L., Engelsma, M. Y., Nowak, B., Hofenk, M. G., and Weyts, F. A. A. (2000). Neuroendocrine-immune interactions in fish: Differential regulation of immune responses by cortisol. *Dev. Comp. Immunol.* **24,** 63–64.

Verburg-Van Kemenade, B. M., Saeij, J. P., Flik, G., and Willems, P. H. (1998). Ca2+ signals during early lymphocyte activation in carp *Cyprinus carpio* L. *J. Exp. Biol.* **201,** 591–598.

Walker, C. H. (1998). Biomarker strategies to evaluate the environmental effects of chemicals. *Environ. Health Perspect.* **106**(Suppl), 613–620.

Wilkund, T., and Bylund, G. (1996). Fin abnormalities of pikeperch on coastal areas of the Finnish south coast. *J. Fish Biol.* **48,** 652–657.

Yoder, C. O., and Rankin, E. T. (1998). The role of biological indicators in a state water quality management process. *Environ. Monit. Assess.* **51,** 61–88.

Zapata, A. G., Varas, A., and Torroba, M. (1992). Seasonal variations in the immune system of lower vertebrates. *Immunol. Today.* **13,** 142–147.

Zapata, A., and Amemiya, C. T. (2000). Phylogeny of lower vertebrates and their immunological structures. *Curr. Top. Microbiol. Immunol.* **248,** 67–107.

Zelikoff, J. T., Bowser, D., Squibb, K. S., and Frenkel, K. (1995). Immunotoxicity of low level cadmium exposure in fish: An alternative animal model for immunotoxicological studies. *J. Toxicol. Environ. Health.* **45,** 235–248.

Zelikoff, J. T. (1998). Biomarkers of immunotoxicity in fish and other non-mammalian sentinel species: Predictive value for mammals? *Toxicology* **129,** 63–71.

Zelikoff, J. T., Raymond, A., Carlson, E., Li, Y., Beaman, J. R., and Anderson, M. (2000). Biomarkers of immunotoxicity in fish: From the lab to the ocean. *Toxicol. Lett.* **112–113,** 325–331.

Zhou, H., Stuge, T. B., Miller, N. W., Bengten, E., Naftel, J. P., Bernanke, J. M., Chinchar, V. G., Clem, L. W., and Wilson, M. (2001). Heterogeneity of channel catfish CTL with respect to target recognition and cytotoxic mechanisms employed. *J. Immunol.* **167,** 1325–1332.

Bryan W. Brooks,[*] Sean M. Richards,[†] James J. Weston,[‡] Philip K. Turner,[§] Jacob K. Stanley,[§] Thomas W. La Point,[§] Richard Brain,[||] Elizabeth A. Glidewell,[*] A. Rene D. Massengale,[¶] Whitney Smith,[#] C. LeRoy Blank,[#] Keith R. Solomon,[||] Marc Slattery,[‡] and Christy M. Foran[**]

[*]Department of Environmental Studies
Baylor University, Waco, Texas

[†]Department of Biological and Environmental Sciences
University of Tennessee at Chattanooga
Chattanooga, Tennessee

[‡]Environmental Toxicology Research Program
School of Pharmacy
University of Mississippi, University, Mississippi

[§]Institute of Applied Sciences
University of North Texas, Denton, Texas

[||]Centre for Toxicology, University of Guelph,
Guelph, Ontario, Canada

[¶]Department of Biology
Baylor University, Waco, Texas

[#]Department of Chemistry and Biochemistry
University of Oklahoma, Norman, Oklahoma

[**]Department of Biology, West Virginia University
Morgantown, West Virginia

Aquatic Ecotoxicology of Fluoxetine: A Review of Recent Research

I. Introduction

Kolpin *et al.* (2002) recently identified widespread occurrence of multiple pharmaceuticals in United States surface waters. Included among these contaminants was fluoxetine (Fig. 1, Table I), a selective serotonin reuptake inhibitor (SSRI). SSRIs are primarily indicated for depression, but are also prescribed to treat compulsive behavior, and eating and personality disorders. SSRIs are preferred to monoamine oxidase inhibitors and tricyclic antidepressants for treatment of affective disorders, due to a lack of receptor

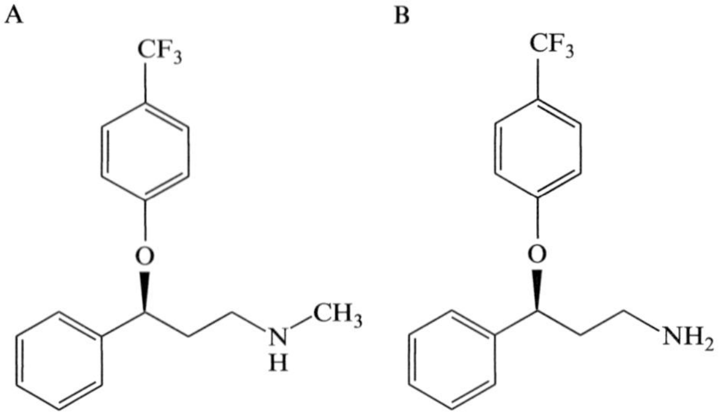

FIGURE 1 Chemical structures of (A) fluoxetine and (B) norfluoxetine.

TABLE 1 Physiochemical and Environmental Fate Parameters of Fluoxetine and Norfluoxetine

Parameter	*Fluoxetine*			*Norfluoxetine*		
Physiochemical Parameter						
Empirical formula	$C_{17}H_{18}F_3NO$			$C_{16}H_{16}F_3NO$		
Molecular weight	309.33			295.3		
PKa	10.06+/−0.10			9.05 +/− 0.13		
Environmental Fate Parameter						
PH	2.0	7.0	11.0	2.0	7.0	11.0
log K_{ow}	1.25	1.57	4.30	0.97	2.05	4.06
BCF	~1	2.00	1071.52	~1	6.97	716.12
log K_{oc}	0.64	0.97	3.70	0.49	1.57	3.58

Values calculated by ACD/Labs Software Version 5.0 (Toronto, Ontario, Canada).

antagonism and few anticholinergic and cardiovascular side effects (Rang *et al.*, 1995). Fluoxetine, the prototype SSRI and a highly prescribed antidepressant (NDC Health, 2003), blocks serotonin reuptake from the presynaptic nerve cleft (Ranganathan *et al.*, 2001). A racemic mixture of two enantiomers, fluoxetine is metabolized by cytochrome P450 isoenzymes to norfluoxetine, its active metabolite, and is primarily excreted as less than 10% unchanged parent compound in urine (Hiemke and Härtter, 2000).

Although the occurrence and estrogenicity of steroid therapeutics in municipal effluents has received attention (Harries *et al.*, 1997; Hemming *et al.*, 2001, 2002; Nichols *et al.*, 1999), environmental hazard and exposure information is sparse for nonsteroid pharmaceuticals (Huggett *et al.*, 2002). Limited data available for pharmaceuticals in aquatic environments pertains mostly to detection (Boyd *et al.*, 2003; Buser and Muller, 1998; Buser *et al.*, 1999; Golet *et al.*, 2001; Hirsch *et al.*, 1999; Huggett *et al.*, 2003; Kolpin *et al.*, 2002; Rossknecht *et al.*, 2001; Stan and Heberer, 1997; Stumpf *et al.*, 1996, 1999; Suter and Giger, 2000; Ternes, 1998) and degradation (Guarino and Lech, 1986; Richardson and Bowron, 1985; Velagaleti and Robinson, 2000) in rivers and lakes. Studies that have examined pharmaceutical effects primarily used single-species, acute laboratory toxicity tests (Fong, 1998; Guarino and Lech, 1986; Honkoop *et al.*, 1999; Huber and Delago, 1998; Lanzky and Halling-Sorensen, 1997; Uhler *et al.*, 2000). Fewer studies have evaluated fish biochemical and reproduction responses to nonsteroid therapeutics (Huggett *et al.*, 2002). Pharmaceutical effects on higher levels of biological organization are not reported in the peer-reviewed literature.

Very little information is also available for fluoxetine exposure (Boyd *et al.*, 2003; Kolpin *et al.*, 2002; Weston *et al.*, 2001) and effects (Brain *et al.*, 2004; Brooks *et al.*, 2003a; Fong, 2001; Foran *et al.*, 2004; Richards *et al.*, 2004) in the aquatic environment. Therefore, the objective of this paper is to: (1) summarize available data on the occurrence and detection of fluoxetine in surface waters; (2) summarize our research on aquatic biota and community responses to fluoxetine; and (3) provide a preliminary aquatic risk characterization for fluoxetine.

II. Fluoxetine Exposure and Detection in Surface Waters

Environmental exposure to norfluoxetine has not been reported, but several investigators detected fluoxetine in water bodies and municipal effluents (Jones-Lepp *et al.*, 2001; Kolpin *et al.*, 2002; Weston *et al.*, 2001). In surface waters, Kolpin *et al.* (2002) estimated maximum fluoxetine concentrations at 0.012 μg/L. Recently, Metcalfe *et al.* (2003) found mean fluoxetine levels to range from 38 ng/L to 99 ng/L and 13 ng/L to 46 ng/L in Canadian effluents and surface waters, respectively. Weston *et al.* (2001) indicated that effluent fluoxetine concentrations may reach as high as 0.54 μg/L. However, the magnitude, frequency, and duration of fluoxetine exposure have not been fully explored. Further, detection of fluoxetine in sediments has not been reported. Whereas the ability to reliably detect and quantify fluoxetine in aqueous and sediment matrices is essential to assessing environmental exposure, Weston *et al.* (2001) identified that extraction and recovery of fluoxetine and norfluoxetine in aqueous samples requires further development. Several techniques for fluoxetine and norfluoxetine detection in water matrices have been reported. Methods of Kolpin *et al.* (2002), Jones-Lepp *et al.* (2001), and Weston *et al.* (2001) follow a general scheme: pre-filtration, extraction, concentration, and detection and quantification.

Kolpin *et al.* (2002) collected, pre-filtered, and extracted analytes from 1-liter surface water samples using SPE (solid phase extraction) cartridges. Cartridges were eluted, evaporated to near dryness, and then brought to a final volume of 1 ml, resulting in a 1000:1 concentration ratio. High-performance liquid chromatography (HPLC) with a reverse phase octylsilane (C_8) column was used to detect and quantify analytes. Jones-Lepp *et al.* (2001) collected effluent samples from nine wastewater treatment plants (WWTP); whether samples were pre-filtered prior to solid phase extraction was not indicated. Two-liter samples were adjusted to a pH of 2.5 and subsequently extracted using SPE C_{18} discs. Analytes adsorbed to discs were eluted and then concentrated to 0.3 mL, resulting in a 6667:1 concentration ratio. Liquid-chromatography electron spray ion trap mass spectrometry (LC-ES/ITMS) was used for detection and quantification. Weston *et al.* (2001) sampled municipal effluent discharge from two WWTPs. One-liter samples were pre-filtered, adjusted to a pH of 9, and extracted with C_{18} SPE discs. Weston *et al.* (2001) selected a pH adjustment to 9.0 because the ionization state and lipophilicity of fluoxetine changes with increasing pH (Table I). Adsorbed analytes were eluted, evaporated to dryness, and reconstituted to a final volume of 0.1 ml, resulting in a 100,000:1 concentration ratio. LC-ES/MS was utilized to detect and quantify effluent fluoxetine and norfluoxetine levels.

Percent recovery of fluoxetine varied between methods: less than 60% in Kolpin *et al.* (2002), 88% (triplicate extractions) in Jones-Lepp *et al.* (2001), and 79–82% in Weston *et al.* (2001). In addition, Weston *et al.* (2001) observed recoveries of 67–77% for norfluoxetine matrix spikes, based on triplicate extractions. Kolpin *et al.* (2002) reported that out of 84 streams sampled, fluoxetine concentrations did not exceed an estimated 0.012 μg/L. Jones-Lepp *et al.* (2001) sampled nine WWTP effluents for fluoxetine; however, detection frequency and specific effluent fluoxetine concentrations were not reported. Weston *et al.* (2001) reported fluoxetine levels up to 0.54 μg/L; however, norfluoxetine was not detected in two WWTP effluents.

III. Single Species Toxicity Test Organism Responses to Fluoxetine

Standardized single species toxicity tests are used to screen for potential hazards of aquatic contaminants, to develop water quality criteria, and to monitor whole effluent toxicity (WET) in the United States. Environmental assessments of pharmaceutical compounds also rely on single species responses if an expected environmental introduction concentration (EIC) exceeds 1 μg/L (FDA-CDER, 1998). Such laboratory studies are attractive because they use clean water or sediments and are less expensive to perform than field studies. Further, laboratory responses are often less variable than data collected in bioassessments or mesocosm studies (Dickson *et al.*, 1996).

Whereas fluoxetine exposure may affect pelagic organisms (Fong, 2001), fluoxetine also may bind to sediments and affect benthic organisms.

Responses of aquatic organisms were assessed with a series of standardized aqueous and sediment toxicity tests (Brain *et al.*, 2004; Brooks *et al.*, 2003a). A green algae (*Pseudokirchneriella subcapitata*), a photoluminescent marine bacterium (*Vibrio fischeri*), a floating emergent macrophyte (*Lemna gibbo*); two cladocerans (*Ceriodaphnia dubia* and *Daphnia magna*), and the fathead minnow (*Pimephales promelas*) were chosen for aquatic toxicity tests. For sediment toxicity tests, the midge (*Chironomus tentans*), and the scud (*Hyalella aztecc*), were utilized.

A. Aqueous Toxicity Tests

L. gibba (ASTM, 1998) and *P. subcapitata* (USEPA, 1989; 1991) toxicity tests followed recommended procedures. Following a 7-day laboratory exposure at 10, 30, 100, 300 and 1000 μg/L fluoxetine, no significant effects were detected on *L. gibba* frond number, wet weight, or chlorophyll-*a*, chlorophyll-*b*, and carotenoid pigments (Brain *et al.*, 2004). An EC_{50} for *P. subcapitata* growth was estimated by nonlinear regression (Bruce and Versteeg, 1992) at 24 μg/L (Table II). This value is almost identical to a previously reported EC_{50} of 28 μg/L for an unnamed green algae (FDA-CDER, 1996). *P. subcapitata* growth was also evaluated for treatment level effects using ANOVA with a Dunnett's post hoc test. A lowest observed effect concentration (LOEC) was observed at 13.6 μg/L (Table II), which was also the lowest treatment level tested. *P. subcapitata* cell deformities were observed at 13.6 and 27.2 μg/L treatment levels. Cells also appeared smaller than untreated controls at these concentrations. The mechanism by which fluoxetine may induce deformations in algal cells is unknown. However, fluoxetine has antimicrobial properties and potentially exerts its

TABLE II Standardized Toxicity Test Organism Responses to Fluoxetine

Organism	*EC_{50}**	*NOEC*†	*LOEC*†	*Matrix*
Pseudokirchneriella subcapitata	24 μg/L	ND	13.6 μg/L	AAP
Ceriodaphnia dubia	234 μg/L	56 μg/L	112 μg/L	RHW
Daphnia magna	820 μg/L	89 μg/L	178 μg/L	RHW
Pimephales promelas	705 μg/L			RHW
Vibrio fischeri	724 μg/L			RHW
Hyalella azteca	>43 mg/kg	ND	5.4 mg/kg	Sediment
Chironomus tentans	15.2 mg/kg	ND	1.3 mg/kg	Sediment

* *P. subcapitata* EC_{50} = growth; *C. dubia, D. magna, P. promelas, H. azteca, and C. tentans* EC_{50} = mortality; *V. fischeri* EC_{50} = photoluminescence.

† *C. dubia, D. magna* NOEC/LOEC = neonates female^{-1}; *H. azteca, C. tentans* NOEC/LOEC = growth.

AAP, AAP media, RHW, reconstituted hard water, Sediment, University of North Texas Water Research Field Station reference sediment; ND, not determined.

toxicity by inhibiting cellular efflux pumps (Munoz-Bellido *et al.*, 2000). Although cell deformities and biovolumes were not quantified, a possible bacteriostatic mechanism of algal and microbial fluoxetine toxicity deserves further investigation. For example, current research in our laboratories is assessing microbial community responses to environmentally relevant fluoxetine levels (Massengale *et al.*, 2003).

C. dubia, D. magna, and *P. promelas* 48-hour acute toxicity tests and *Vibrio fischeri* acute tests were performed in reconstituted hard water (USEPA, 1991). Each test was repeated, and LC_{50} estimated by Trimmed Spearman Karber (Hamilton *et al.*, 1977). Average LC_{50} for *C. dubia, D. magna, P. promelas*, and *V. fischeri* were 234, 820, 705, and 724 μg/L, respectively. A *D. magna* LC_{50} of 820 μg/L is similar to a nominal value of 940 μg/L reported for a *Daphnia* spp. (FDA-CDER, 1996). The LC_{50} of 705 μg/L for *P. promelas* is lower than a previously reported 48-hour LC_{50} of 2 mg/L for rainbow trout (*Oncorhynchus mykiss*; FDA-CDER, 1996).

In addition, 7-day *C. dubia* and 21-day *D. magna* static-renewal studies were performed to evaluate potential fluoxetine effects on cladoceran reproduction. These tests also followed standard methods (USEPA, 1989); however, organisms were fed an algae-Cerophyll suspension according to methods of Hemming *et al.* (2002) following daily renewals. *C. dubia* NOEC and LOEC were determined to be 56 and 112 μg/L, respectively, by ANOVA with a Bonferroni's adjustment. Although a treatment level of 112 μg/L was statistically different from control organisms ($\alpha = 0.05$), the observed difference may or may not be of ecological relevance because the difference was an average of 2.1 neonatesfemale. *D. magna* reproduction NOEC and LOEC were 89 μg/L and 178 μg/L, respectively; growth (dry weight) NOEC was determined at 178 μg/L and the LOEC was 356 μg/L. In addition, a two-fold increase in the number of undeveloped eggs was observed at the 356 μg/L treatment level.

B. Sediment Toxicity Tests

Ten day *C. tentans* and *H. azteca* sediment toxicity tests followed standard methods and were performed using a Zumwalt testing system (USEPA, 2000). Reference sediments were obtained from pond mesocosms at the University of North Texas Water Research Field Station. Sediments were characterized for total organic carbon (22340 mg/kg), percent moisture (60%), and grain size distribution (41.2% sand, 39.2% silt, 19.6% clay). In addition to physical characterization, sediments were evaluated for 17 metals, 44 volatile organics, 56 semi-volatile organics, 4 triazine herbicides, 6 organophosphorus insecticides, 3 organochlorine herbicides, 20 organochlorine pesticides, 2 carbamate pesticides, and 7 PCB congeners (La Point *et al.*, unpublished data). Sediments were considered "clean" and were spiked with fluoxetine according to Suedel and Rodgers (1996). Following preliminary range finding toxicity tests, *C. tentans* and *H. azteca* treatment levels were selected at 0, 1.4, 2.8, 5.6, 11.2, and 22.4 mg/kg, and 0, 5.4, 10.8, 21.6, and 43.2 mg/kg, respectively.

C. tentans survival was reduced by fluoxetine treatments; an LC_{50} of 15.2 mg/kg was estimated (Table II). In addition, each fluoxetine treatment level significantly reduced *C. tentans* growth such that a LOEC of 1.3 mg/kg was observed (Table II). *H. azteca* survival was not affected by the highest treatment level tested (43 mg/kg; Table II). However, *H. azteca* growth was also significantly reduced by all treatment levels (LOEC = 5.6 mg/kg; Table II).

C. Fluoxetine Effects on Invertebrate Reproduction

In addition to 10-day tests with *H. azteca*, a 42-day study was performed to evaluate potential fluoxetine effects on *H. azteca* reproduction (USEPA, 2000). *H. azteca* fecundity (young per female) was not significantly affected by fluoxetine treatment levels. Fluoxetine treatments stimulated *H. azteca* reproduction, though not significantly. An increase in *C. dubia* fecundity was also observed with 56 μg/L fluoxetine treatment. Flaherty *et al.* (2001) observed a comparable reproductive stimulation when *D. magna* were exposed to 36 μg/L fluoxetine for 30 days. However, our 21-day study with *D. magna* did not observe such fecundity stimulation by fluoxetine. Similarly to Flaherty *et al.* (2001), Fong *et al.* (1998) observed fluoxetine to induce mussel spawning. In invertebrates, serotonin may stimulate ecdysteroids, ecdysone, and juvenile hormone, which are responsible for controlling oogenesis and vitellogenesis (Nation, 2002). In some fish species (see further discussion below) serotonin may stimulate the release of gonadotropin. Gonadotropin stimulates sex steroid synthesis and controls oogenesis development, including vitellogenesis (Arcand-Hoy and Benson, 2001). Although serotonergic effects on ecdysteroids, ecdysone, and juvenile hormone are less understood (LeBlanc *et al.*, 1999), observed stimulation in fecundity may result from increased synaptic serotonin levels. However, because invertebrate reproduction is energy intensive, such an increase in *C. dubia* or *H. azteca* reproduction should not necessarily be associated with maintenance of offspring viability or fitness.

IV. Medaka Reproduction and Endocrine Function Responses to Fluoxetine

Among the issues raised by environmental detection of fluoxetine are concerns over potential sub-lethal effects on aquatic organisms, including behavioral responses. Considering the potential for environmental SSRIs to act as they do in humans, to alter or increase serotonin concentrations, the potential disruptive effects of low-level, chronic exposure must be considered. Serotonin is likely to be one of the most potent and ubiquitous neuromodulators in vertebrates (Azmitia, 1999). It is synthesized in cells lining the gut, in neurons of the hypothalamus that regulate pituitary activity, and in the brainstem of vertebrates. Many of these neurons release serotonin into the synaptic cleft, where it acts as a neurotransmitter. In addition, cerbrospinal-fluid contacting neurons in the hypothalamus and cells in the periphery release serotonin into general circulation, where it acts on more distant target tissues in the central nervous system or vascular and gastrointestinal muscle, T cells, and platelets.

Because of the critical nature of the functions regulated by serotonin, there is a potential for environmental SSRIs to alter appetite, the immune system, and reproduction, as well as other behavioral functions (Fong, 2001; Meguid *et al.*, 2000; Mosner and Lesch, 1998). Serotonin acts directly on the immune system by modulating cellular function and indirectly through actions on the central nervous system (Mossner and Lesch, 1998). In studies with lymphocytes from HIV-positive patients, treatment with a serotonin receptor agonist resulted in increased T cell counts (Hofmann *et al.*, 1996), whereas serotonin itself was found to increase their proliferative capacity (Eugen-Olsen *et al.*, 1997). A similar relationship between serotonin and immune function has also been described for fish (Khan and Deschaux, 1997). To the extent that serotonin alters immune function, an increase in serotonin may produce beneficial changes in the immune response, but also may elevate the rate of negative impacts such as autoimmune disease.

Serotonin is an important neuromodulator of sexual function in vertebrates and invertebrates. Changes in serotonin metabolism or concentration are correlated with reproductive phases of human females (Hindberg and Naesh, 1992) and other animals, including female fish (Hernandez-Rauda *et al.*, 1999). In fact, studies in fish indicated that serotonin potentiates effects of gonadotropin-releasing hormone on gonadotropin release from the pituitary (Khan and Thomas, 1994). In some seasonally reproductive animals, serotonin concentration varies with reproductive potential and gonadal recrudescence (Hernandez-Rauda *et al.*, 1999). Literature relating serotonin and SSRIs to reproductive function has been recently reviewed for many groups of invertebrates and vertebrates, including fish (Fong, 2001). The role of serotonin in reproduction, and therefore the potential for SSRIs to disrupt normal serotonin function, varies across family groups. Serotonin and SSRIs potentiate spawning and oocyte maturation in some bivalves and crustaceans (Fong, 2001). Serotonin also induces oocyte maturation in Japanese medaka (*Oryzias latipes*; Iwamatsu *et al.*, 1993) but inhibits this process in another teleost, the mummichog (*Fundulus heteroclitus*; Cerda *et al.*, 1998). Few studies in vertebrates have looked for a correlation between changes in reproductive and endocrine function and aqueous exposure to SSRIs.

To assess potential fish endocrine function and reproduction responses to environmental SSRIs, Foran *et al.* (2004) exposed Japanese medaka for four weeks to fluoxetine treatments of 0, 0.1, 0.5, 1.0, and 5.0 μg/L. Japanese medaka were chosen because this species is a widely used model organism for the study of contaminant-induced developmental effects (Metcalfe *et al.*, 1999) and reproductive impairment (Arcand-Hoy *et al.*, 1998). Reproduction, including fecundity, rate of fertilization, egg hatching success, abnormal development, and endocrine function (including vitellogenin and circulating plasma steroids) were assessed following the exposure period. Methods for reproduction endpoint assessment and vitellogenin, plasma estradiol (E2) and testosterone (T), and *ex vivo* gonadal steroid release followed those reported elsewhere (Foran *et al.*, 2002; Zhang *et al.*, 2002). Nominal exposure concentrations were verified following Weston *et al.* (2001).

Fluoxetine exposure for four weeks resulted in few changes in medaka reproductive success. Medaka pairs produced an average of 158 $\pm$ 41 eggs over the two-week reproductive assessment period. Fecundity was unaffected by fluoxetine treatments (ANOVA; $F = 0.776$, $p = 0.55$); however, statistical power was limited by treatment-level replication and response variability. More than 87% of all eggs were fertilized in each group. The percent of fertilized eggs successfully hatched within 30 days of fertilization ranged from 84–94% in all treatment groups. Thus, fecundity and fertility were unaffected by fluoxetine exposure. During observations of developing embryos, several abnormalities were noted. These included edema, curved spine, incomplete development (no pectoral fins, reduced eyes), and nonresponsiveness. Whereas few abnormalities were noted in untreated organisms (4 of 820, or 0.49%), developmental abnormalities were observed more frequently at all fluoxetine treatment levels. The number and percent of developmental alterations for each treatment level were: 0.1 μg/L, 21/863 or 2.43%; 0.5 μg/L, 17/637 or 2.53%; 1.0 μg/L, 18/913 or 1.97%; 5.0 μg/L, 17/758 or 2.24%. These observations indicate that developmental abnormalities are 4.02–5.18 times more frequent in fluoxetine treatments.

Most adult physiological measurements were unaffected by fluoxetine exposure. However, female circulating E2 levels were increased by exposure to two fluoxetine treatment levels. Condition factor [weight (g) / length $(mm)^3$] did not differ for animals across treatment levels (female ANOVA, $F = 0.351$ and $p = 0.84$; male ANOVA, $F = 0.863$ and $p = 0.50$). Gonadal somatic index, measured as the profile area of the gonad normalized by the animal's weight, was also unchanged with exposure (female ANOVA, $F = 1.252$ and $p = 0.32$; male ANOVA, $F = 1.524$ and $p = 0.23$). Hepatic VTG content and circulating T concentrations were not affected by fluoxetine treatment levels (female ANOVA, $F = 0.524$ and $p = 0.72$; male ANOVA, $F = 0.102$ and $p = 0.98$). Release of E2 and T from *ex vivo* gonadal tissue incubated with 25-hyroxycholesterol did not change with treatment for either ovarian tissue (E2 ANOVA, $F = 0.711$ and $p = 0.59$; T ANOVA, $F = 0.484$ and $p = 0.75$) or testes (E2 ANOVA, $F = 0.401$ and $p = 0.80$; T ANOVA, $F = 0.881$ and $p = 0.50$). Although plasma E2 concentrations were unaffected in males, female circulating E2 was significantly increased by 0.1 μg/L and 0.5 μg/L fluoxetine treatments ($p = 0.01$ and $p = 0.054$, respectively).

These results provide some early information to associate physiological change with environmental fluoxetine exposure, and demonstrate the limitation of assessing reproductive impacts using only one model species. A 4-week exposure of environmentally relevant concentrations of fluoxetine did not affect medaka fecundity, rate of egg fertilization, or hatching success. However, developmental abnormalities were noted at all fluoxetine exposure levels. Further, a complex response was noted among the endocrine endpoints; female circulating steroid concentrations were elevated at the 0.1 μg/L and 0.5 μg/L exposure levels. Because of the small plasma volumes collected from medaka, blood from two animals of the same sex was pooled, leaving three tissue samples for plasma steroid analysis in each treatment group. However, the statistically significant response with limited sample numbers may indicate a dramatic effect of fluoxetine with 0.1 μg/L and 0.5 μg/L exposure. Absence of a concentration-response relationship in the change of circulating E2 highlights the potential for different factors to affect the response to SSRIs, including basal circulating steroid levels, sexual dimorphisms in cytochrome P450 enzyme activity, and potential sexual dimorphisms in serotonin systems (Hernandez-Rauda *et al.*, 1999). A wide range of impacts and the potential for regulatory biofeedback to counteract elevations in serotonin raises an issue as to whether a traditional concentration-response relationship would be expected with a long-term aquatic exposure to SSRIs.

Although Japanese medaka are a commonly used model organism in reproductive assessments of contaminant effects, the response of oocytes to serotonin is known to vary between species (Cerda *et al.*, 1998; Fong, 2001; Iwamatsu *et al.*, 1993). Therefore, results of Foran *et al.* (2004), which indicate no statistically significant reproduction changes in response to fluoxetine treatment, may not be representative of effects in other teleosts. Clearly, further studies on the sub-lethal consequences of fluoxetine exposure are necessary, and these studies should consider study species sensitivity, behavioral responses, and endpoint selection to serotonin modulation. For example, recent research in our laboratories detected an upregulation of dopamine and a decrease of norepinephrine in brain tissue collected from male medaka exposed to 0.1, 1, and 5 μg/L fluoxetine in the Foran *et al.* (2004) study (Brooks *et al.*, 2003b).

V. Fluoxetine Effects: Community Responses to Fluoxetine, Ibuprofen, and Ciprofloxacin Mixtures

Whereas a direct assessment of fluoxetine effects on lotic or lentic freshwater communities has not been performed, Richards *et al.* (2004) used lentic microcosms to investigate effects of fluoxetine, ibuprofen, and ciprofloxacin mixtures. Eight 12,000-liter microcosms and their aquatic communities were established, maintained, and treated with ibuprofen, ciprofloxacin, and fluoxetine according to methods described in detail by Richards *et al.* (2004). Control microcosms (n = 3) received no treatment, low treatment microcosms (LT, n = 1) received 6, 10, and 10 μg/L, medium treatment (MT, n = 1) received 60, 100, and 100 μg/L, and high-treatment (HT, n = 3) received 600, 1000, and 1000 μg/L of ibuprofen, ciprofloxacin, and fluoxetine, respectively. These pharmaceuticals were selected based on mode of action and frequency of prescription in North America (NDC Health, 2003). Treatment concentrations were based on distributional analyses of upper centiles (95th, 99th, and 99.9th) estimated from actual surface water concentrations for ibuprofen (Buser *et al.*, 1999; Metcalfe and Koenig, 2001; Stumpf *et al.*, 1999) or on centiles of distributions of measured environmental concentrations from similarly prescribed pharmaceuticals in surface water (Daughton and Ternes, 1999; NDC Health, 2003) for ciprofloxacin and fluoxetine. HT concentrations purposely exceeded that of individual pharmaceuticals found in the environment because Richards *et al.* (2004) wanted to account for the possibility of additivity among compounds with the same mode of action *and* provide a high-exposure scenario for future probabilistic risk assessments. Pharmaceuticals within the microcosms were monitored and reintroduced as necessary to maintain nominal concentrations. Resultant 48h time-weighted average concentrations were within 10% of nominal values. Biological samples (phytoplankton, zooplankton, macrophytes, and bacteria) were collected every 7 days; fish were observed daily.

The initial and most obvious response was observed in fish. Juvenile sunfish (*Lepomis gibbosus*, n = 30 per microcosm) were contained in mesh cages; naturally occurring plankton were the primary food source, as no external food was added to the cages or microcosm. Treatment-dependent mortality was observed in sunfish. Within the first 96 hours of HT exposure, all sunfish died (n = 90). The trial was repeated on day 8; within four days, 98.8% of these fish died. After 35 days of exposure in the MT microcosm, 46.6% of sunfish had died (n = 14/30). During the same period, 1.1% of controls died (1/90) and no LT fish (n = 30) died.

The mechanisms of fish toxicity are unclear; treatment levels were not expected to induce mortality because MT concentrations were less than 230, 130, and 11-fold lower than those equivalent to mammalian whole-body therapeutic doses for ibuprofen, ciprofloxacin, and fluoxetine, respectively (Canadian Pharmacists Association, 2000). One plausible hypothesis suggested by Richards *et al.* (2004) for such a mechanism is that fluoxetine exposure led to increased plasma serotonin levels. Serotonin constricts the arterio-arterial branchial vasculature (Nilsson and Sundin, 1998). This would lead to impaired gas exchange and hypoxia, potentially leading to death. However, Khan and Thomas (1992) failed to increase levels or potentiate effects of serotonin by i.p. injection of 10 μg/g fluoxetine. Their dose of 10 μg/g was 10-fold greater than the estimated body dose experienced by fish in HT microcosms (assuming that the concentration in the fish came to equilibrium with the water 1000 μg/L $\approx$ 1 μg/g). Other potential factors that could affect lethality, such as dissolved oxygen and pH, were not significantly different between treatments. Synergistic interactions, wherein the combination of the three drugs may have increased the potency of one or all, may also account for toxicity observed in sunfish; however, preliminary laboratory studies suggest that the observed response was attributable to fluoxetine (Table II; Johnson, personal communication).

Zooplankton, phytoplankton, and macrophytes all responded to treatment with the mixture of pharmaceuticals (Richards *et al.*, 2004); however, these changes were not all attributable to fluoxetine. Zooplankton and phytoplankton communities were characterized by a decrease in species composition but an increase in numbers of some species. The components of the mixture responsible for these responses have not been identified at this time. The macrophytes, *Myriophyllum spicatum*, *Myriophyllum sibiricum*, and *Lemna gibba* L., all declined at the HT, and *L. gibba* showed sub-lethal effects (chlorosis and necrosis) at the MT. Based on subsequent 7-day plant bioassays with the individual pharmaceuticals, these responses were attributed to ciprofloxacin (Brain *et al.*, 2004). Fish mortality, reduced diversity, and community-level effects among plankton populations, along with the macrophyte mortality observed in this study, raise important questions about the potential for similar effects in surface waters. However, the causative agents in the treatment mixture have not all been identified.

VI. Ecological Risk Characterization for Fluoxetine

Ecological risk assessment (ERA) procedures often rely on deterministic hazard quotients to characterize risk to aquatic organisms. Probabilistic risk assessment methods are more attractive than deterministic ratios because risk can be expressed as the probability that adverse effects will occur (Solomon *et al.*, 1996). Further, probabilistic procedures can quantify variability associated with exposure and effect measures and quantify uncertainty inherent to risk assessments (Hart, 2001). However, lack of environmental exposure and hazard information for fluoxetine currently preclude such probabilistic approaches. Data presented here provide a foundation for future probabilistic risk assessments of fluoxetine.

Ecological risk of pharmaceuticals to aquatic organisms is currently characterized with a hazard quotient (HQ); however, alternative approaches have been suggested (Lange and Dietrich, 2002). A HQ is expressed as the relationship between a predicted environmental concentration (PEC) and a predicted no-effect concentration (PNEC). If a HQ derived from a PEC/PNEC ratio is <1, then risk to the environment is considered low. The U.S. Food and Drug Administration requires that an environmental assessment, a modified ERA, be performed for pharmaceuticals if predicted environmental introduction concentrations (EIC) are greater than 1 μg/L (FDA-CDER, 1998). This approach does not address additive effects of therapeutics with similar mechanisms of action, does not consider interaction effects of compounds with different mechanisms of action, and relies on acute toxicity test responses. Also, a 10-fold dilution factor is generally applied to an EIC to predict expected environmental concentrations (EEC or PEC). This technique may be appropriate for many lotic systems; Dorn (1996) indicated that annual mean flows greater than 75% of permitted, effluent dischargers in the United States receive 10-fold dilution. However, such an exercise becomes problematic in regions where effluent discharges do not receive upstream dilution (Brooks *et al.*, 2004a). Perennial municipal effluents influence historically ephemeral streams in the arid southwestern United States (Bradley *et al.*, 1995). For example, flow of the Trinity River south of Dallas/Fort Worth, TX, is often dominated by greater than 90% municipal effluents (Dickson *et al.*, 1989). Because the EIC is approximately equal to the PEC in these streams, fluoxetine concentrations in effluent dominated systems may represent maximal hazard to aquatic organisms (Brooks *et al.*, 2004b; Marsh *et al.*, 2003).

Although default EIC calculations do not consider effluent dominated streams, EICs may be conservative if they are not adjusted for metabolism. Fluoxetine EICs in the United States were estimated using annual consumption data for the year 2000 (Table III; CNN, 2001; McLean, 2001). If instream dilution, degradation, and metabolism are not included in these estimations, an EIC or PEC for fluoxetine is approximately 0.439 μg/L (Table III). Webb (2001) reported a similar PEC of 0.37 μg/L for fluoxetine in the United Kingdom. However, fluoxetine is normally excreted as 10% parent compound or fluoxetine *N*-glucuronide in urine (Hiemke and Härtter, 2000). When such metabolism is included in EIC calculations, a value of 0.0439 μg/L was calculated for systems not receiving dilution. Further, an EIC of 0.00439 μg/L was generated when a 10-fold dilution factor and metabolism were considered (Table III). This is similar to a PEC of 0.003 μg/L reported by Webb (2001), which included WWTP biodegradation and 10-fold dilution factors.

Interestingly, Kolpin *et al.* (2002), Weston *et al.* (2001), and Metcalfe *et al.* (2003) measured fluoxetine concentrations in surface waters and

TABLE III Aquatic predicted environmental concentration (PEC) of fluoxetine in the United States with and without corrections for dilution and metabolism*

	No metabolism	*90% metabolism*
No dilution	0.439 μg/L†	0.0439 μg/L
10 fold dilution	0.0439 μg/L	0.00439 μg/L

* 90% metabolism based on 10% parent compound and glucoronide conjugate excreted in urine (Hiemke and Hartter, 2000).

† EIC aquatic (μg/L) = A × B × C × D, where A = kg/year produced for direct use (active moiety); B = 1/liters per day entering POW; C = year/365 days; and D = 10^9 μg/kg (conversion factor).

municipal effluents at higher levels than those predicted by lowest PEC calculations (0.00439 μg/L; Table III). Weston *et al.* (2001) detected fluoxetine in municipal effluents from 0.32 μg/L to 0.54 μg/L, while Metcalfe *et al.* (2003) found mean effluent fluoxetine levels to range from 0.038 μg/L to 0.099 μg/L. Kolpin *et al.* (2002) reported maximum fluoxetine levels of 0.012 μg/L in surface waters; Metcalfe *et al.* (2003) reported higher mean fluoxetine levels in surface waters (0.013–0.046 μg/L). The lowest fluoxetine effect level, as required by USFDA in environmental assessments of pharmaceuticals, is 13.6 μg/L for *P. subcapitata* growth (Table II). Based on standardized toxicity test data, an HQ for fluoxetine is calculated at <1, suggesting little risk to the aquatic environment. However, when fish physiological and reproductive responses were evaluated, lower fluoxetine exposure levels of 0.1 μg/L and 0.5 μg/L affected female medaka plasma estradiol levels, and the number of developmental abnormalities were elevated at all exposure levels. Further, Brooks *et al.* (2003b) observed modulation of dopamine and norepinephrine levels in medaka brain tissue by 0.1 μg/L fluoxetine. When these nonstandard steroid, developmental, and neurotransmitter data are considered, the lowest observed response level of fluoxetine on aquatic biota occurs at concentrations detected in municipal effluents and at one order of magnitude higher than highest surface-water concentrations reported (Kolpin *et al.*, 2002; Metcalfe *et al.*, 2003).

VII. Conclusions

Herein, we summarized current data on fluoxetine occurrence in surface waters and aquatic organism and community responses to fluoxetine exposure, and a preliminary aquatic risk characterization for fluoxetine was provided. Fluoxetine is reported in effluents and surface waters at low to mid ng/L concentrations (Kolpin *et al.*, 2002; Weston *et al.*, 2001). Adverse effects of fluoxetine are observed in standardized aquatic toxicity tests at

μg/L levels. Little risk to aquatic systems is expected from such fluoxetine exposure levels if a hazard quotient approach is utilized to characterize risk. However, such a deterministic ratio of exposure and effect levels should not preclude fluoxetine from further risk consideration.

Cairns (1983) previously considered reliance on aquatic toxicity test endpoints for regulatory contaminant decisions. Standardized toxicity tests are not intended to predict structural or functional responses to contaminants (Dickson *et al.*, 1992; La Point and Waller, 2000) and may not represent most sensitive species responses (Cairns, 1986). Further, standardized test endpoints do not provide information on biochemical, developmental, behavioral, or transgenerational responses to fluoxetine exposure. For example, Foran *et al.* (2004) observed ng/L treatment levels of fluoxetine to affect Japanese medaka plasma estradiol levels and increase developmental abnormalities, and Brooks *et al.* (2003b) detected medaka brain neurotransmitter modulation at the same concentrations. Although the mechanism(s) by which fluoxetine induced these responses and whether such responses may impact population viability are not fully understood, potential fluoxetine effects on fish individuals and populations warrant further study.

Daughton and Ternes (1999) suggested that chronic studies with environmental concentrations of pharmaceuticals are necessary to assess aquatic ecosystem responses. Because pharmaceuticals are continuously released into the environment, it is relevant to perform such chronic life-cycle-type tests, which encompass sensitive stages of organism development. Extrapolation from single-species toxicity tests alone does not assess the potential for pharmaceuticals to affect aquatic communities. Aquatic microcosms and mesocosms are well suited for evaluation of multitrophic-level stressor responses (Boudou and Ribeyre, 1997; Brooks *et al.*, 1997; Halling-Sorensen *et al.*, 1998; Kennedy *et al.*, 2002) and are ideal for assessment of direct and indirect effects of parent compounds and metabolites on complex aquatic communities (Hill *et al.*, 1994). Richards *et al.* (2004) found that a mixture of fluoxetine with other pharmaceuticals impacted aquatic microcosms over a 35-day study period. Fluoxetine treatment levels used by Richards *et al.* (2004) are higher than reported environmental concentrations; however, their study clearly identified the importance of evaluating contaminant effects on multiple levels of biological organization.

Based on recent Kolpin *et al.* data (2002) and limited data from other sources, the probability of fluoxetine occurring individually at concentrations high enough to affect aquatic communities is judged to be low; however, potential additivity of action must also be considered. In typical surface waters receiving wastewater effluent, there could be hundreds of pharmaceuticals; those with similar modes of action could have additive effects on indigenous "nontarget" aquatic organisms. In the UK and Canada alone, over 3000 active pharmaceuticals are licensed for use (Pfluger and Dietrich,

2001; Servos *et al.*, 2002), few of which have even been analyzed for in surface water. Many unaccounted pharmaceuticals share the same mode of action. For example, in their review of pharmaceuticals in the environment, Daughton and Ternes (1999) discussed 50 pharmaceuticals of concern; of these, three were SSRIs. Consequently, when surface-water concentrations of pharmaceuticals sharing a common mechanism of action are further elucidated, the effective (additive) environmental concentrations could be more substantial (Daughton and Ternes, 1999; Massengale *et al.*, 2003). The types and quantities of pharmaceuticals present in surface waters will obviously vary by region; however, it is clear that there is a high potential for a large number of pharmaceuticals to simultaneously occur in the environment. Further, results of Richards *et al.* (2004) suggest that a more definitive assessment of risks posed by parent compounds and mixtures of pharmaceuticals in surface waters should be conducted.

Daughton and Ternes (1999) also suggested that bioassays or biomarkers should be developed that focus on specific mechanisms of pharmaceutical action on nontarget biota. This is decidedly critical because environmental pharmaceuticals, unlike pesticides, are most likely not acutely toxic to aquatic life. For example, the beta-adrenergic receptor blocking therapeutics propranolol and metaprolol reduce cladoceran heart rate and respiration at levels lower than those affecting survival, growth, and fecundity (Brooks *et al.*, 2003c). Brooks *et al.* (2003a) observed that fluoxetine adversely affected growth of a green algae, *P. subcapitata* (Table II); the mechanism(s) by which fluoxetine exerts its toxicity on algae has not been reported in the peer-reviewed literature. However, Munoz-Bellido *et al.* (2000) identified that fluoxetine has antibacterial properties, potentially interfering with efflux pumps. An average EC_{50} value for *V. fischeri*, a model marine bacterium, was estimated at 724 μg/L (Table II; Massengale *et al.*, 2003). Richards *et al.* (2004) evaluated effects of fluoxetine, ibuprofen, and ciprofloxacin mixtures on bacterial communities in aquatic microcosms. Whereas initial measures of bacterial cell numbers were not affected by treatments, phylotypes of bacterial community samples are currently being evaluated for treatment effects. In addition, Massengale *et al.* (2003) recently utilized a Biolog ECOplate method, which performs community-level physiological profiling through carbon-utilization analysis, to screen for microbial community responses to environmentally realistic fluoxetine exposures.

Existing data clearly indicate a need for a greater understanding of fluoxetine effects on nontarget biota, of pharmaceutical interactions and effects on multiple levels of biological organization, and of environmental fate of fluoxetine, including bioaccumulation. Such information is required before more definitive assessments of pharmaceuticals in the environment may be performed. Similarly, lack of information on temporal and spatial occurrence of fluoxetine in aquatic systems presently limits predicted environmental concentration estimates.

Acknowledgments

This research was supported by a U.S. Congressional Environmental Sensors and Signals grant, a Texas Water Resources Institute/United States Geological Survey grant, the Baylor University Research Committee, the Institute of Applied Sciences at the University of North Texas, the University of Oklahoma, the Environmental Toxicology Research Program at the University of Mississippi, and the Canadian Network of Toxicology Centres and Canada R&D. Erica March, Bethany Peterson, Lindsey Odom, John Rimoldi, David Johnson, Monica Lam, Scott Mabury, Christian Wilson, and Temidayo Fadelu were vital for sample collection, analysis, and data processing.

References

American Society for Testing and Materials (1998). Standard guide for conducting static toxicity tests with *Lemna gibba* G-3. Annual Book of ASTM Standards 14.02, 784–793.

Arcand-Hoy, L. D., Nimrod, A. C., and Benson, W. H. (1998). Endocrine-modulating substances in the environment: Estrogenic effects of pharmaceutical products. *Int. J. Toxicol.* **17**, 139–158.

Arcand-Hoy, L. D., and Benson, W. H. (2001). Toxic responses of the reproductive system. *In* "Target Organ Toxicity in Marine and Freshwater Teleosts, Volume II: Systems" (D. Schlenk and W. H. Benson, Eds.), pp. 175–202. Taylor and Francis, New York.

Azmitia, E. C. (1999). Serotonin neurons, neuroplasticity, and homeostasis of neural tissue. *Neuropsychopharmacology* **21**, 33S–45S.

Boudou, A., and Ribeyre, F. (1997). Aquatic ecotoxicology from the ecosystem to the cellular and molecular levels. *Environ. Health Perspect.* **105**(Suppl. 1), 21–35.

Boyd, G. R., Reemtsma, H., Grimm, D. A., and Mitra, S. (2003). Pharmaceuticals and personal care products (PPCPs) in surface and treated waters of Louisiana, USA and Ontario, Canada. *Sci. Total Environ.* **311**, 135–149.

Bradley, P. M., McMahon, P. B., and Chapell, F. H. (1995). Effects of carbon and nitrate on denitrification in bottom sediments of an effluent-dominated river. *Wat. Resources Res.* **31**, 1063–1068.

Brain, R., Johnson, D., Richards, S. M., Sanderson, H., Sibley, P. K., and Solomon, K. R. (2004). Effects of 25 pharmaceutical compounds to *Lemna gibba* using a 7-day static renewal test. *Environ. Toxicol. Chem.* **23**, 371–382.

Brooks, B. W., Chambers, J. A., Libman, B. S., and Threlkeld, S. T. (1997). Main and interactive effects in a wetland mesocosm experiment: Algal community responses to agrichemical runoff. *Proc. Miss. Water Resour.* **27**, 111–118.

Brooks, B. W., Turner, P. K., Stanley, J. K., Weston, J., Glidewell, E. A., Foran, C. M., Slattery, M., La Point, T. W., and Huggett, D. B. (2003a). Waterborne and sediment toxicity of fluoxetine to select organisms. *Chemosphere* **52**, 135–142.

Brooks, B. W., Smith, W., Blank, C. L., Weston, J. J., Slattery, M., and Foran, C. M. (2003b). Pharmaceutical neuromodulation of teleost catecholamines. Annual Meeting of the

Society of Environmental Toxicology and Chemistry, Austin, Texas. November 9–13, 2003.

Brooks, B. W., Dzialowski, E. M., Turner, P. K., Stanley, J. K., and Glidewell, E. A. (2003c). Freshwater invertebrate responses to pharmaceuticals. Annual Meeting of the American Society of Limnology and Oceanography, Salt Lake City, Utah. Feburary 9–14, 2003.

Brooks, B. W., Stanley, J. K., White, J. C., Turner, P. K., Wu, K. B., and La Point, T. W. (2004a). Laboratory and field responses to cadmium in effluent-dominated stream mesocosms. *Environ. Toxicol. Chem.* **23,** 1057–1064.

Brooks, B. W., Riley, T., and Taylor, R. D. (2004b). Water quality of effluent dominated stream ecosystems: Ecotoxicological, hydrological, and management considerations. *Hydrobiol.* In press.

Bruce, R. D., and Versteeg, D. J. (1992). A statistical procedure for modeling continuous toxicity data. *Environ. Toxicol. Chem.* **11,** 1485–1494.

Buser, H.-R., and Muller, M. D. (1998). Occurrence of the pharmaceutical drug clofibric acid and the herbicide mecoprop in various Swiss lakes and in the North Sea. *Environ. Sci. Technol.* **32,** 188–192.

Buser, H.-R., Poiger, T., and Muller, M. D. (1999). Occurence and environmental behavior of the chiral pharmaceutical drug ibuprofen in surface waters and in wastewater. *Environ. Sci. Technol.* **33,** 2529–2535.

Cairns, J., Jr. (1983). Are single species toxicity tests alone adequate for estimating environmental hazard? *Hydrobiol.* **100,** 47–57.

Cairns, J., Jr. (1986). The myth of the most sensitive species. *BioSci.* **36,** 670–672.

Canadian Pharmacists Association (2000). *In* "Compendium of Pharmaceuticals and Specialties" (L. Welbanks, Ed.). Canadian Pharmacists Association, Toronto.

Cerda, J., Subhoder, N., Reich, G., Wallace, R. A., and Selman, K. (1998). Oocyte sensitivity to serotonergic regulation during the follicular cycle of the teleost *Fundulus heteroclitus*. *Biol. Reprod.* **59,** 53–61.

CNN (2001). FDA clears 'generic Prozac' for sale. http://www.cnn.com/2001/HEALTH/08/01/prozac.barr/ (Accessed 23 June 2002).

Daughton, C. G., and Ternes, T. A. (1999). Pharmaceuticals and personal care products in the environment: Agents of subtle change? *Environ. Health Perspect.* **107,** 907–938.

Dickson, K. L., Waller, W. T., Kennedy, J. H., Arnold, W. R., Desmond, W. P., Dyer, S. D., Hall, J. F., Knight, J. T., Jr., Malas, D., Martinez, M. L., and Matzner, S. L. (1989). A water quality and ecological survey of the Trinity Rive, Vols I and II, City of Dallas Water Utilities, Dallas, TX.

Dickson, K. L., Waller, W. T., Kennedy, J. H., and Ammann, L. P. (1992). Assessing the relationship between ambient toxicity and instream biological response. *Environ. Toxicol. Chem.* **11,** 1307–1322.

Dickson, K. L., Waller, W. T., Kennedy, J. H., Ammann, L. P., Guinn, R., and Norberg-King, T. J. (1996). Relationships between effluent toxicity, ambient toxicity, and receiving systems impacts: Trinity River dechlorination case study. *In* "Whole Effluent Toxicity Testing: An Evaluation of Methods and Prediction of Receiving System Impacts" (D. R. Grothe, K. L. Dickson, and D. K. Reed-Judkins, Eds.), pp. 287–308. SETAC Press, Pensacola, Florida.

Dorn, P. B. (1996). An industrial perspective on whole effluent toxicity testing. *In* "Whole Effluent Toxicity Testing: An Evaluation of Methods and Prediction of Receiving System Impacts" (D. R. Grothe, K. L. Dickson, and D. K. Reed-Judkins, Eds.), pp. 16–37. SETAC Press, Pensacola, Florida.

Eugen-Olsen, J., Afzelius, P., Andresen, L., Iversen, J., Kronborg, G., Aabech, P., Nielsen, J. O., and Hofmann, B. (1997). Serotonin modulates immune function in T cells from HIV-seropositive subjects. *Clinical Immunol. Immunopathol.* **84,** 115–121.

FDA-CDER (1996). Retrospective review of ecotoxicity data submitted in environmental assessments, Docket no. 96N-0057. FDA Center for Drug Evaluation and Research, Rockville, Maryland.

FDA-CDER (1998). Guidance for industry – environmental assessment of human drugs and biologics applications, Revision 1. FDA Center for Drug Evaluation and Research, Rockville, Maryland. http://www.fda.gov/cder/guidance/1730fnl.pdf (Accessed 23 June 2002).

Flaherty, C. M., Kashian, D. R., and Dodson, S. I. (2001). Ecological impacts of pharmaceuticals on zooplankton: The effects of three medications on *Daphnia magna*. Annual Meeting of the Society of Environmental Toxicology and Chemistry, Baltimore, Maryland, November 11–15, 2001.

Fong, P. P. (1998). Zebra mussel spawning is induced in low concentrations of putative serotonin reuptake inhibitors. *Biol. Bull.* **194,** 143–149.

Fong, P., Huminski, P., and D'Urso, L. (1998). Induction and potentiation of parturition in fingernail clams by selective serotonin re-uptake inhibitors. *J. Exp. Zool.* **280,** 260–264.

Fong, P. P. (2001). Antidepressants in aquatic organisms: A wide range of effects. *In* "Pharmaceuticals and Personal Care Products in the Environment: Scientific and Regulatory Issues" (C. G. Daughton and T. L. Jones-Lepp, Eds.), pp. 264–281. American Chemical Society, Washington, D.C.

Foran, C. M., Peterson, B. N., and Benson, W. H. (2002). Transgenerational and developmental exposure of Japanese medaka (*Oryzias latipes*) to ethinylestradiol results in endocrine and reproductive differences in the response to ethinylestradiol as adults. *Toxicol. Sci.* **68,** 389–402.

Foran, C. M., Weston, J. J., Slattery, M., Brooks, B. W., and Huggett, D. B. (2004). Reproductive assessment of Japanese medaka (*Oryzias latipes*) following a four-week fluoxetine exposure. *Arch. Environ. Contam. Toxicol.* **46,** 511–517.

Golet, E. M., Alder, A. C., Hartmann, A., Ternes, T. A., and Giger, W. (2001). Trace determination of fluoroquinolone antibacterial agents in solid-phase extraction urban wastewater by and liquid chromatography with fluorescence detection. *Anal. Chem.* **73,** 3632–3638.

Guarino, A. M., and Lech, J. J. (1986). Metabolism, disposition, and toxicity of drugs and other xenobiotics in aquatic species. *Vet. Hum. Toxicol.* **28**(Suppl. 1), 38–44.

Halling-Sorensen, B., Nors-Nielsen, S., Lanzky, P. F., Ingerslev, F., Holten-Lutzhoft, H. C., and Jorgensen, S. E. (1998). Occurrence, fate, and effects of pharmaceutical substances in the environment—a review. *Chemosphere* **36,** 357–393.

Hamilton, M. A., Russo, R. C., and Thurston, R. V. (1977). Trimmed Spearman-Karber method for estimating median lethal concentrations in toxicity bioassays. *Environ. Sci. Technol.* **11,** 714–719.

Harries, J., Sheahan, D., Jobling, S., Matthiessen, P., Neall, P., Sumpter, J., Tylor, T., and Zaman, N. (1997). Estrogenic activity in five United Kingdom rivers detected by measurement of vitellogenesis in caged male trout. *Environ. Toxicol. Chem.* **16,** 534–542.

Hart, A. (2001). Probabilistic Risk Assessment for Pesticides in Europe: Implementation and Research Needs. Central Science Laboratory, Sand Hutton, York, United Kingdom.

Hemming, J. M., Waller, W. T., Chow, M. C., Denslow, N. D., and Venables, B. (2001). Assessment of the estrogenicity and toxicity of a domestic wastewater effluent flowing through a constructed wetland system using biomarkers in male fathead minnow (*Pimephales promelas* Rafinesque, 1820). *Environ. Toxicol. Chem.* **20,** 2268–2275.

Hemming, J. M., Turner, P. K., Brooks, B. W., Waller, W. T., and La Point, T. W. (2002). Assessment of toxicity reduction in wastewater effluent flowing through a treatment wetland using *Pimephales promelas, Ceriodaphnia dubia*, and *Vibrio fischeri*. *Arch. Environ. Contam. Toxicol.* **42,** 9–16.

Hernandez-Rauda, R., Rozas, G., Rozas, G., Rey, P., Otero, J., and Aldegunde, M. (1999). Changes in the pituitary metabolism of monoamines (dopamine, norepinephrine, and serotonin) in female and male rainbow trout (*Oncorhynchus mykiss*) during gonadal recrudescence. *Physiological Biochem. Zool.* **72**, 352–359.

Hiemke, C., and Härtter, S. (2000). Pharmacokinetics of selective serotonin reuptake inhibitors. *Pharmacol. Ther.* **85**, 11–28.

Hill, I. R., Heimbach, F., Leeuwangh, P., and Matthiessen, P. (1994). *Freshwater Field Tests for Hazard Assessment of Chemicals.* CRC Press, Boca Raton, Florida.

Hindberg, I., and Naesh, O. (1992). Serotonin concentrations in plasma and variations during the menstrual cycle. *Clinical Chem.* **38**, 2087–2089.

Hirsch, R., Ternes, T., Haberer, K., and Kratz, K. L. (1999). Occurrence of antibiotics in the aquatic environment. *Sci. Total Environ.* **225**, 109–118.

Hofmann, B., Afzelius, P., Iversen, J., Kronborg, G., Aabech, P., Benfield, T., Dybkjaer, E., and Nielsen, J. O. (1996). Buspirone, a serotonin receptor agonist, increases CD4 T-cell counts and modulates the immune system in HIV-seropositive subjects. *AIDS* **10**, 1339–1347.

Honkoop, P. J. C., Luttikhuizen, P. C., and Piersma, T. (1999). Experimentally extending the spawning season of a marine bivalve using temperature change and fluoxetine as synergistic triggers. *Mar. Ecol. Prog. Ser.* **180**, 297–300.

Huber, R., and Delago, A. (1998). Serotonin alters decisions to withdraw in fighting crayfish, *Astacus astacus*: The motivational concept revisited. *J. Comp. Phys.* **182A**, 573–583.

Huggett, D., Khan, I., Foran, C., and Schlenk, D. (2003). Determination of beta-adrenergic receptor blocking pharmaceuticals in United States wastewater effluent. *Environ. Pollut.* **121**, 199–205.

Huggett, D. B., Brooks, B. W., Peterson, B., Foran, C. M., and Schlenk, D. (2002). Toxicity of select beta-adrenergic receptor blocking pharmacueticals (β-blockers) on aquatic organisms. *Arch. Environ. Contam. Toxicol.* **43**, 229–235.

Iwamatsu, T., Toya, Y., Sakai, N., Yasutaka, T., Nagata, R., and Nagahama, Y. (1993). Effect of 5-hyroxytryptamine on steroidogenesis and oocytye maturation in pre-ovulatory follicles of the medaka *Oryzias latipes. Develop. Growth Differ.* **35**, 625–630.

Jones-Lepp, T. L., Alvarez, D. A., Petty, J. D., Osemwengie, L. I., and Daughton, C. G. (2001). Analytical chemistry for mapping trends of pharmaceutical and personal care product pollution from personal use: Some current research and future needs. 10th Symposium on Handling of Environmental and Biological Samples in Chromatography, Mainz/ Wiesbaden, Germany, April 1–4, 2001.

Kennedy, J. H., La Point, T. W., Balci, P., Stanley, J., and Johnson, Z. B. (2002). Model Aquatic Ecosystems in Ecotoxicological Research: Considerations of Design, Implementation, and Analysis. *In* "Handbook of Ecotoxicology" (D. J. Hoffman, B. A. Rattner, G. A. Burton, Jr., and J. Cairns, Jr., Eds.), Vol. 2, pp. 45–74. Lewis Publishers, Boca Raton, Florida.

Khan, N., and Deschaux, P. (1997). Role of serotonin in fish immunomodulation. *J. Exp. Biol.* **200**, 1833–1838.

Khan, I. A., and Thomas, P. (1992). Stimulatory effects of serotonin on maturational gonadotropin release in the Atlantic croaker, *Micropogonias undulatus. Gen. Comp. Endocrin.* **88**, 388–396.

Khan, I. A., and Thomas, P. (1994). Seasonal and daily variations in the plasma gonadotropin II response to a LHRH analog and serotonin in Atlantic craoker (*Micropogonias undulatus*): Evidence for mediation by 5-HT2 receptors. *J. Exp. Zool.* **269**, 531–537.

Kolpin, D. W., Furlong, E. T., Meyer, M. T., Thurman, E. M., Zaugg, S. D., Barber, L. B., and Buxton, H. T. (2002). Pharmaceuticals, hormones, and other organic wastewater contaminants in US streams, 1999–2000: A national reconnaissance. *Environ. Sci. Tech.* **36,** 1202–1211.

Lange, R., and Dietrich, D. (2002). Environmental risk assessment of pharmaceutical drug substances—conceptual considerations. *Toxicol. Lett.* **31,** 97–104.

Lanzky, P. F., and Halling-Sorensen, B. (1997). The toxic effect of the antibiotic metronidazole on aquatic organisms. *Chemosphere* **35,** 2553–2561.

LeBlanc, G. A., Campbell, P. M., den Besten, P., Brown, R. P., Chang, E. S., Coats, J. R., deFur, P. L., Dhadialla, T., Edwards, J., Riddiford, L. M., Simpson, M. G., Snell, T. W., Thorndyke, M., and Matsumura, F. (1999). The endocrinology of invertebrates. *In* "Endocrine Disruption in Invertebrates: Endocrinology, Testing, and Assessment" (P. L. deFur, M. Crane, C. G. Ingersoll, and L. Tattersfield, Eds.), pp. 23–107. SETAC Press, Pensacola, Florida.

La Point, T. W., and Waller, W. T. (2000). Field assessment in conjunction with whole effluent toxicity testing. *Environ. Toxicol. Chem.* **19,** 14–24.

Marsh, K. E., Foran, C. M., Willet, K., and Brooks, B. W. (2003). *In* "Aquatic resources and human health. Achieving Sustainable Freshwater Systems: A Web of Connections" (M. M. Holland, E. R. Blood, and L. Shaffer, Eds.), pp. 65–83. Island Press, New York.

Massengale, A. R. D., Glidewell, E. A., and Brooks, B. W. (2003). A novel method for assessing pharmaceutical effects on aquatic microbial communities. Annual Meeting of the Society of Environmental Toxicology and Chemistry, Austin, Texas, November 9–13, 2003.

McLean, B. (2001). Prozac: A Bitter Pill. http://www.fortune.com/indexw.jhtml?channel=artcol.jhtml&doc_id=203499(Accessed 23 June 2002).

Meguid, M. M., Fetissov, S. O., Varma, M., Sato, T., Zhang, L., Laviano, A., and Rossi-Fanelli, F. (2000). Hypothalamic dopamine and serotonin in the regultion of food intake. *Nutrition* **16,** 843–857.

Metcalfe, C., and Koenig, B. (2001). A survey of levels of ibuprofen in surface water in the lower Great Lakes. Water Quality Centre, Trent University, Peterborough, United Kingdom.

Metcalfe, C. D., Gray, M. A., and Kiparssos, Y. (1999). The Japanese medaka (*Oryzias latipes*): An *in vivo* model for assessing the impacts of aquatic contaminants on the reproductive success of fish. *In* "Impact Assessment of Hazardous Aquatic Contaminants" (S. S. Rao, Ed.), pp. 29–52. Lewis Publishers, Boca Raton, Florida.

Metcalfe, C. D., Miao, X., Koenig, B. G., and Struger, I. (2003). Distribution of acidic and neutral drugs in surface waters near sewage treatment plants in the lower Great Lakes, Canada. *Environ. Toxicol. Chem.* **22,** 2881–2889.

Mossner, R., and Lesch, K. P. (1998). Role of serotonin in the immune system and in neuroimmune interactions. *Brain Behav. Immunity* **12,** 249–271.

Munoz-Bellido, J. L., Munoz-Criado, S., and Garcìa-Rodrìguez, J. A. (2000). Antimicrobial activity of psychotropic drugs: Selective serotonin reuptake inhibitors. *Inter. J. Antimicrob. Ag.* **14,** 177–180.

Nation, J. L. (2002). Insect physiology and biochemistry. CRC Press, Boca Raton, Florida.

Nichols, K., Snyder, E., Miles-Richardson, S., Pierens, S., Snyder, S., and Giesy, J. P. (1999). Effects of municipal wastewater exposure in situ on the reproductive physiology of the fathead minnow (*Pimephales promelas*). *Environ. Toxicol. Chem.* **18,** 2001–2012.

Nilsson, S. N., and Sundin, L. (1998). Gill Blood Flow Control. *Comp. Biochem. Physiol.* **119A,** 137–147.

NDC Health (2003). The internet drug index (The top 200 prescriptions of 2002: 2002 US prescriptions based on more than 3.05 billion U.S. prescriptions). http://www.rxlist.com/top200.htm (Accessed 30 July 2003).

Pfluger, P., and Dietrich, D. R. (2001). Effects on pharmaceuticals in the environment – an overview and principle considerations. *In* "Pharmaceuticals in the Environment" (K. Kummerer, Ed.), pp. 11–17. Springer-Verlag, Berlin, Germany.

Rang, H. P., Dale, M. M., Ritter, J. M., and Gardner, P. (1995). *Pharmacology* 590–592. Churchill Livingstone Inc., New York.

Ranganathan, R., Sawin, E. R., Trent, C., and Horvitz, H. R. (2001). Mutations in the *Caenorhabditis elegans* serotonin reuptake transporter MOD-5 reveal serotonin-dependent and independent activities of fluoxetine. *J. Neurosci.* **21**, 5871–5884.

Richards, S. M., Wilson, C. W., Johnson, D. J., Castle, D. M., Lam, M., Mabury, S. A., Sibley, P. K., and Solomon, K. (2004). Effects of pharmaceutical mixtures in aquatic microcosms. *Environ. Toxicol. Chem.* **23**, 1035–1042.

Richardson, M. L., and Bowron, J. M. (1985). The fate of pharmaceutical chemicals in the aquatic environment. *J. Pharmacokin. Pharmacol.* **37**, 1–12.

Rossknecht, H., Hetzenauer, H., and Ternes, T. A. (2001). Pharmaceuticals in Lake Constance. *Nachrichten Aus Der Chemie* **49**, 145–149.

Servos, M. R., Bennie, D. T., Starodub, M. E., and Orr, J. C. (2002). *In* Pharmaceuticals and personal care products in the environment: A summary of published literature. Report No. 02-309: National Water Research Institute, Burlington, Ontario.

Solomon, K. R., Baker, D. B., Richards, R. P., Dixon, K. R., Klaine, S. J., La Point, T. W., Kendall, R. J., Giddings, J. M., Giesy, J. P., Hall, L. J., Jr., and Williams, W. M. (1996). Ecological risk assessment of atrazine in North American surface waters. *Environ. Toxicol. Chem.* **15**, 31–76.

Stan, H. J., and Heberer, T. (1997). Pharmaceuticals in the aquatic environment. *Analysis* **25**, 20–23.

Stumpf, M., Ternes, T., Haberer, K., Seel, P., and Baumann, W. (1996). Determination of drugs in sewage treatment plants and river water. *Vom Wasser.* **86**, 291–303.

Stumpf, M., Ternes, T. A., Wilken, R-D., Rodrigues, S. V., and Baumann, W. (1999). Polar drug residues in sewage and natural waters in the state of Rio de Janeiro, Brazil. *Sci. Total Environ.* **225**, 135–141.

Suedel, B., and Rodgers, J. H., Jr. (1996). Toxicity of fluoranthene to *Daphnia magna, Hyalella azteca, Chironomus tentans*, and *Stylaria lacustris* in water-only and whole sediment exposures. *Bull. Environ. Contam. Toxicol.* **57**, 132–138.

Suter, M. J. F., and Giger, W. (2000). Trace determinants of emerging water pollutants: Endocrine disruptors, pharmaceuticals, and specialty chemicals. *Chimia.* **54**, 13–16.

Ternes, T. (1998). Occurrence of drugs in German sewage treatment plants and rivers. *Water Res.* **32**, 3245–3260.

Uhler, G. C., Huminski, P. T., Les, F. T., and Fong, P. P. (2000). Cilia-driven rotational behavior in gastropod (*Physa elliptica*) embryos induced by serotonin and putative serotonin reuptake inhibitors (SSRIs). *J. Exp. Zool.* **286**, 414–421.

USEPA (1989). *Short-Term Methods for Estimating the Chronic Toxicity of Effluents and Receiving Waters to Freshwater Organisms*, 2nd Edition. Environmental Monitoring Systems Laboratory, Cincinnati, Ohio.

USEPA (1991). *Methods for Measuring the Acute Toxicity of Effluents Receiving waters to Freshwater and Marine Organisms*, 4th Edition. EPA/600/4-90/027.

USEPA (2000). *Methods for Measuring the Toxicity and Bioaccumulation of Sediment-Associated Contaminants with Freshwater Invertebrates*, 2nd Edition. EPA 600/R-99/064.

Velagaleti, R., and Robinson, J. (2000). Degradation and depletion of pharmaceutical chemicals in the environment. Abstracts of Papers of the American Chemical Society. 219, 51-ENVR.

Webb, S. F. (2001). A data based perspective on the environmental risk assessment of human pharmaceuticals II – aquatic risk characterization. *In* "Pharmaceuticals in the Environement: Sources, Fate, Effects and Risks" (K. Kummerer, Ed.), pp. 203–219. Springer-Verlag, Berlin.

Weston, J. J., Huggett, D. B., Rimoldi, J., Foran, C. M., and Stattery, M. (2001). Determination of fluoxetine (Prozac^{TM}) and norfluoxetine in the aquatic environment. Annual Meeting of the Society of Environmental Toxicology and Chemistry, Baltimore, Maryland. November 11–15, 2001.

Zhang, L., Khan, I. A., and Foran, C. M. (2002). Characterization of the estrogenic response to genistein in Japanese medaka (*Oryzias latipes*). *Comp. Biochem. Phys. Part C.* **132**, 203–211.

Michael Cleuvers

Department of General Biology
Aachen University of Technology
D-52056 Aachen

Aquatic Ecotoxicity of Pharmaceuticals Including the Assessment of Combination Effects

I. Introduction

A. Background

In recent years, reports about residues of pharmaceuticals in surface and drinking waters have increased in scientific publications (Daughton and Jones-Lepp, 2001; Daughton and Ternes, 1999; Kümmerer, 2001; Ternes *et al.*, 1998, 2001). Among the detected substances in rivers were beta blockers (metoprolol up to 1.54 μg/L), beta-sympathomimetics (Hirsch *et al.*, 1996; Sedlak and Pinkston, 2001), analgesic and anti-inflammatory drugs (diclofenac up to 1.2 μg/L; Buser *et al.*, 1998a, 1999; Stumpf *et al.*, 1998; Ternes, 1998), estrogens (17-estradiol up to 0.013 μg/L; Adler *et al.*, 2001; Huang and Sedlak, 2001; Kuch and Ballschmitter, 2000), and antibiotics (erythromycin up to 1.7 μg/L; Adler *et al.*, 2001; Hirsch *et al.*, 1999; Lindsey *et al.*, 2001), as well as lipid lowering agents (clofibrinic acid up to 0.2 μg/L;

Ahrer *et al.*, 2001; Buser *et al.*, 1998b; Öllers *et al.*, 2001; Stan *et al.*, 1994) and antiepileptic drugs (carbamazepine up to 2.1 μg/L; Möhle *et al.*, 1999; Seiler *et al.*, 1999; Ternes, 1998). Particularly, some small streams receiving a relatively large amount of water from sewage water treatment plants are found to be considerably polluted, with peak concentrations of several pharmaceuticals greater than 1 μg/L. Due to their specific mode of action and the fact that these compounds are intentionally designed to exert an effect on humans, mammals, or other vertebrates, residues of pharmaceuticals could be as, or even more, detrimental to human health than pesticides, which are created to affect weeds, fungi, and invertebrate varmints. The mode of action of pharmaceuticals is not well enough understood to make general statements about potential environmental effects caused by these substances. Drinking water treatment diminishes residues, but is not able to remove these substances completely. Thus even in tap water, some pharmaceuticals like clofibrinic acid can be detected in concentrations up to 270 ng/L (Heberer *et al.*, 1997, 2001a,b; Ternes *et al.*, 2001). In contrast to the amount of analytical data, information about the ecotoxicological effects of drug residues is sparse (Cleuvers, 2002; Webb, 2001). To create a broader basis for the evaluation of the ecotoxicological relevance of pharmaceutical compounds and mixtures, biotests with *Desmodesmus subspicatus, Daphnia magna*, and *Lemna minor* were performed.

B. Concepts for the Prediction of Mixture Toxicity

Drug residues found in the aquatic environment usually occur as mixtures, not as single contaminants. Thus, scientific assessment of risk to aquatic life should consider this complex exposure situation. By analyzing combination effects, ecotoxicologists try to elucidate the problem of risk assessment for complex mixtures of various substances through two decades of excellent studies (Altenburger *et al.*, 2000; Backhaus *et al.*, 2000; Faust *et al.*, 2001). Basically, two different concepts are in use for the prediction of mixture toxicity, generally termed *concentration addition* and *independent action*.

The concept of concentration addition can be traced back to the early works of the pharmacologists Loewe and Muischnek (Loewe, 1927, 1953; Loewe and Muischnek, 1926). Concentration addition can be described mathematically for a mixture of *n* substances by the equation (Berenbaum, 1985):

$$\sum_{i=1}^{n} \frac{c_i}{ECx_i} = 1. \qquad (1)$$

In this equation, c_i is the individual concentration of the single substance present in a mixture with a total effect of $x\%$, and ECx_i is the concentration of the single substance that would alone cause the same effect x as observed for the mixture. As an important point, concentration addition means that substances applied below their individual no effect concentration (NOEC) can nevertheless contribute to the total effect of the mixture. Concentration addition is based on the idea of a *similar action* of chemicals, whereas interpretations of this term can differ considerably. From a mechanistic point of view, similar action means in a strict sense that substances should have the same specific interaction with a molecular target site in the observed test organism (Pöch, 1993). In contrast, used in a more common sense, a similar action could be observed for all substances that are able to cause the same toxicological response under consideration, such as death of the test organisms. For example, concentration addition is able to predict mixture toxicities of inert chemicals, or chemicals that are not reactive when considering overall acute effects and that do not interact with specific receptors in the organism (Broderius *et al*, 1995; van Loon *et al.*, 1997). The mode of action of such compounds is called narcosis (van Leeuwen *et al.*, 1992; Verhaar *et al.*, 1992). Narcosis type toxicity is considered to be caused by an absolutely nonspecific mode of action, in that the potency of a chemical to induce narcosis is entirely dependent on its hydrophobicity, generally expressed by its n-octanol-water partition coefficient (log K_{ow}). As a result, in the absence of any specific mechanism of toxicity, a chemical will, within certain boundaries, always be as toxic as its log K_{ow} indicates. Thus, the narcosis type of action is also called baseline toxicity (Verhaar *et al.*, 1992).

The alternative concept of independent action was already formulated by Bliss (1939). It is based on the idea of dissimilar action of compounds in a mixture, such that the compounds have different molecular target sites and modes of action. As a result of such a dissimilar action, the relative effect of one of the toxicants in a mixture should remain unchanged in the presence of another one. For a binary mixture, the combination effect can be calculated by the equation:

$$E(c_{mix}) = 1 - [(1 - E(c_1)) \leftrightarrow (1 - E(c_2))] \tag{2}$$

or in general,

$$E(c_{mix}) = 1 - \subseteq_{i=1}^{n} (1 - E(c_i)), \tag{3}$$

in which $E(c_1)$ and $E(c_2)$ are the effects of the single substance and $E(c_{mix})$ is the total effect of the mixture. Following this equation, a substance applied in a concentration below its individual NOEC will not contribute to the total effect of the mixture, such that there will be no mixture toxicity if the concentrations of all used single substances are below their NOEC. At given concentrations of the single compounds in a mixture, the combination effect will in general be higher if the substances follow the concept of concentration addition. Thus misleadingly, the different concepts were sometimes brought in correlation with the terms *synergism* and *antagonism*. But synergisms or antagonisms between the used substances and their effects can occur independently of a similar or dissimilar mode of action.

II. Methods

A. *Daphnia* Acute Immobilization Test

Daphnia tests were conducted following the European Guidelines. (Commission of the European Communities, 1992) using the water flea *Daphnia magna* Strauss. Daphnids were bred in ADaM, a culture medium imitating natural fresh water (Klüttgen *et al.*, 1994). Experiments were run at temperatures of $20 \pm 1^{\circ}C$ and photoperiods of 16 hours light, 8 hours dark (about 20 $\mu Es^{-1}\ m^{-2}$). Twenty daphnids less than 24 h old were used for the controls; each treatment was subdivided into four replicates each containing five daphnids. Culture volume was 50 ml. Immobility was observed after 24 and 48 hours, with the latter being the endpoint for effect calculation.

B. *Algae* Growth Inhibition Test

Algae tests were conducted following the European Guideline. (Commission of the European Communities, 1993). Planktonic chlorococcale green alga *Desmodesmus subspicatus* (Chodat) were used; they were obtained from Sammlung von Algenkulturen at the University of Göttingen, Germany. Initial cell densities were adjusted at 10^4 cells/ml using a calibration curve of chlorophyll fluorescence (Ex, 460; Em, 685 nm) versus cell number and appropriate dilution of preculture. Medium was prepared according to the protocol using deionized water and analytical grade chemicals. Algae were incubated at $23 \pm 2^{\circ}C$ under continuous white light (120 $\mu Es^{-1}\ m^{-2}$) and were kept in suspension by continuous shaking (approx. 80 rpm). Results were quantified in terms of average growth rates calculated from cell numbers based on measurements of chlorophyll fluorescence.

C. *Lemna* Growth Inhibition Test

Lemna tests with the duckweed *Lemna minor* were performed according to the International Standards Organization (ISO, 2001) using the Steinberg medium (Steinberg, 1943, 1946) with the modification that two species of phosphate were added. Biotests were carried out in 400 ml beakers with an outer diameter of 80 mm, filled with 150 ml medium. Inoculum for each beaker was 12 fronds. Only plants with two or three fronds were chosen. Six control replicates and three treatment replicates were used. Tests were carried out in a climatic exposure test cabinet, calibrated at 25 ± 2°C, with fluorescent tubes mounted on the top. Light intensity was adjusted at 100 $\mu Es^{-1}\ m^{-2}$. Test duration was seven days (168 h). Number and area of the fronds were determined at days 0, 3, 5, and 7. Measurements and evaluations were performed by means of the digital image analysis system Scanalyzer (LemnaTec, Wuerselen, Germany). At days 0, 3, 5, and 7, images of the beakers were taken for analysis. Based on total frond area, the specific average growth rates were calculated following the ISO Standard (ISO, 2001).

D. Test Substances and Calculation of Effect Concentrations

Substances used for the biotests were clofibrinic acid (metabolite of several lipid lowering drugs), carbamazepine (antiepileptic), propranolol and metoprolol (beta-blockers), ibuprofen sodium, diclofenac sodium and naproxen sodium (analgesics/anti-inflammatory drugs), captopril (anti-hypertensive), and metformin (anti-diabetic). All were supplied in analytical grade by Sigma Aldrich, Taufkirchen, Germany. Generally, substances were applied in at least five concentrations (1, 3.2, 10, 32, and 100 mg/L). Depending on the results, some tests were repeated, including lower (0.1 and 0.32 mg/L) or higher (320 mg/L) concentrations.

Effective concentrations (EC_x) were determined with the program Sigma Plot 2000 using nonlinear curve fitting based on a sigmoid model (4-parameter logistic function). For the assessment of mixture toxicities, half the calculated effect concentrations ($EC_5/2$, $EC_{10}/2$, $EC_{20}/2$, $EC_{50}/2$, $EC_{80}/2$, or $EC_{90}/2$) were used. If the substances follow the concept of concentration addition, theoretically the combination effect of the mixtures would add up to a total effect of about 5, 10, 20, 50, and 80% or 90%, respectively.

E. Analysis of the Mode of Action

To find out something about the mode of action of the observed substances, we used quantitative structure activity relationships (QSARs; Lipnick, 1995; Schüürmann and Markert, 1998; Verhaar *et al.* 1992) for nonpolar narcotic chemicals proposed by Verhaar *et al.* (1995) for *Daphnia magna* (equation 4) and by van Leeuwen *et al.* (1992) for the chlorophyte *Pseudokirchneriella subcapitata* (equation 5), the sensitivity of which is seen to be comparable to *Desmodesmus subspicatus*.

$$\text{Log } EC_{50}[\text{mmol/L}] = -0.95\ \log K_{ow} - 1.32 \qquad (4)$$

$$\text{Log } EC_{50}[\text{mmol/L}] = -1.00 \log K_{ow} - 1.23 \tag{5}$$

III. Results

The EU-Directive 93/67/EEC (Commission of the European Communities, 1996) classifies substances according to their EC_{50} value in different classes: <1 mg/L (very toxic to aquatic organisms); 1–10 mg/L (toxic to aquatic organisms); and 10–100 mg/L (harmful to aquatic organisms). Substances with an EC_{50} above 100 mg/L would not be classified. The toxicity of the tested pharmaceuticals was very heterogeneous, with EC_{50} values ranging from 7.5 mg/L (propranolol) to 174 mg/L (naproxen) in the *Daphnia* test; from 5.8 mg/L (propranolol) to >320 mg/L (metformin and naproxen) in the algal test; and from 7.5 mg/L (diclofenac) to >320 mg/L (metoprolol) in the *Lemna* test (Table I). Most EC_{50} values were in the range of 10–100 mg/L, or even greater. Values below 10 mg/L were measured only for propranolol (*Daphnia* and *Desmodesmus*), metoprolol (*Desmodesmus*), and diclofenac (*Lemna*). With the exception of metformin, metoprolol, and propranolol, *Lemna* was the most sensitive species to all tested substances.

To evaluate the mixture toxicity, we performed *Daphnia* and algal tests with combinations of two substances. Figures 1 and 2 show the results of a test with the two frequently detected drugs, clofibrinic acid (lipid lowering agent) and carbamazepine (anti-epileptic compound). In the *Daphnia* test, these substances followed the concept of concentration addition. As a result, the measured effects were much stronger than expected based on the very weak effects measured singly. For example, at the $EC_{90}/2$ level, the concentrations responsible for the singly measured effects of clofibrinic acid

TABLE I EC_{50} of the Tested Pharmaceuticals as Obtained in the Biotests with *Daphnia magna*, *Desmodesmus subspicatus* and *Lemna minor**

	EC_{50} [mg/L]		
Test substance	Daphnia	Desmodesmus	Lemna
Clofibrinic acid	72	115	12.5
Carbamazepine	>100	74	25.5
Ibuprofen sodium	108	315	22
Diclofenac sodium	68	72	7.5
Naproxen sodium	174	>320	24.2
Captopril	>100	168	25
Metformin	64	>320	110
Propranolol	7.5	5.8	114
Metoprolol	>100	7.3	>320

* Endpoint for *Daphnia* was immobilization after 48 hours, and for *Desmodesmus* and *Lemna* inhibition of the average growth rate after three and seven days, respectively.

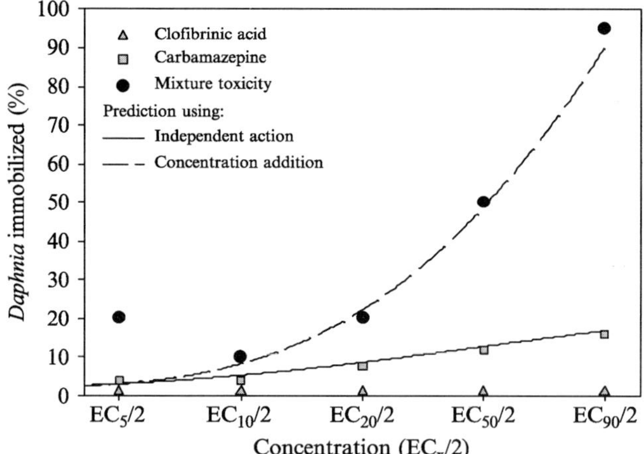

FIGURE 1 Measured mixture toxicity of clofibrinic acid and carbamazepine as obtained in the acute *Daphnia* test in comparison to the singly measured toxicities and the mixture toxicity predicted by the concepts of concentration addition and independent action.

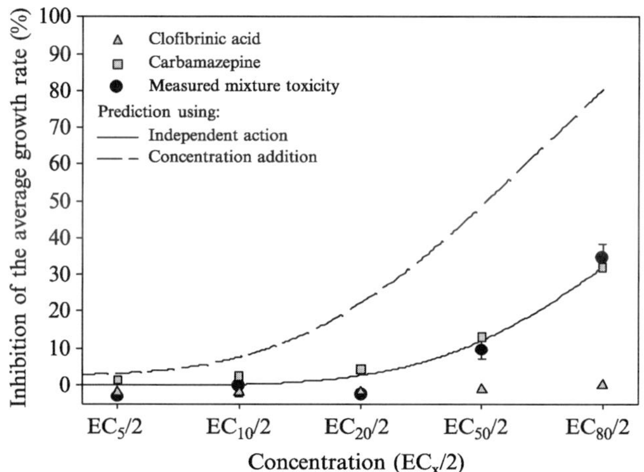

FIGURE 2 Measured mixture toxicity of clofibrinic acid and carbamazepine as obtained in the algal test in comparison to the singly measured toxicities and the mixture toxicity predicted by the concepts of concentration addition and independent action.

(1% immobilized daphnids) and carbamazepine (16% immobilized daphnids) cause a strong effect in the mixture of about 95% immobilization of the daphnids. In the algal tests (Fig. 2), the mixture effect could be well calculated using the concept of independent action, which predicts much lower combination effects than concentration addition. Regarding the mode of action for both substances, the measured EC_{50} is greater than predicted by

TABLE II Log K_{ow}, Molecular Weight, and the Predicted and Measured EC_{50}s for the Substances Used in the Tests with *Daphnia* and *Desmodesmus* with Binary Mixtures of Chemicals*

			EC_{50} [mg/L]			
			Predicted for narcosis		*Measured*	
Test substance	*Log K_{ow}*	*Molecular weight [g/mol]*	Daphnia	*Algae*	Daphnia	*Algae*
Clofibrinic acid	3.1	214.63	11.7	10.0	72	115
Carbamazepine	2.45	236.27	53.2	49.4	>100	74
Diclofenac sodium	1.56	318.1	50.2	51.6	68	72
Ibuprofen sodium	~1	218.3	1.2	1.3	108	315

* The log K_{ow} of Ibuprofen-Na was not available in literature, but because of its free solubility in water it was estimated to be not higher than 1.

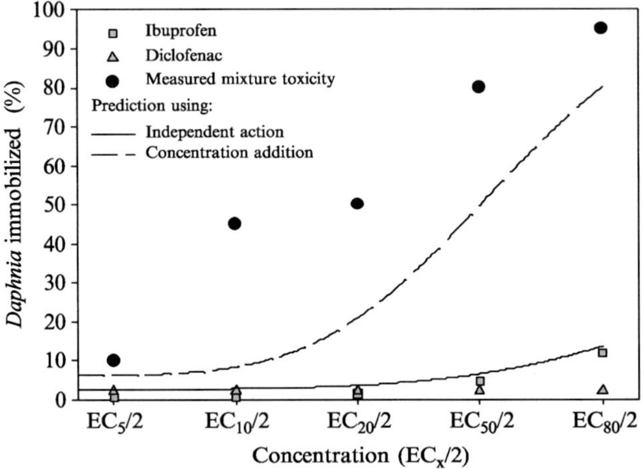

FIGURE 3 Measured mixture toxicity of diclofenac and ibuprofen as obtained in the acute *Daphnia* test in comparison to the singly measured toxicities and the mixture toxicity predicted by the concepts of concentration addition and independent action.

the QSAR for non-polar narcosis (Table II), demonstrating that no specific mechanism is present.

In the second mixture test, ibuprofen and diclofenac were used, both belonging to the non-steroidal anti-inflammatory drug (NSAID) group. In the *Daphnia* test, the mixture effect was even somewhat stronger than predicted by concentration addition (Fig. 3). The combination effect in the algal test could be predicted well by that concept, which one would

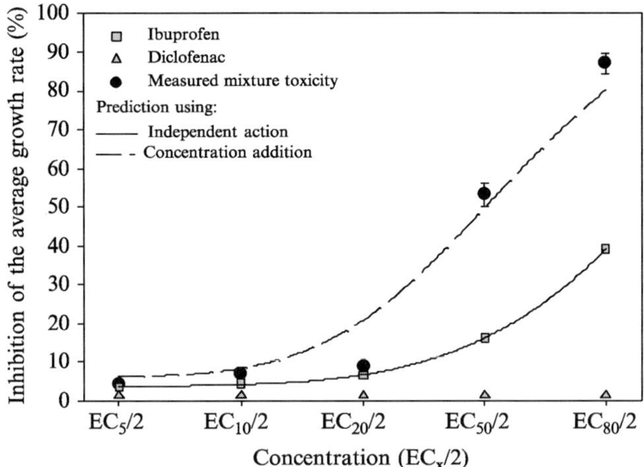

FIGURE 4 Measured mixture toxicity of diclofenac and ibuprofen as obtained in the algal test in comparison to the singly measured toxicities and the mixture toxicity predicted by the concepts of concentration addition and independent action.

expect for substances with a similar mode of action (Fig. 4). But as shown in Table II, that similarity is not due to a specific toxic action, but is a result of non-polar narcosis as in the first combination study. The log K_{ow} of ibuprofen sodium was not available in literature. Due to the fact that ibuprofen sodium is freely soluble in water, its log K_{ow} was estimated to be less than 1. But in any case, the measured EC_{50} is so high that any specific toxic action can be excluded.

IV. Discussion

The measured toxicity of the substances was heterogeneous and (for most substances) moderate, with EC_{50} between 10 and 100 mg/L or even greater. In the acute *Daphnia* and algal tests only the tested beta-blockers showed EC_{50} values below 10 mg/L (propranolol in both tests and metoprolol in the algal test only). Bearing in mind that environmental concentrations of pharmaceuticals are in most cases below 1 μg/L, acute effects of such compounds in the aquatic environment are very unlikely. Nevertheless, further studies with other test species, such as fish and benthic macroinvertebrates (as well as chronic studies with daphnids), should be performed to evaluate the toxic potential of these compounds. In the *Lemna* test, only the EC_{50} of diclofenac was less than 10 mg/L, but in six of nine cases, *Lemna* was the most sensitive organism. This is not surprising, however, because this is a chronic test with the possibility to record several sensitive, sublethal parameters such as frond size (Cleuvers and Ratte,

2002). In the case of diclofenac, the EC_{50} was nearly 10 times lower than measured in the algae and *Daphnia* test. Thus, *Lemna* growth inhibition tests are very useful and serve as an additional source of information about phytotoxicity for higher plants, which may differ greatly from that to algae (Lewis, 1995). Unfortunately, no special QSARs for narcosis are available for the *Lemna* test as it was performed in this study, but due to the lower EC_{50} values, it is not unlikely that specific toxic actions of some tested pharmaceuticals have occurred in the *Lemna* tests.

Due to the fact that the selected compounds belong to different classes of pharmaceuticals with diverse modes of action, no general trend exists in sensitivity of *Daphnia* and *Desmodesmus*. But the three studied anti-inflammatory drugs diclofenac, ibuprofen, and naproxen interestingly show the same descending order of sensitivity of *Lemna* → *Daphnia* → *Desmodesmus*. Moreover, with all used test organisms, diclofenac was the most toxic and naproxen was the least toxic compound. This leads to the question of the mode of action of these substances, which is normally known only for the target organisms, namely humans and a few laboratory mammals. Diclofenac, ibuprofen, and naproxen all inhibit the cyclooxygenases, the key enzymes catalyzing the biosynthesis of prostaglandins, which are *inter alia* responsible for the genesis of pain and inflammations (Vane, 1971; Vane and Botting, 1998). This inhibition is responsible for the analgesic and anti-inflammatory effect of these drugs. But as shown above, in the tests with *Daphnia* as well as with *Desmodesmus*, these compounds act unspecifically by nonpolar narcosis. Thus, the toxicity may be associated with the log K_{ow} of the drugs rather than with any specific toxic action. As a result, the combination effect of diclofenac and ibuprofen could be calculated with the concept of concentration addition, whereas in the *Daphnia* test, the observed mixture toxicity was somewhat stronger than predicted. As a result, we should keep in mind that toxicity of a single substance could increase strongly in combination with other similarly acting substances. For example, in the *Daphnia* test the used concentrations $EC_{20}/2$ and $EC_{50}/2$ were below the NOEC, but together showed an effect of about 50% and 80% immobilization of daphnids, respectively. Additionally, synergisms cannot be excluded between various substances independent of a similar or dissimilar mode of action. This shows that the use of individual NOECs is inapplicable regarding the assessment of mixture toxicities.

In the combination test with clofibrinic acid and carbamazepine, species-dependent results were obtained with regard to the best concept to calculate the combination effect of these compounds. In the *Daphnia* test, these substances follow the concept of concentration addition, which is generally seen to be appropriate only for similar acting substances (Pöch, 1993). This, of course, raises the question—which similar mode of action can be found for a lipid lowering agent and an anti-epileptic drug belonging to utterly different classes of chemicals? Again, it could be shown that in both test species, the measured toxicity was not higher than the baselir predicted by the QSAR for non-polar narcosis. But in contrast tc found with *Daphnia*, the algal test's combination effect was less than predicted by concentration addition; it could be well predicted using the concept of independent action, which is unusual for substances acting by nonpolar narcosis.

V. Conclusions

The toxicity of the tested pharmaceuticals was very heterogeneous. In the majority of cases, based upon the obtained EC_{50}s, *Lemna* was the most sensitive species; it is recommended to perform such tests routinely in addition to the other standard tests. Regarding the assessment of the environmental risk of pharmaceuticals, acute effects seems to be unlikely, but we need to bear in mind that considerable combination effects of substances can occur, even if the toxicity of a single substance is low. Thus, the combination effect of pharmaceuticals detected together in water samples should be determined to achieve a better assessment of the ecotoxicological potential of drug residues in the aquatic environment.

Acknowledgments

The author thanks Stephanie Esser for technical support and Arnd Weyers for critically reading the manuscript.

References

Adler, P., Steger-Hartmann, T., and Kalbfus, W. (2001). Distribution of natural and synthetic estrogenic steroid hormones in water samples from southern and middle Germany. *Acta Hydrochim. Hydrobiol.* **29,** 227–241.

Ahrer, W., Scherwenk, E., and Buchberger, W. (2001). Occurrence and fate of fluoroquinolone, macrolide, and sulfonamide antibiotics during wastewater treatment and in ambient waters in Switzerland. *In* "Pharmaceuticals and Personal Care Products in the Environment: Scientific and Regulatory Issues" (C. G. Daughton and T. Jones-Lepp, Eds.), pp. 56–69. Symposium Series 791, American Chemical Society, Washington, D.C.

Altenburger, R., Backhaus, T., Boedeker, W., Faust, M., Scholze, M., and Grimme, L. H. (2000). Predictability of the toxicity of multiple chemical mixtures to *Vibrio fischeri*: Mixtures composed of similarly acting chemicals. *Environ. Toxicol. Chem.* **19,** 2341–2347.

Backhaus, T., Altenburger, R., Boedeker, W., Faust, M., Scholze, M., and Grimme, L. H. (2000). Predictability of the toxicity of a multiple mixture of dissimilarly acting chemicals to *Vibrio fisch. Environ. Toxicol. Chem.* **19,** 2348–2356.

Berenbaum, M. C. (1985). The expected effect of a combination of agents: The general solution. *J. Theoret. Biol.* **114,** 413–431.

Bliss, C. I. (1939). The toxicity of poisons applied jointly. *Ann. Rev. Appl. Biol.* **26**, 585–615.

Broderius, S. J., Kahl, M. D., and Hoglund, M. D. (1995). Use of joint toxic response to define the primary mode of action for diverse industrial organic chemicals. *Environ. Toxicol. Chem.* **14**, 1591–1605.

Buser, H. R., Poiger, T., and Müller, M. D. (1998a). Occurrence and fate of the pharmaceutical drug diclofenac in surface waters: rapid photodegradation in a lake. *Environ. Sci. Technol.* **33**, 2529–2535.

Buser, H., Muller, M. D., and Theobald, N. (1998b). Occurrence of the pharmaceutical drug clofibric acid and the herbicide mecoprop in various Swiss lakes and in the North Sea. *Environ. Sci. Technol.* **32**, 188–192.

Buser, H.-R., Poiger, T., and Müller, M. D. (1999). Occurrence and experimental behaviour of the pharmaceutical drug ibuprofen in surface waters and in wastewater. *Environ. Sci. Technol.* **33**, 2529–2535.

Cleuvers, M. (2002). Aquatic ecotoxicology of selected pharmaceutical agents – algal and acute *Daphnia* tests. *UWSF – Z Umweltchem Ökotox* **14**, 85–89.

Cleuvers, M., and Ratte, H. T. (2002). Phytotoxicity of coloured substances: Is Lemna duckweed an alternative to the algal growth inhibition test? *Chemosphere* **49**, 9–15.

Commission of the European Communities (1992). Methods for determination of ecotoxicity; Annex V, C.2, *Daphnia*, acute toxicity to *Daphnia* L 383A, 172–178. EEC Directive 92/69/EEC.

Commission of the European Communities (1993). Methods for determination of ecotoxicity; Annex V, C.3, Algal inhibition test. L 383A, 179–186. EEC Directive 92/69/EEC.

Commission of the European Communities (1996). Technical guidance document in support of commission directive 93/67/EEC on risk assessment for new notified substances and commission regulation (EC) No 1488/94 on risk assessment for existing substances. Part II: Environmental Risk Assessment. Office for official publications of the European Communities, Luxembourg.

Daughton, C. G., and Ternes, T. A. (1999). Pharmaceuticals and personal care products in the environment: agents of subtle change? *Environ. Health Perspect.* **107**, 907–938.

Daughton, C. G., and Jones-Lepp, T., Eds. (2001). Pharmaceuticals and personal care products in the environment: Scientific and regulatory issues. pp. 56–69. Symposium Series 791, American Chemical Society, Washington, D.C.

Faust, M., Altenburger, R., Backhaus, T., Blanck, H., Boedeker, W., Gramatica, P., Hamer, V., Scholze, M., Vighi, M., and Grimme, L. H. (2001). Predicting the joint algal toxicity of multi-component s-triazine mixtures at low-effect concentrations of individual toxicants. *Aquatic Toxicol.* **56**, 13–32.

Heberer, Th., Dünnbier, U., Reilich, Ch., and Stan, H. J. (1997). Detection of drugs and drug metabolites in ground water samples of a drinking water treatment plant. *Fresenius Environ. Bull.* **6**, 438–443.

Heberer, Th., Fuhrmann, B., Schmidt-Bäumler, K., Tsipi, D., Koutsouba, V., and Hiskia, A. (2001a). Occurrence of pharmaceutical residues in sewage, river, ground and drinking water in Greece and Germany. *In* "Pharmaceuticals and Personal Care Products in the Environment: Scientific and Regulatory Issues" (C. G. Daughton and T. Jones-Lepp, Eds.), pp. 56–69. Symposium Series 791, American Chemical Society, Washington, D.C.

Heberer, Th., Verstraten, I. M., Meyer, M. T., Mechlinski, A., and Reddersen, K. (2001b). Occurrence and fate of pharmaceuticals during bank filtration—preliminary results from investigations in Germany and the United States. *Water Res. Update* **120**, 4–17.

Hirsch, R., Ternes, T. A., Haberer, K., and Kratz, K. L. (1999). Occurrence of antibiotics in the aquatic environment. *Sci. Total Environ.* **225**, 109–118.

Hirsch, R., Ternes, T. A., Haberer, K., and Kratz, K. L. (1996). Nachweis von Betablockern und Bronchospasmolytika in der aquatischen Umwelt. *Vom Wasser* **87**, 263–274.

Huang, C. H., and Sedlak, D. L. (2001). Analysis of estrogenic hormones in municipal wastewater effluent and surface water using enzyme-linked immunoadsorbent assay and gas chromatography/tandem mass spectrometry. *Environ. Toxicol. Chem.* **20**, 133–139.

ISO (International Standards Organisation) (2001). Standard ISO/WD 20079. Water quality – Duckweed growth inhibition; determination of the toxic effect of water constituents and waste water to duckweed (*Lemna minor*). Draft document.

Klüttgen, B., Dülmer, U., Engels, M., and Ratte, H. T. (1994). ADaM, an artificial freshwater for the culture of zooplankton. *Water Res.* **28**, 743–746.

Kuch, H. M., and Ballschmitter, K. (2000). Determination of endogenous and exogenous estrogens in effluents from sewage treatment plants at the ng/l-level. *Fresenius J. Anal. Chem.* **366**, 392–395.

Kümmerer, K., Ed. (2001). *Pharmaceuticals in the environment. Sources, fate, effects and risk.* Springer, Berlin, Germany.

Lewis, M. A. (1995). Use of freshwater plants for phytotoxicity testing—a review. *Environ. Poll.* **87**, 319–336.

Lindsey, M. E., Meyer, M., and Thurman, E. M. (2001). Analysis of trace levels of sulfonamide and tetracycline antimicrobials in groundwater and surface water using solid-phase extraction and liquid chromatography/mass spectrometry. *Anal. Chem.* **73**, 4640–4646.

Lipnick, R. L. (1995). Structure-activity-relationships. *In* "Fundamentals of Aquatic Toxicology" (G. M. Rand, Ed.), pp. 609–655. Taylor & Francis, London, United Kingdom.

Loewe, S. (1927). Die Mischarznei. Versuch einer allgemeinen Pharmakologie der Arzneikombination. *Klinische Wochenschrift* **6**, 1077–1085.

Loewe, S. (1953). The problem of synergism and antagonism of combined drugs. *Arzneimittel-Forschung/Drug Research* **3**, 285–290.

Loewe, S., and Muischnek, H. (1926). Über Kombinationswirkungen. 1. Mitteilung: Hilfsmittel der Fragestellung. *Naunyn-Schmiedebergs Archiv für experimentelle Pathologie und Pharmakologie* **114**, 313–326.

Möhle, E., Horvath, S., Merz, W., and Metzger, J. W. (1999). Determination of hardly degradable organic compounds in sewage water – identification of pharmaceutical residues. *Vom Wasser* **92**, 207–223.

Öllers, S., Singer, H. P., Fässler, P., and Müller, R. S. (2001). Simultaneous quantification of neutral and acidic pharmaceuticals and pesticides at the low-ng/l level in surface and waste water. *J. Chromatogr. A* **911**, 225–234.

Pöch, G. (1993). *Combined effects of drugs and toxic agents. Modern evaluation in theory and practice.* Springer, Berlin, Germany.

Schüürmann, G., and Markert, B. (1998). Ecotoxicology: Ecological Fundamentals, Chemical Exposure, and Biological Effects. pp. 665–749. John Wiley & Sons, New York.

Sedlak, D. L., and Pinkston, K. E. (2001). Factors affecting the concentrations of pharmaceuticals released to the aquatic environment. *Wat. Res. Update* **120**, 56–64.

Seiler, R. L., Zaugg, S. D., Thomas, J. M., and Howcroft, D. L. (1999). Caffeine and pharmaceuticals as indicators of waste water contamination in wells. *Ground Water* **37**, 405–410.

Stan, H. J., Heberer, T., and Linkerhägner, M. (1994). Vorkommen von Clofibrinsäure in aquatischen System—führt die therapeutische Anwendung zu einer Belastung von Oberflächen-, Grund-und Trinkwasser? *Vom Wasser* **83**, 57–68.

Steinberg, R. A. (1943). Use of *Lemna* as test organism. *Chronica Botanica* **7**, 420–424.

Steinberg, R. A. (1946). Mineral requirements of *Lemna minor. Plant Physiol.* **21**, 42–48.

Stumpf, M., Ternes, T. A., Haberer, K., and Baumann, W. (1998). Isolierung von Ibuprofen-Metaboliten und deren Bedeutung als Kontaminanten der aquatischen Umwelt. *Vom Wasser* **91**, 291–303.

Ternes, T. (1998). Occurrence of drugs in German sewage treatment plants and rivers. *Water Res.* **32,** 3245–3260.

Ternes, T. A., Bonerz, M., and Schmidt, T. (2001). Determination of neutral pharmaceuticals in wastewater and rivers by liquid chromatography – electrospray tandem mass spectrometry. *J. Chromatogr. A* **938,** 175–185.

Vane, J. R., and Botting, R. M. (1998). Mechanism of action of nonsteroidal antiinflammatory drugs. *Am. J. Med.* **104,** 2S–8S.

Vane, J. R. (1971). Inhibition of prostaglandin synthesis as a mechanism of action for aspirin like drugs. *Nature* **231,** 232–235.

Van Leeuwen, C. J., van der Zandt, P. T. J., Aldenberg, T., Verhaar, H. J. M., and Hermens, J. L. M. (1992). Application of QSARs, extrapolation and equilibrium partitioning in aquatic assessment: I. Narcotic industrial pollutants. *Environ. Toxicol. Chem.* **11,** 267–282.

Van Loon, W. M. G. M., Verwoerd, M. E., Eijnker, F. G., van Leeuwen, C. J., van Duyn, P., van deGuchte, C., and Hermens, J. L. M. (1997). Estimating total body residues and baseline toxicity of complex organic mixtures in effluents and surface waters. *Environ. Toxicol. Chem.* **16,** 1358–1365.

Verhaar, H. J. M., van Leeuwen, C. J., and Hermens, J. L. M. (1992). Classifying environmental pollutants. 1. Structure-activity relationships for prediction of aquatic toxicology. *Chemosphere* **25,** 471–491.

Verhaar, H. J. M., Mulder, W., and Hermens, J. L. (1995). QSARs for ecotoxicity. *In* "Overview of structure-activity relationships for environmental endpoints, part 1: General outline and procedure." Report prepared within the framework of the project "QSAR for prediction of fate and effect of chemicals in the environment" (J. L. M. Hermens, Ed.). Environmental Technologies RTD Programme, European Commission, contract number EV5V-CT92-0211.

Webb, S. F. (2001). A data-based perspective on the environmental risk assessment of human pharmaceuticals I – collation of available ecotoxicity data. *In* "Pharmaceuticals in the Environment. Sources, Fare, Effects and Risks" (K. Kümmerer, Ed.), pp. 175–201. Springer, Berlin, Germany.

PART IV

Principal Considerations

Reinhard Länge* **and Daniel Dietrich**[†]

*Schering AG, Experimental Toxicology
Research Laboratories
Berlin, Germany

[†]Environmental Toxicology
University of Konstanz
Konstanz, Germany

Environmental Risk Assessment of Pharmaceutical Drug Substances—Conceptual Considerations

I. Introduction

Following the European Union (EU) directive 93/39/EEC (European Union, 1993) on the registration of new medicinal products in the EU, the environmental risk by use and disposal of these compounds have to be evaluated in order to provide appropriate information to the users of these products. Similar regulations are in effect or planned in other countries, such as the U.S. and Canada (Department of Health Canada, 2001; FDA, 1998). In the draft note for guidance (CPMP, 1994) and in a recently published discussion paper by the Committee for Pharmaceuticals and Medicinal Products (CPMP) (EMEA, 2001), a procedure for such a risk assessment is described.

In essence, this procedure follows the general principle of environmental risk assessments as applied to chemicals in the EU under the chemicals legislation, in which the intrinsic properties of the compounds are analyzed by certain experimental studies, and by application of an assessment factor the predicted no effect concentration (PNEC) is estimated. Additionally, the environmental exposure is calculated based on the predicted market volume, the water consumption of the target population, and a dilution factor accounting for dilution of effluent when reaching the surface waters. This estimate is considered to be the predicted environmental concentration (PEC). The ratio of PEC to the PNEC provides a first estimate of the potential risk of the medicinal compound.

Given the nature of active compounds in medicinal products, it was questioned whether the assessment of the intrinsic ecotoxicological properties of such compounds by using simplified, short-term aquatic tests is an adequate basis for an environmental risk assessment. Indeed, the limitations of the small short-term aquatic test battery were revealed by a study of Henschel *et al.* (1997), whereby four pharmaceuticals and their respective metabolites were tested with different standard and nonstandard ecotoxicity tests. The most sensitive tests were the nonstandard tests, such as the BF-2 fish cell line (cytotoxicity and proliferation inhibition of fish cells) for three of the four substances. The non-standard ciliate test, fish embryo test, and the Daphnia acute test were sensitive to an additional three compounds. The algal growth inhibition test was relatively insensitive for all four compounds tested. This demonstrates that the classical short-term aquatic tests may underestimate the toxicity of the four compounds tested. Although the CPMP discussion paper (EMEA, 2001) stated that the use of short-term tests for three different aquatic species (algae, *Daphnia spec.*, and fish) were recommended, it is recognized that for certain groups of compounds a different approach may be needed. No further guidance is given in the discussion paper or in the earlier draft guideline on what criteria should be used to determine whether additional species, endpoints, and test durations should be envisaged for specific active compounds.

In the following, considerations are presented as guidance on what criteria could be useful tools to pre-evaluate the potential for the ecotoxicological effects based on the large database typically developed during the development of a drug substance. These criteria should help to determine an adequate testing strategy for the ecotoxicological effects of medicinal compounds.

II. Available Relevant Pharmacological, Pharmacodynamic, and Toxicological Information

Common procedures in the research and development of medicinal compounds provide a huge database that should be used for a rationale of an adequate testing strategy for the ecotoxicological properties.

Preliminary information on the activity of an active substance is gained in research, where a certain pharmacodynamic property is evaluated for treatment of human diseases. This desired property in a patient might provide for an unwanted adverse effect in a nontarget aquatic species. Thus the specific mode of action is a very relevant basis for further considerations of a test strategy in environmental organisms. It must be emphasized, however, that although the intrinsic activity properties of a pharmaceutical compound and mode of action may be similar across species, the actual affected endpoints (downstream of a receptor interaction) may differ dramatically among classes of organisms, and species within the same classes.

For example, serotonin (5-hydroxy-tryptamine, 5-HT) is a neurotransmitter not only found in humans and mammalians, but also in all other phyla so far examined, including invertebrates. However, the 5-HT induced effects downstream of the 5-HT receptor differ among the various species (human regulation of appetite, sleep, sexual arousal, and depression; bivalves: regulation of reproductive processes such as spawning, oocyte maturation, germinal vesicle breakdown, sperm reactivation, and parturition; *freshwater gastropods (Lymnaea stagnalis* and *Biomphalaria glabrata)*: regulation of egg laying and induction of penile erection; *protozoans*: cilia regeneration, nudibranches, ciliary reaction; *Aplysia*: regulation of muscle contraction; *crustaceans*: regulation of ovarian growth) (Fong, 2001; Fong *et al.*, 1998; Muschamp and Fong, 2001). This also implies that 5-HT analogs (methiothepin) or 5-HT reuptake inhibitors [SSRI; fluoxetine (prozac), fluvoxamine (luvox) or paroxetine (paxil)] used in patients with depression or obsessive-compulsive behaviors associated with Tourette's Syndrome, may adversely influence the normal function of various other species. However, whether or not aquatic organisms can be adversely affected at the concentrations of antidepressants released into the environment is at present difficult to determine. The mere fact that similar, if not identical, pharmacological targets are present in species other than the human should drive the environmental testing and risk assessment of tailored pharmaceuticals with highly specific activities. Consequently, the pharmaceutical database (containing compound class, type of pharmacological activity, efficacy, treatment doses, and physico-chemical parameters) could serve as a primary source of information of preliminary risk estimation. The combined information on the pharmacodynamical properties of such a compound, and the distribution of identical or similar physiological targets in environmental organisms, should lead to directed and tailored environmental testing.

In contrast to the desired pharmacodynamical properties discussed above, adverse reactions during pharmacological and toxicological testing of a pharmaceutical are a completely different entity. Important pieces of information on the adverse effects of a compound that may be useful for targeting an ecotoxicological test strategy can be obtained from the mammalian toxicology database, as well as from patient information. This database can provide an indication of the ratio of acute and chronic toxicity, target organs, and specific effects regarding genotoxicity, reproduction, and developmental toxicity or immunotoxicity.

Particularly relevant for further ecotoxicological testing considerations are characteristic parameters on the relationship of pharmacological and toxicological responses to the plasma levels and information regarding the metabolism and excretion of the compound in question. The potential metabolites may even form a complete ecotoxicologically relevant entity of their own. For example, while the lipid regulator clofibrate may not be detected, its major metabolite, clofibric acid, appears in significant amounts in the environment (Ternes, 1998). The question arises whether clofibric acid can adversely affect cholesterol synthesis and steroid genesis, consequently influencing endocrine regulation in aquatic species (Pfluger and Dietrich, 2001).

Pharmacological and toxicological investigations, in conjunction with a more broadened biological/phylogenetic understanding, will demonstrate species-specific responses, thus providing additional information regarding the requirement for further ecotoxicological considerations and testing. For example, if a response is relatively unspecific in functional systems that are preserved over a large variety of species as discussed for serotonin above, additional and extensive testing should be discussed. On the other hand, if the compound is tailored for such specific conditions that it is unlikely that this response can occur in environmental organisms, no additional testing should be envisioned.

III. Use of Pharmacodynamic Information from Mammalian Species in Ecotoxicological Test Strategies

Typically, information on pharmacodynamic activity contains observations made in *in vitro* studies and *in vivo* studies. *In vitro* studies demonstrate the activity of a drug substance on a variety of systems, such as receptors, specific tissues, and organs. *In vivo* studies, on the other hand, demonstrate how specific activities cause responses in the whole organism. This information is a useful tool for selecting test organisms in an ecotoxicological test strategy. If the pharmacological effect is based on a specific receptor-mediated reaction (see serotonin above), it has to be considered whether this receptor is likely to exist in the suite of test organisms available for standard laboratory testing, and what endpoints may suitably demonstrate adverse effects causal to exposure by this pharmacological entity.

Juvenile fathead minnows were exposed to the potential model androgen methyl testosterone (Zerulla *et al.*, 2002). Instead of androgenization as expected, a feminization response of the fish was observed. This feminization response, as determined via vitellogenin protein and mRNA, was clearly coupled with the presence of an aromatase, enabling the aromatization of testosterone to an estrogen. Thus, in addition to the absolute requirement of understanding the physiology, underlying kinetics, and reliability of the endpoints such as vitellogenin employed (Schmid *et al.*, 2002), the mechanism of compound interaction (and in this case specifically, the metabolism) must be investigated in order to allow proper risk assessment. Indeed, generalizations can provide false-positive answers, as recently shown by Hutchinson (2002), whereby the classic synthetic nonsteroidal estrogen or "endocrine disrupter" diethylstilbestrol (DES) demonstrated interaction with the ecdysteroid receptor of the marine copepod *Tisbe battagliai*; however, many known endocrine disrupters and related active pharmaceuticals did not interact, thus severely questioning the presence of an endocrine-mediated effect in this species.

In most cases, it is unlikely that scientific knowledge on environmental organisms is so detailed that the existence of certain receptors can be confirmed. The pharmacological response, though, may be an indicator for an adequate test strategy. It may be worthwhile to consider the use of an early life stage test in fish, when a compound is known to affect the development and growth of blood vessels, an endocrine-controlled mechanism in higher organisms, or developing organs and tissues. In contrast, *Daphnia magna* is unlikely to show any response to this particular pharmacological effect, since these organisms have no vascular system.

Compounds such as antibiotics, known to interfere particularly with the metabolic system in the microbial cell, may show distinct effects in microbial test systems, while other test organisms may be less sensitive under acute exposure situations. Thus, the antimycotic compound metronidazole was tested in fish, crustaceans, and algae. EC_{50} values for the algal tests (*Selenastrum capricornutum, Chlorella spec.*) were between 12 and 45 mg/L; in fish (*Brachydanio rerio*) and crustaceans (*Acatia tonsa*) there were no effects at 500 and 100 mg/L, respectively, which were the highest concentrations tested (Lanzky and Halling-Sørensen, 1997). The antibiotic ciprofloxacin had an EC_{50} of 0.005 mg/L in the cyanobacterium *Microcystis aeruginosa*, 0.6 mg/L in activated sludge, and 2.97 mg/L in the alga *Selenastrum capricornutum*; in fish and *Daphnia magna*, there was no effect up to the highest test concentration of 100 and 60 mg/L, respectively (Halling-Sorensen *et al.*, 2000). These examples indicate the relative sensitivity of microorganisms to antibiotics in comparison to higher organisms. However, these observations should not come as a surprise, as the cyanobacterium *M. aeruginosa* and the green algae *S. capricornutum* are phylogenetically much closer to the pathogenic bacteria; their specific type of gyrases is inhibited more by fluorquinolone antibiotics (Backhaus *et al.*, 2000) than organisms of higher phylogenetic class (crustaceans and fish).

IV. Use of Pharmacological and Toxicological Information from Mammalian Species in Ecotoxicological Test Strategies

Pharmacological and toxicological information becomes available during the development process of a drug substance and comprises a large amount of data on adverse effects. Information on the acute and repeated-dose effects, including the target organs, disturbance of fertility in the male or female gender, embryotoxicity or teratogenicity, and genotoxicity are typically available. For the selection of an ecotoxicological test strategy, the types of effects displayed in mammalian toxicity tests should be considered. Vertebrates have many physiological functions in common; few examples are reported where endpoints studied in fish were similarly affected as those in mammalian species. Thus, the natural estrogen estradiol as well as the synthetic steroid ethinylestradiol have a strong feminization effect in fish, such as affected reproduction and normal development in goldfish (Bjerselius *et al.*, 2001) and fathead minnows (Laenge *et al.*, 2001), respectively.

In addition to the direct effects elicited by a drug in a given species, secondary adverse effects as a consequence of a primary drug interaction must also be considered. For example, cyclosporine and staurosporin selectively inhibit P-glycoprotein (Pgp), a member of the ABC family of transport proteins, in freshwater mussels (*D. polymorpha* and *C. fluminea*) and thus reduce the active excretion of reactive metabolic products from these organisms, potentially leading to a progressive self-intoxication or to an accumulation of other toxic xenobiotics. Other compounds with Pgp modulating or inhibiting activity or compounds that could serve as substrates for Pgp are calcium channel blockers, calmodulin antagonists, anti-hypertensives, vinca alkaloids, steroids, anti-arrhythmics, anti-parasitics and anti-estrogens (Epel and Smital, 2001). Indeed, the presence and degree of expression of Pgp in aquatic organisms may directly influence uptake and retention of active compounds such as steroids, anti-steroids, and possibly quinolone antibiotics (Backhaus *et al.*, 2000), and therefore modulate the physiologically relevant concentrations within the exposed organisms. This example suggests that knowledge of the environmentally present and biologically available concentration of a pharmaceutical compound may not suffice for a risk assessment.

V. Use of Pharmacokinetic Information in Ecotoxicological Test Strategies

Pharmacokinetic investigations in mammalian species demonstrate the availability of the administered substance, distribution, metabolism, and excretion. It further shows, in combination with toxicological or pharmacological studies, at which endogenous level a drug substance exerts activity.

In ecotoxicological studies, the exposure is typically determined on the basis of the dissolved fraction of the test compound. Although the bioavailability of the substance in the organism is not further assessed by blood plasma analysis, it might be useful for preliminary evaluation of the ecotoxicological potency to compare mammalian blood plasma levels with exposure levels in ambient waters of aquatic organisms. The simplified assumption is made that similar (same order of magnitude) levels in plasma of mammalians and in the ambient environment of aquatic organisms are comparably effective. Under this assumption, effective plasma levels, or plasma levels at the recommended human dose in comparison to the predicted environmental concentration, may indicate whether chronic exposure of wildlife organisms needs to be assessed in specific long-term tests.

It has to be stated, however, that there are not many examples supporting this hypothesis yet. In the case of the compound ethinylestradiol (EE2), it was found that the concentration of 4 pg/ml in water caused developmental and reproductive disturbances in fish (Laenge *et al.*, 2001). In comparison, pharmacological effects (contraceptive) are found at plasma levels of 25–100 pg/ml (Hümpel *et al.*, 1990) given a recommended daily dose of 30 μg EE2 per person. This example indicates that fish seemed to be even more sensitive to EE2 than man, and a factor of at least 10 had to be applied if the ratio of the effective environmental concentration is compared to effective plasma levels. Future studies should confirm whether there is a general similarity between mammalian and other vertebrate species such as fish in pharmacologically effective levels.

VI. Criteria for the Development of an Ecotoxicological Test Strategy

The aforementioned considerations may be used to select a science-based test strategy for the environmental hazard and risk assessment of drug substances. Apparently, the information on the pharmacodynamics and toxicology of a drug substance mainly supports ecotoxicity testing specifically in fish, because common functions between mammalian species and invertebrates are not very well understood and cannot *a priori* be assumed (Hutchinson, 2002).

Therefore, the following example of a testing strategy is mainly focused on whether ecotoxicity testing for a candidate substance should go beyond the standard acute tests in algae, daphnia, and fish by performing a longer-term test in fish, employing relevant endpoints.

A. Two-Tiered Strategy

1. Tier 1: Algae, Daphnia, Fish

Information

- General level of acute toxicity and pharmacological activity (underlying mechanism) of the compound in question in the above wildlife species.
- Availability of mammalian pharmacology and toxicology, as well as ecotoxicology datasets for benchmark compound(s) with comparable pharmaco/toxicokinetic and dynamic properties as the compound in question.

Questions

- Is the observed ecotoxicity, in comparison to the mammalian toxicity, unexpected?
- Are the test organisms appropriate (presence/absence of enzymes, receptors, transporters, etc.) for detecting effects of the compound in question? Which aquatic organism(s) are most likely affected?

Preliminary risk assessment

- Determination of the order of magnitudes between expected environmental exposure concentrations and observed effect concentrations.

2. Tier 2: Criteria for Further Testing

Questions

- Is the acute/chronic ratio (the LD_{50} vs. the lowest NOEL in the most sensitive species and the most sensitive endpoint) in mammalian species greater than 1000?
- Is the substance genotoxic?
- What chronic effects are known (longevity or mortality, reproductive toxicity, embryo toxicity, organ toxicity, carcinogenicity)?
- Is the pharmacological/toxicological active plasma level in the mammalian test system within the same order of magnitude as the PEC?
- Are the pharmaco/toxicokinetics driven by a specific uptake/excretion mechanism (transporters, metabolism/conjugation)?

Experimentation

- Identification of the most suitable (susceptible) species, test systems, test duration, and endpoints.

Risk assessment

- Standard environmental risk assessment based on a comparison of environmental concentrations (EC) or PEC and observed no-effect concentrations in the test system(s).

VII. Conclusions and Recommendations

Most of the available literature on pharmaceuticals in the environment deals with the analytical detection of these compounds in the aquatic environment. For very few of these chemicals, the environmental fate is known. Data on the effects of these compounds in environmentally relevant organisms are, with a few exceptions (Fong, 2001; Hutchinson, 2002; Pfluger and Dietrich, 2001), rare. Indeed, most currently available data on the effects of pharmaceuticals in the environment were obtained using standard test systems/procedures, such as algal growth inhibition tests (OECD, 1984a), the *Daphnia magna* acute and chronic toxicity test (OECD, 1984b; Wollenberger *et al.*, 2000) or embryotoxicity tests with zebrafish (DRETA, *Danio rerio* embryo teratogenesis assay) (Dietrich and Prietz, 1999; Dietrich *et al.*, 1998), the African clawed frog *Xenopus laevis* (FETAX, frog embryo teratogenesis assay *Xenopus*) (Kiamos *et al.*, 1998; Neeser *et al.*, 1996), or with the fathead minnow (*Pimephales promelas*) (Laenge *et al.*, 2001; Schmid *et al.*, 2002; Zerulla *et al.*, 2002).

This lack of information is not necessarily surprising, in view of the tailored testing schemes and specifically defined endpoints required to detect potential adverse effects (Epel and Smital, 2001; Fong, 2001; Pfluger and Dietrich, 2001). Indeed, it is unlikely that testing for potential endocrine-modulating chemicals (also known as "endocrine disrupters") with the standard short-term acute test systems as proposed by the CPMP (EMEA, 2001) for pharmaceuticals would reveal their chronic toxicity potency. Thus, an approach tailored to the pharmacological function and activity, rather than a uniform approach, is of essence for the ecotoxicological testing of pharmaceuticals.

The tiered approach as suggested above demands a thorough understanding of the compound in question and an extrapolation and interpretation of the potential environmental activities, in addition to the standard short-term testing systems very early on in the risk assessment process. The combination of present knowledge from the base set of pharmaceuticals registration, environmental biology, and rudimentary toxicology should provide a basis to decide whether a compound presents with a potential for low, medium, or high environmental risk. The second tier should provide for a testing approach specifically tailored to the type of pharmacological activity inherent to the compound in question, relevant for those compounds that have been selected preliminary as potential medium or high environmental concern. Consequently, no "check-the-box" type of testing system should be implemented at this stage, but rather the experimentation should be knowledge and experience driven. Although it asks for a closer interaction of industry, academia, and regulatory agencies, this approach will provide for a sound and robust risk assessment.

The use of validated and standardized methods is essential in order to provide valid and reproducible results. The methods published under the OECD testing guidelines scheme are useful tools; however, specific issues of ecotoxicological relevance may require the need for adaptation and modification of those guidelines to assess additional relevant endpoints.

These considerations are thought to be a start for further research in this area. Suggestions need to be put in more concrete terms, when more data are available that may show similarities in the toxicological and pharmacological responses in mammalian and wildlife organisms when treated with medicinal compounds.

References

Backhaus, T., Scholze, M., and Grimme, L. H. (2000). The single substance and mixture toxicity of quinolones to the bioluminescent bacterium. *Vibrio fischeri*. *Aquat. Tox.* **49,** 49–61.

Bjerselius, R., Liundstedt-Enkel, K., Olsen, H., Mayer, I., and Dimberg, K. (2001). Male goldfish reproductive behavior and physiology are severely affected by exogenous exposure to 17β-estradiol. *Aquat. Tox.* **53,** 139–152.

CPMP (1994). Assessment of potential risks to the environment posed by medicinal products for human use (excluding products containing live genetically modified organisms. Phase 1: Environmental risk assessment. Ad hoc working group on environmental risk assessment for non-GMO containing medicinal products. III/5504/94, draft 6, version 4, European Commission, Brussels.

Department of Health Canada (2001). Notice to interested parties – Intent to develop environmental assessment regulations for products regulate under the food and drug act. *Canada Gazette* **135,** 35.

Dietrich, D. R., and Prietz, A. (1999). Fish embryotoxicity and teratogenicity of pharmaceuticals, detergents and pesticides regularly detected in sewage treatment plant effluents and surface waters. *Tox. Sci.* **48**(Suppl. 1), 151.

Dietrich, D. R., Prietz, A., and Kiamos, M. A. (1998). *Danio rerio* embryotoxicity and teratogenicity assay (DRETA) for detecting waterborne embryo-toxicants and teratogens. *Tox. Sci.* **48**(Suppl. 1), 259.

EMEA (2001). Draft CPMP discussion paper on environmental risk assessment of non-genetically modified organism (non-GMO) containing medicinal products for human use. CPMP/SWP/4447/00, 25 January 2001. European Agency for the Evaluation of Medicinal Products, London.

Epel, D., and Smital, T. (2001). Multidrug-multixenobiotic transporters and their significance with respect to environmental levels of pharmaceuticals and personal care products. *In* "Pharmaceuticals and Personal Care Products in the Environment: Scientific and Regulatory

Issues" (C. G. Daughton and T. Jones-Lepp, Eds.), pp. 244–263. American Chemical Society, Washington, D.C.

European Union (1993). Directive 93/39/EEC of the Council for changing Directive 65/65/EEEC, 75/318/EEC and 75/319/EEC related to pharmaceuticals. *Official Journal of the European Commission* **L214,** 22–32.

FDA (1998). Guidance for Industry for the submission of an environmental assessment in human drug applications and supplements. U.S. Food and Drug Administration, Center for Drug Evaluation and Research (CDER) and Center for Biologics Evaluation and Research, July 1998 (Revision 1), CMC6. *http://www.fda.gov/cder/guidance/index.htm.*

Fong, P. P. (2001). Antidepressants in aquatic organisms: A wide range of effects. *In* "Pharmaceuticals and Personal Care Products in the Environment: Scientific and Regulatory Issues" (C. G. Daughton and T. Jones-Lepp, Eds.), pp. 264–281. American Chemical Society, Washington, D.C.

Fong, P. P., Huminski, P. T., and D'Urso, L. M. (1998). Induction and potentiation of parturition in fingernail clams (*Sphaerium striatinum*) by selective serotonin re-uptake inhibitors (SSRIs). *J. Exp. Zool.* **280,** 260–264.

Halling-Sørensen, B., Holten, B., Luetzoeft, H.-C., Andersen, H. R., and Ingerslev, F. (2000). Environmental risk assessment of antibiotics: Mecillinam, trimethoprim and ciprofloxacin. *J. Antimicrobial Chemotherapy* **46**(Suppl. 1), 53–58.

Henschel, K.-P., Wenzel, A., Diedrich, M., and Fliedner, A. (1997). Environmental hazard assessment of pharmaceuticals. *Reg. Tox. Pharm.* **25,** 220–225.

Hümpel, M., Täuber, U., Kuhnz, W., Pfeffer, M., Brill, K., Heithecker, R., Louton, T., and Steinberg, B. (1990). Comparison of serum ethinyl estradiol, sex-hormone-binding globulin, corticoid-binding globulin and cortisol levels in women using two low-dose combined contraceptives. *Hormone Res.* **33,** 35–39.

Hutchinson, T. (2002). Reproductive and developmental effects of endocrine disrupters in invertebrates: *In vitro* and *in vivo* approaches. *Tox. Lett.* **131**(1–2), 75–81.

Kiamos, M. A., Prietz, A., and Dietrich, D. R. (1998). Optimizing the experimental design of the FETAX assay. *Tox. Sci.* **42**(Suppl. 1), 1280.

Laenge, R., Hutchinson, T. H., Croudace, C. P., Siegmund, F., Schweinfurth, H., Hampe, P., Panter, G. H., and Sumpter, J. P. (2001). Effects of the synthetic estrogen 17α-ethinylestradiol on the life-cycle of the fathead minnow (*Pimephales promelas*). *Env. Toxicol. Chem.* **20,** 1216–1227.

Lanzky, P. F., and Halling-Sørensen, B. (1997). The toxic effect of the antibiotic metronidazole on aquatic systems. *Chemosphere* **35,** 2553–2561.

Muschamp, J. W., and Fong, P. P. (2001). Effects of the serotonin receptor ligand methiothepin on reproductive behavior of the freshwater snail *Biomphōlaria glabrata*: Reduction of egg laying and induction of penile erection. *J. Exp. Zool.* **289,** 202–207.

Neeser, S., Prietz, A., and Dietrich, D. R. (1996). Automatization of the developmental toxicity test with Xenopus embryos (FETAX). Proceedings of the 1st International Statuskolloquium in Ecotoxicology within the EUREGIO Bodensee, Konstanz, Germany, November 26, 1996.

OECD (1984a). *Alga*, Growth Inhibition Test. OECD Guidelines For Testing of Chemicals, 201.

OECD (1984b). *Daphnia sp.*, Acute Immobilisation and Reproduction Test. OECD Guidelines for Testing of Chemicals, 202.

Pfluger, P., and Dietrich, D. R. (2001). Pharmaceuticals in the environment – an overview and principle considerations. *In* "Pharmaceuticals in the Environment" (K. Kümmerer, Ed.), pp. 11–17. Springer Verlag, Heidelberg, Germany.

Schmid, T., Gonzalez-Valero, J., Rufli, H., and Dietrich, D. R. (2002). Determination of the half-life of vitellogenin in male fathead minnows as an indicator for endocrine modulation. *Tox. Lett.* **131,** 65–74.

Ternes, T. A. (1998). Occurrence of drugs in German sewage treatment plants and rivers. *Water Res.* **32**, 3245–3260.

Wollenberger, L., Halling-Sorensen, B., and Kusk, K. O. (2000). Acute and chronic toxicity of veterinary antibiotics to *Daphnia magna*. *Chemosphere* **40**, 723–730.

Zerulla, M., Länge, R., Steger-Hartmann, T., Panter, G., Hutchinson, T., and Dietrich, D. R. (2002). Morphological sex reversal upon short-term exposure to endocrine modulators in juvenile fathead minnow (*Pimephales promelas*). *Tox. Let.* **131**, 51–63.

Jürg P. Seiler
ToxiConSeil
CH-3475 Riedtwil, Switzerland

Pharmacodynamic Activity of Drugs and Ecotoxicology—Can the Two Be Connected?

I. Introduction

Pharmaceutical drugs may be the best investigated and characterized man-made chemicals, for their toxic effects as well as pharmacodynamic activities and pharmacokinetic properties. Within the collected information required for their registration and marketing, ecotoxicology data form a conspicuous exception. In the past, the simplistic opinion was that drugs and their metabolites would be excreted postapplication in a diluted form, only to be further diluted in wastewater canals and sewage treatment plants, and would never be detected in appreciable concentrations in the environment, let alone to pose an environmental hazard. This has changed with the refined analytical techniques that allow for the detection, identification, and quantitation of pharmaceuticals in environmental compartments, notably in the aqueous one.

While the first findings of environmental contaminations by pharmaceutical compounds were thought to constitute "special cases" due to specific, rare, local conditions, this point of view was soon abandoned. It was recognized that a number of drug residues could be found in sewage treatment plant effluents, surface waters, and even in ground water and drinking water supplies. This insight evoked—at least for the last cited contamination—increased attention to, and speculation on, possible consequences (Potera, 2000). This situation also led to the further insight that such contaminations should be assessed with respect to the possible environmental risks involved. It was further recognized that pharmaceuticals, as biologically active substances, might be able to exert certain effects not only in humans (or domestic animals in the case of veterinary medicines), but also in environmental species. The discussion was further fueled by the recognition that environmental contamination with hormonally active substances could lead to "endocrine disruption," with estradiol and ethinylestradiol as the most active compounds in this respect, both estrogens used in medicinal applications and indications.

Regulatory guidance in this area developed, but more slowly in human than veterinary pharmaceuticals. The approach for an experimental hazard identification—if one was indeed deemed necessary (Straub, 2002)—was seen by many as unsatisfactory (Halling-Sorensen *et al.*, 1998). Therefore, drawing on the wealth of information available on the biological activities of pharmaceutical substances in order to obtain better information on possible environmental effects and toxicities, as well as to design more intelligent and rational testing programs, was considered to be a possibility.

II. Problem Statement

Apart from technical sources of environmental contamination such as manufacturing and disposal, the main source of entry of pharmaceuticals and their metabolites into the environment is by excretion from patients treated for medical reasons. Such treatments are intended to maintain, improve, or restore the health of a human being and hence have an absolute benefit, at least if the possible toxicological risks to humans are considered to be outweighed by the potential health benefits; for ethical reasons, this precludes banning or severely restricting their use and application. Therefore, a potential risk to the environment, should one indeed exist, has to be managed in other ways.

Risk management necessitates a knowledge of the hazards incurred, as well as the quantitation of the exposure incurred by the organisms or systems subject to these hazards. The exposure side of this equation is now fairly well-developed, since a number of publications have already shown that pharmaceuticals can be detected in up to μg/l concentrations in various aqueous compartments of the environment (Daughton and Ternes, 1999; Stumpf *et al.*, 1996; Ternes, 1998; Zuccato *et al.*, 2000). Less well known are the actual hazards to environmental species or systems, with the notable exception of estrogenic compounds, where effects on homeostasis of fish sex hormones, as evidenced by vitellogenin production in males, have been detected at environmental concentrations (Sumpter, 1998). Since toxicological, pharmacodynamic, and pharmacokinetic properties of drug substances are well known, it is certainly tempting to try and use this extensive information to derive clues as to what kinds of hazard to expect. The simple question would thus be: Can we actually derive the existence and nature of potential ecotoxicological effects from this information? A look at the conventionally assumed dose or concentration relationships in pharmaceutical toxicology and pharmacology, and the relation of therapeutically active doses, concentrations, or exposures to the concentrations found in the environment may, at least at first sight, suggest otherwise (Fig. 1). Since the toxicological no-observed-effect-level (NOEL) should ideally be situated at doses higher than those needed to elicit the desired pharmacodynamic response, it would be obvious that the toxicological properties of a drug substance (such as those observed in the conventional routine test species) could not be indicative of any prospective ecotoxicological effects. This of course assumes that the respective sensitivities, not only in mammalian (including man) but also in ecological species, would lie within a factor of about 10, which has conventionally been utilized to cover species differences in sensitivity for the determination of an acceptable daily intake (ADI) with respect to the exposure to pesticides and other chemicals. Another case, analogous to the one for the toxicological properties, could also be made for the pharmacodynamic properties of pharmaceuticals, such that at environmental concentrations, the exposure of eco-species would be far too low as to result in any measurable or relevant pharmacodynamically mediated, adverse effects.

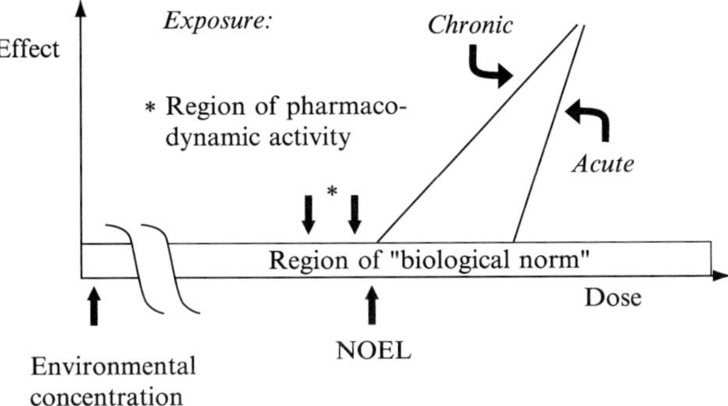

FIGURE 1 Schematic representation of dose/concentration-effect relationships in toxicology and pharmaco-dynamics.

III. Mammalian Pharmacodynamics and Ecospecies

A. Target Specificity

In a preliminary approach, one has to consider that a general prerequisite needs to be fulfilled in order to obtain a respective response from an eco-organism. The eco-organism in question must possess the structure or function that is targeted in the therapeutic indication, meaning it must express the respective receptor or use the respective biosynthetic pathway with enzymes exhibiting structures nearly identical to their human counterparts. Thus, any organism lacking beta-adrenoceptors could fairly confidently be expected to be uninfluenced pharmacodynamically by beta-blockers, and many of the conventional pharmacodynamic targets might therefore be considered to play a role in mammals only. Whether it can indeed be assumed that most of the mammalian pharmacodynamic targets would also be restricted to mammals remains to be discussed.

In many cases (at least with the more recently introduced drugs), the pharmacodynamic activity of a compound is strictly tailored and optimized to its human target, sometimes even to a specific subset of this potential target, intentionally making humans the most sensitive of all tested species to this particular compound. This, in turn, may lead to the assumption that other species would probably need higher exposures in order to have the respective response elicited. The early non-steroidal anti-inflammatory drugs (NSAIDs), for instance, targeted both isoforms of the enzyme cyclooxygenase (COX) in a relatively unspecific way. The more recent drugs in this class, however, are much more specific in their inhibition of the inducible form of COX-2 responsible for the inflammatory reactions, while leaving uninhibited the constitutive form of COX-1 that protects the gastrointestinal mucosa. Another example in this respect could be given by the "-statin" class of anti-hyperlipidemic drugs. While the first such compounds were inhibiting their target enzyme hydroxymethyl-glutaryl-CoA reductase (HMG-CoA reductase) in a relatively nonspecific way with regard to the different tissue localizations of the enzyme, the more recent compounds are exhibiting higher specificities toward the human liver. This combines higher activity in the essential target organ with potentially lower peripheral effects, reducing the potential for adverse events (Arzneimittelkompendium, 2002). Again, there is no information on the respective sensitivities of the HMG-CoA reductase in ecological species (with the obvious exception of rodents), and therefore some species in the environment might be as sensitive (or more so) than the human liver to the inhibitory action of these compounds. The appearance of ecological effects would, however, depend on further factors, such as whether the HMG-CoA reductase-catalyzed step is also the rate-limiting step in these organisms.

B. Quantitative Considerations

Also from a quantitative (but rather simplistic) point of view, it might seem apparent that there cannot be a direct connection between the primary pharmacodynamic properties of an active drug substance and its potential ecotoxicological effects. Considering the analgetically active doses and concentrations of indomethacin, an NSAID, it can be calculated that a rat would have to drink about 250,000 liters of sewage plant effluent contaminated with the highest measured concentration of indomethacin (1.3 μg/l) in order to profit from the pharmacodynamic effects of this drug; in experimental paradigms of analgesia in rats, effects are observed at or above doses of 100 mg/kg.

The same arguments could be employed when considering the possibilities of chronic effects on humans induced by contaminants in drinking water. It has been estimated that, even under worst-case assumptions, the lifetime consumption of any pharmaceutical drug through ingestion of an average volume of 2 liters of contaminated drinking water per day would lead to an exposure equivalent to only one or two therapeutic daily doses. Indeed, model calculations have shown that exposures to ethinylestradiol, phenoxymethylpenicillin, or cyclophosphamide through environmental contamination caused by their excretion from treated patients should not cause any significant risks to the human population at large (Christensen, 1998). In this light, expressions of concern such as, "Concentrations measured in water may give rise to human exposure in the ng per day range, at least three to four orders of magnitude lower than those producing a pharmacological effect. Risks arising from acute exposure can therefore be regarded as unlikely. *However, possible effects of life-long exposures have still to be determined*," (Zuccato *et al.*, 2000), may seem to be exaggerated. The investigations in the conventional set of chronic toxicity studies should provide sufficient information on doses and exposure levels without any observable toxic or pharmacodynamic effects, since the distance between these experimentally determined NOELs and the analytically measured environmental concentrations again would span several orders of magnitude (Fig. 1).

C. Individual (Sub-Threshold) Sensitivities

Even though the separation between contamination levels and NOEL seems to sufficiently cover inter-individual sensitivity levels, recent investigations have shown that exposures normally believed to be harmless (or at least not leading to overt toxicities) could result in certain damage to the organism. An investigation on changes of certain neuropsychological parameters in vineyard workers in the region of Bordeaux, France who were either directly (pesticide mixing and spraying) or indirectly (harvesting) exposed to pesticides revealed that subtle influences on these parameters could be detected, even in those workers who had no direct contact with the pesticides but only with the treated vines and grapes during harvesting (Baldi *et al.*, 2001). Since the differences observed between the exposed persons and the unexposed controls disappeared, when the tasks used in determining these parameters were slowed, the neuropsychological damage in the exposed persons went undetected in normal life. At yet another level, it should be mentioned that the existence (although very much disputed) of the syndrome termed "multiple chemical sensitivity" may prove the possibility that negative effects may be induced by extremely low levels of contaminants (Labarge and McCaffrey, 2000). We shall return to this aspect of imperceptible changes and the possible later appearance of overt consequences again.

When considering that, even for humans, certain "sub-threshold" toxicological phenomena could exist, it should not be *a priori* declared impossible that low environmental concentrations of active drug substances might have certain activities related to their pharmacodynamic properties toward ecological species. The calculation cited above for the analgesic action of indomethacin on rats shows only one side of the coin, the "illuminated" one. Experiments performed on rodents have shown by what doses these species are pharmacodynamically affected; this information is not available for other organisms.

It is not known whether a drug that is optimized for action on a "human target," be it an enzyme or a receptor, will exhibit similar (or even greater) activity against an analogous target in one or several eco-species. Similar targets, however, could govern different processes in different species, as the example of prostaglandins shows. In birds, these compounds play a role in egg shell formation; inhibition of their biosynthesis by the treatment of birds with the COX-inhibitor indomethacin in doses of 50–100 mg/kg has indeed resulted in egg shell thinning, comparable to the effects reported with environmental contamination of DDT (Lundholm, 1997). The doses needed to produce this effect are far above environmentally achievable exposures, but the example serves to show that eco-organisms may use analogous pathways for different endpoints.

Analogous signaling pathways are used in different groups of organisms (phyla) for a variety of purposes, whereby specific substances might "pharmacodynamically-mediate" the most diverse outcomes. A striking example is the symbiotic signalling used by plants in their relationship with nitrogen-fixing rhizobial bacteria. The symbiotic relationship between the rhizobial bacterium *Sinorhizobium meliloti* and alfalfa is initiated by the transcriptional activation of the bacterial nodulation (*nod*) genes through the agonistic actions of phytochemicals produced by alfalfa. Since a possible homology between nodD (the product of the constitutively expressed gene *nodD*) and the estrogen receptor ER-α has been identified, it should come as no surprise that "endocrine-disrupting" environmental chemicals, such as DDT, pentachlorophenol, or bisphenol A, would also interfere with the plant-*Rhizobium* signalling pathway, similar to their effects on endocrine signalling in mammalian cells (Fox *et al.*, 2001).

Another example is the utilization of serotonergic signalling in different species. While in humans serotonin reuptake inhibitors (SSRIs) are used therapeutically to treat depression (or just as mood improvers), the same compounds can influence mussels in quite a different way. SSRIs will not make mussels happier (as far as can be ascertained with this species), but they have been shown to induce parturition in fingernail clams at concentrations as low as 10 nM for fluvoxamine, up to 10 μM for paroxetine (Fong *et al.*, 1998).

D. Secondary Effects

A direct extrapolation from effects on endpoints conventionally investigated in laboratory species to potential effects on unrelated species or taxa may not be relevant for the characterization of environmental hazards. This caveat is more important for pharmaceuticals developed for their specific properties; they can exhibit additional activities in areas that are completely different from those intended to be therapeutically exploited, although possibly still (indirectly) related to their primary effects. An example is the previously cited "-statins," which have been demonstrated to affect more than the cholesterol biosynthesis in mammals. They may also exhibit effects on gene expression, as well as on immunologic parameters, by decreasing expression of the major histocompatibility complex II and consequent effects on T cells (Kwak *et al.*, 2000; Palinski, 2000). They have also been shown to sensitize smooth muscle cells to the apoptotic effect of the Fas-ligand (Knapp *et al.*, 2000). Furthermore, epidemiologic studies have shown that previous treatment with "-statins," but not with other lipid-lowering agents like "-fibrates," has a beneficial influence on the prevalence of Alzheimer's disease and other dementias. This influence seems to be mediated through cholesterol signalling, but is independent from the presence or extent of hyperlipidemia (Jick *et al.*, 2000; Wolozin *et al.*, 2000).

In still another area, effects on multicellular organisms might be possible. It has been repeatedly shown that inhibition of the intracellular gap-junctional communication may lead to adverse effects. Since the intracellular communication through these gap-junctions, formed by a variety of connexins (highly conserved proteins), is necessary for the concerted development of tissues, its inhibition has been shown to be related to effects such as neoplasia or teratogenicity (Trosko *et al.*, 2002; Upham *et al.*, 1998). Again, gap-junctional intercellular communication is a well-conserved property throughout biological multicellular systems, as are the respective proteins that form these gap junctions (the connexins). Pharmaceuticals exhibiting inhibitory properties might be regarded as posing an ecotoxicological risk, too. While this property of chemical substances is not yet exploited therapeutically and can thus be seen as outside the scope of this paper, a potential use of this target in chemoprevention and chemotherapy of human cancer has been suggested (Trosko, 2003; Trosko and Ruch, 2002).

With regard to the possibility of predicting environmental effects from the primary (or secondary) pharmacodynamic effects of a drug substance, the "anthropocentric" view may yield insufficient clues or evidence for potential hazards. This view has to be expanded in order to use available information on these compounds. However, it is necessary to figure out the purpose that these data should serve, taking into account the fundamental difference in the assessment between mammalian and ecological toxicology.

IV. Ecotoxicology *vs.* Mammalian (Human) Toxicology

A. Effects on Individuals *vs.* Population Effects

When questioning whether some substances might harm the environment (eco-species as single entities or ecological systems as a whole), and whether such influences can be predicted by the toxicological or pharmacological properties of a substance, a fundamental difference between ecotoxicology assessments and the conventional "anthropocentric" evaluation of the pharmaco-toxicological profile of a drug has to be considered. In the human situation, the concern is directed toward the protection of the individual against untoward, toxic effects; in the environmental situation, it is not the fate of the individual (mouse, bird, frog, fish, Daphnia, or individual cells of blue-green algae) that is of concern. The purpose of an ecotoxicology assessment is to protect whole species, populations, or ecosystems against harmful effects. An acute, point contamination, even if it may lead to a massive kill-off of individuals, could be regarded as ecologically inconsequential when a rapid return to "normal" conditions would subsequently lead to a full recovery of the affected community. The fire at the Sandoz plant in Schweizerhalle and its short- and long-term consequences for the affected part of the Rhine river has demonstrated the above. At the other end of the spectrum, it has to be questioned whether a continuous contamination with a mutagenically active substance, such as the antineoplastic agent cyclophosphamide found in sewage water in concentrations of 7 ng/l–143 ng/l (Steger-Hartmann *et al.*, 1997), would lead to environmental effects induced by enhanced mutation frequencies in eco-organisms. While the occurrence of genotoxic effects (even if induced in single individuals only) through the action of environmental genotoxins is considered an unacceptable risk to humans, the situation is less clear for eco-species and eco-systems, where presumably selection pressure from changes in external conditions will play a much more important role in the way population stability is maintained (or changed).

B. Accumulation of Substances

Long-term exposure to a low concentration of a contaminant that does not produce immediate effects can lead to a massive endangerment of one or more species. One example is DDT: The exposure to DDT, aggravated by the accumulation in the food chain as well as in exposed individuals, has led to the near extinction of certain birds of prey due to its eggshell-thinning effects.

C. Accumulation of Effects

An accumulation of a substance finally resulting in dangerously high exposures is not always necessary. An accumulation of slight effects may also lead to sudden, toxic consequences. An example of such an "effect-accumulation" can be seen in the recently reported hyperlactatemia (up to fulminant lactacidosis) in HIV-infected patients taking antiretroviral therapy (John *et al.*, 2001). The incriminated nucleoside reverse transcriptase inhibitors have a high specificity for the viral reverse transcriptase in comparison to the DNA polymerases α and β, but their specificity toward the mitochondrial polymerase γ is generally less pronounced. Although an inhibition of the replication of mitochondrial DNA (mtDNA) or its transcriptional inactivation may not become immediately apparent, it will result in a progressive inability of the mitochondrion to sustain a sufficient supply of respiratory chain proteins. This progressive mitochondrial dysfunction will ultimately result in the accumulation of lactate as the end-product of glycolysis, resulting in hyperlactatemia (and eventually lactacidosis), as well as other symptoms of mitochondrial dysfunction such as hepatic steatosis, myopathy, nephrotoxicity, peripheral neuropathy, and pancreatitis (Kakuda, 2001).

This example from human toxicology illustrates a possibility that may also exist in environmental situations. Model considerations have demonstrated that, while an ecosystem may be able to absorb changes in environmental conditions and while the consequences on that ecosystem could remain imperceptible over a large range of such changes, the system may at some time point (or upon the influence of an additional factor, itself without deleterious effects) collapse into a completely different state (Scheffer *et al.*, 2001). It may also be a possibility that effects produced by the action of chemical contaminants could accumulate over a certain time period, without first showing any appreciable consequences for the ecosystem; the influence of an additional, new factor might then seal the fate of the system.

D. Metabonomics as a Sensitive Tool

Some clues to the above possibility might also be gleaned from metabonomic information. This methodology allows for the detection of alterations in the metabolic state of organisms indicative for early toxic changes in physiology and metabolism; such alterations are already detectable at doses or concentrations that do not yet lead to overt toxicity and histopathological damage (Nicholson *et al.*, 1999; Robertson *et al.*, 2000). Bundy *et al.* (2001) have applied this methodology to investigate such early changes of intoxication in earthworms (*Eisenia veneta*), and the authors stress that this approach should not only be able to provide an early warning of toxicity, but may also be useful in the assessment of integrated effects of cumulative exposures to mixtures of contaminants. Metabonomic changes may also be seen as exactly the kind of changes described above, namely changes that are seemingly innocuous (or at least not immediately harmful) but that upon exposure to another factor might have deleterious consequences. An example is the slight, normally imperceptible disturbance of the immunological homeostasis in North Sea mammals induced by certain contaminants (PCBs, alkyl-tin compounds) that was at the bottom of the despite massive mortality of seals upon viral infections, these events that would not have led to this catastrophic outcome in immunologically fully competent animals (Ross *et al.*, 1996).

E. How Can the "Human Experience" Be Used?

In summarizing these considerations, it can be stated that when contamination with a pharmacologically active substance influences the physiology or behavior of individuals in eco-species, those effects remaining at the individual level should be of lesser (or non-relevant) concern in an ecotoxicology assessment than those leading to untoward consequences on whole species or systems. An expanded view, drawing from the knowledge of primary mechanisms of pharmacodynamic actions, might eventually be able to establish leads for more specific and accurately directed investigations into the potential for environmental effects induced by low concentrations of such substances. The following section will explore where such leads can be detected, and what questions should be asked in order to make use of the information on mechanisms of action for the determination of potential hazards to the environment.

F. Evolutionary Conserved Targets of Pharmacological Intervention

While some of the above considerations about the practical impossibility of extrapolating effects that have been determined in the conventional pharmacodynamic and toxicological assays at considerably high doses and exposures to the environmental situation with very low exposures may be applied to most of the older and less specific pharmaceuticals, the development of pharmaceuticals targeting more basic mechanisms of cellular functions may result in additional possibilities of influencing eco-organisms. Such basic mechanisms, like those connected with signal transduction or with cell division, are generally well conserved in evolution and can thus be identified throughout the living world from unicellular to mammalian organisms. For instance, structurally conserved mitogen-activated protein kinases (MAPK) have been demonstrated to be present in yeast as well as in human cells (Widmann *et al.*, 1999). A number of cellular events are regulated by fluctuations in intracellular Ca^{2+} ($[Ca^{2+}]_i$) concentrations, which are themselves governed through the activation of voltage-gated ion channels and/or the inositol 1,4,5-triphosphate receptor; this pathway of producing such $[Ca^{2+}]_i$ transients is also highly conserved and can be found to function in the events following fertilization of eggs in sea urchins to mammals (Deguchi *et al.*, 1996).

Signal transduction by tyrosine protein kinases is a general mechanism that can be observed to function in mammalian cells as well as in mussels (Canesi *et al.*, 1999). Therefore, it might be tempting to ask whether the new antineoplastic drug Gleevec that targets and inhibits the abl-kinase, constitutively expressed through the recombinant form of BCR-abl in chronic myelocytic leukemia, can also influence cellular events dependent on the c-abl (or c-kit) mediated signal transduction pathway in eco-organisms. Although in mammalian cells there is a redundancy in signal transduction pathways that reduces the cytotoxic effects of this compound on normal cells, thus resulting in its specificity of action toward the targeted neoplastic cells, this redundancy might not be as developed in non-mammalian organisms, making them potentially more vulnerable to the inhibitory actions of such compounds.

The advances in genome sequencing have further provided evidence for the evolutionary development of the extended family of G-protein coupled receptors (GPCRs), whereby similar genes for gastrin/cholecystokinin and other receptors have been discovered in the Drosophila genome (Hewes and Taggert, 2001). Not only similar DNA sequences, but similar receptors with similar functions, can be observed throughout the world of organisms, as evolutionary conservation of structures and functions can be found repeatedly. Receptors for insulin, insulin-like growth factor-I (IGF-I), and glucagon are present in fish, amphibians, reptiles, birds, and mammals, as well as in invertebrates, with a structural homology of about 85% between the human and the chicken IGF-I receptor cDNA (Navarro *et al.*, 1999). Calcitonin plays a similar role in fish and in mammals, a fact that is amply demonstrated by the therapeutic utilizability of salmon calcitonin (Miacalcic) in humans, where the "foreign" form of the peptide hormone is even more active than the endogenous substance, due to an increased binding to its receptor (Arzneimittelkompendium, 2002). Mussels (*Mytilus galloprovincialis*) express peripheral benzodiazepine receptors, the density of which (maximal number of benzodiazepine binding sites) is significantly increased in the presence of pollution (Betti *et al.*, 2003). ACTH receptor-like mRNA is present not only in immunocytes from humans, but also from mollusks (Ottaviani *et al.*, 1998). Furthermore, opiate receptors of the μ_3 subtype can be found in neural tissues of invertebrates (Stefano and Scharrer, 1996), with the μ opiate receptor cDNA of the mussel *Mytilus edulis* exhibiting a 95% sequence identity to the human neuronal $\mu 1$ receptor (Cadet and Stefano, 1999). Metabotropic glutamate receptors found in *Drosophila* seem to exhibit similar pharmacological properties as their mammalian counterparts (Panneels *et al.*, 2003), and nitric oxide release is stimulated by the action of 2-arachidonyl glycerol on a cannabinoid receptor in invertebrate immunocytes in the same way as in human immune tissues (Stefano *et al.*, 2000).

In a general way, it can be stated that most of the fundamental pathways of cellular signalling, due to their success in the selection process, have been conserved throughout evolution in one form or another. Supporting this view, the increasing knowledge about DNA sequences of whole genomes has shown quite a number of similar sequences in many species from very different parts of the phylogenetic tree. Many of these sequences will give rise to homologous gene products, although their function may differ widely between species. These findings are increasingly discussed for their potential use in target identification for human diseases, as well as for model definition in the drug development process; the application of "phylogenomics" to the search for suitable human targets has been advocated (Searls, 2003).

In most of these cases it is not known, however, whether human pharmaceuticals may induce effects at these environmental targets in concentrations similar to the exposures needed to influence their human counterparts, although from the extent of sequence homology an educated guess may be possible. At least for the insulin receptor, comparative data on binding affinities of human insulin for the receptors of different species are available (Navarro *et al.*, 1999), which show that these affinities remain more or less constant throughout the species tested and are about equivalent to the EC_{50} values obtained for the action of insulin at the human receptor. On the other hand, it has been stated above that the salmon calcitonin exhibits greater activity in humans than the human peptide; it may well be assumed that the converse might also take place, such that some pharmaceutical targeted to a human effector affects a similar structure of an eco-species at relatively small concentrations, due to higher affinity or efficacy.

G. Target Concentrations and "Eco-Kinetics"

When considering the conserved structure of many receptors and enzymes targeted in human therapeutic situations, it would seem rational to suppose that, generally, the EC_{50} values for the environmental targets should not differ dramatically from those determined for the human target. However, this would mean that, at least for those drug substances that can influence their targets at *in vitro* EC_{50} values of around 1–10 nM, environmental concentrations in the range of 300–5 μg/L should already constitute effective target concentrations. Therefore, another important aspect to consider would certainly be pharmacokinetics (or "eco-kinetics"). For the patient, or generally for any mammalian organism, the applied dose of a pharmaceutical drug has to be sufficient to deliver an appropriate concentration of the active moiety to the target. The absolute dose to be administered depends therefore on the pharmacokinetic properties of the drug, especially on its rate and extent of absorption, its distribution in the body, and its rate of biotransformation to inactive metabolites (first-pass effect). It is obvious that these same parameters would govern the extent of activity of other mammals, too; the rough estimates on "water doses" needed to evoke an analgesic effect in rats or mice given above will certainly be applicable for concentrations of any environmental substance.

For a Daphnia immersed in contaminated water, the situation may, however, be completely different. Biotransformation to inactive metabolites may, in this case, not play a major role anymore, because this organism should ultimately reach an equilibrium between the (constant) external concentration of the substance and its internal concentration unless it is endowed with a very efficient transport system for the removal of xenobiotics. Therefore, it could be conceivable to reason that one might be able to see effects in certain eco-organisms already at environmental concentrations not far above the EC_{50} values determined in *in vitro* assays with the isolated target(s). This is, however, an area where even less is known than with respect to pharmacodynamics, and where information would most probably be very difficult to obtain experimentally.

H. Considerations on Ecotoxicology Assays

In the consideration of the potential for ecotoxicological hazards posed by pharmaceuticals as environmental contaminants, a broadening of the angle of view is needed. In this respect, the sole conduct of acute toxicity studies on a few selected eco-organisms to derive some information about the potential hazards of the tested substance will have to be regarded indeed as insufficient, except in a few exceptional cases for which the high acute toxicity to fish of pyrethroid insecticides may serve as an example. It would probably be more worthwhile to look at the pharmacodynamic properties of the drug, at the range of activities and the necessary target concentrations, and derive a potential activity profile for ecological effects from this information.

In order to judge the environmental relevance of a contamination with any chemical substance ("xenobiotic"), its potential toxicity needs to be characterized by defining the ecotoxicological hazards and the possible exposure situations. The hazard identification is conventionally performed by a standardized set of ecotoxicity tests, which may or may not yield information relevant to the specific substance under investigation. This is especially true for pharmaceuticals, where the potential effects may not be detected by, for example, acute toxicity testing.

I. The Mechanistic Approach

A more promising approach for compounds where the properties are known to a very great extent, is constituted by a critical assessment of the pharmacodynamic mechanism(s) of action in relation to potential environmental targets. Thus, any substance that is acting as an antagonist at one of the peroxisome-proliferator activated receptors (PPAR-α, -β, or -γ) should then be assessed for its potential of disturbing related cellular events that depend on the activity of nuclear receptors. This assessment should not however, remain restricted to the human pharmacodynamic effect, but should attempt to investigate all possibilities that originate from its primary activities. Since PPAR-γ has an important role in cellular differentiation (Rosen and Spiegelman, 2001), an inhibition of its activity may have consequences for embryonic development of sensitive species. While a number of structurally unrelated chemicals have been demonstrated to disturb gene activation mediated by this receptor (Maloney and Waxman, 1999), it can certainly be estimated that the possibility for such an activity may indeed be more pronounced for pharmaceuticals that derive their usefulness from this very interaction.

As another example, the utilization of the inositide- and Wnt-signalling pathways in species as diverse as humans and slime molds may be cited. Lithium is thought to exert its mood-stabilizing effects through its influence on the inositide-signalling pathway in humans, while its side effects (especially its teratogenic effects) may be related to its influence on Wnt-signalling. It has been shown recently that, in the slime mold *Dictyostelium discoideum*, the same two pathways are expressed and that lithium is able to influence both of them. However, the two pathways regulate distinct functions in this mold, with inositide signalling responsible for the aggregation of the organism when submitted to stress, and with the Wnt pathway responsible for the formation of fruiting bodies under conditions of low nutrients. Low lithium concentrations have been found to inhibit the latter, and high concentrations to inhibit the former, pathway. Lithium may thus not influence the "mood" of this slime mold, but rather the way it is able to react to additional changes in the environment (Spinney, 2003).

In an analogous way, those pharmacodynamic properties should also be considered as potential mechanisms for ecotoxicological effects that would not contribute to the primary (human) efficacy of a pharmaceutical drug; the possibility that some eco-organisms might be more sensitive to such a "secondary" mechanism should certainly be kept in mind. Another aspect to be examined is the "absolute" activity of the drug substance. The lower the target concentration needed to elicit a pharmacodynamic response, the greater the chance that (given a fixed contamination level) some respective effects might be induced in an environmentally exposed organism. These effects need not lead to acute, dramatic consequences, but they might "pave the way" for additional disturbances to obtain higher relevance and to result, in combination, in detectable adverse events.

V. Conclusion

The application of the entire pharmacodynamic knowledge about pharmaceuticals to an assessment of where to start looking for potential ecotoxicities might confer an advantage to human and veterinary pharmaceuticals over general chemicals with respect to the environmental risk assessment. While it will probably be in rare cases only that the primary pharmacodynamic mechanism can be applied one-to-one from the therapeutic situation to an environmental danger (ecto-parasiticides in veterinary medicine being one of the obvious exceptions), the possibilities presented by this knowledge for obtaining clues about potential targets for environmental toxicity should not be underestimated.

Subsequent to these improvements in defining potential targets, the possibility of devising scientifically-based testing strategies for the detection, identification, and quantitation of ecotoxicological hazards (instead of having to use the "shotgun approach" of putting drug substances through a general testing battery, yielding results of questionable relevance), should allow a much better risk estimation for the environmental component of the drug assessment process, and ultimately lead to a better understanding of the potential influence of drugs on the environment.

References

Arzneimittelkompendium der Schweiz (2002). Lescol (Fluvastatin), p. 1438; Miacalcic (Calcitonin), p. 1635; Selipran (Pravastatin), p. 2353; Sortis (Atorvastatin), p. 2444; Zocor (Simvastatin), p. 2929 (J. Morant and H. Ruppanner, Eds.). Documed, Basel, Switzerland.

Baldi, I., Filleul, L., Mohammed-Brahim, M., Fabrigoule, C., Dartigues, J.-F., Schwall, S., Drevet, J.-Ph., Salamon, R., and Brochard, P. (2001). Neuropsychologic effects of long-term exposure to pesticides: Results from the French Phytoner study. *Environ. Health Persp.* **109,** 839–844.

Betti, L., Giannaccini, G., Nigro, M., Dianda, S., Gremigni, V., and Lucacchini, A. (2003). Studies of peripheral benzodiazepine receptors in mussels: Comparison between a polluted and a nonpolluted site. *Ecotoxicol. Environ. Saf.* **54,** 36–42.

Bundy, J. G., Osborn, D., Weeks, J. M., Lindon, J. C., and Nicholson, J. K. (2001). An NMR-based metabonomic approach to the investigation of coelomic fluid biochemistry in earthworms under toxic stress. *FEBS Letters* **500,** 31–35.

Cadet, P., and Stefano, G. B. (1999). *Mytilus edulis* pedal ganglia express μ opiate receptor transcripts exhibiting high sequence identity with human neuronal μ1. *Mol. Brain Res.* **74,** 242–246.

Canesi, L., Ciacci, C., Betti, M., Malatesta, M., Gazzanelli, G., and Gallo, G. (1999). Growth factors stimulate the activity of key glycolytic enzymes in isolated digestive gland cells from mussels (*Mytilus galloprovincialis* Lam.) through tyrosine kinase mediated signal transduction. *Gen. Comp. Endocrinol.* **116,** 241–248.

Christensen, F. M. (1998). Pharmaceuticals in the environment—a human risk? *Reg. Toxicol. Pharmacol.* **28,** 212–221.

Daughton, C. G., and Ternes, T. A. (1999). Pharmaceuticals and personal care products in the environment: Agents of subtle change? *Environ. Health Persp.* **107**(Suppl. 6), 907–938.

Deguchi, R., Osanai, K., and Morisawa, M. (1996). Extracellular Ca^{2+} entry and Ca^{2+} release from inositol 1,4,5-triphosphate-sensitive stores function at fertilization in oocytes of the marine bivalve *Mytilus edulis. Development* **122,** 3651–3660.

Fong, P. P., Huminski, P. T., and D'Urso, L. M. (1998). Induction and potentiation of parturition in fingernail clams (Sphaerium striatinum) by selective serotonin reuptake inhibitors (SSRIs). *J. Exper. Zool.* **280,** 260–264.

Fox, J. E., Starcevic, M., Kow, K. Y., Burow, M. E., and McLachlan, J. A. (2001). Endocrine disrupters and flavonoid signalling. *Nature* **413,** 128–129.

Halling-Sørensen, B., Nors-Nielsen, S., Lanzky, P. F., Ingerslev, F., Holten-Lützhøft, H. C., and Jørgensen, S. E. (1998). Occurrence, fate and effects of pharmaceutical substances in the environment—a review. *Chemosphere* **36,** 357–393.

Hewes, R. S., and Taggart, P. H. (2001). Neuropeptides and neuropeptide receptors in the Drosophila melanogaster genome. *Genome Res.* **11,** 1126–1142.

Jick, H., Zornberg, G. L., Jick, S. S., Seshadri, S., and Drachman, D. A. (2000). Statins and the risk of dementia. *Lancet* **356,** 1627–1631.

John, M., Moore, C. B., James, I. R., Nolan, D., Upton, R. P., McKinnon, E. J., and Mallal, S. A. (2001). Chronic hyperlactatemia in HIV-infected patients taking antiretroviral therapy. *AIDS* **15,** 717–723.

Kakuda, T. N. (2001). Pharmacology of nucleoside and nucleotide reverse transcriptase inhibitor-induced mitochondrial toxicity. *Clin. Therapeutics* **22,** 685–708.

Knapp, A. C., Huang, J., Starling, G., and Kiener, P. A. (2000). Inhibitors of HMG-CoA reductase sensitize smooth muscle cells to Fas-ligand and cytokine-induced cell death. *Atherosclerosis* **152,** 217–227.

Kwak, B., Mulhaupt, F., Myit, S., and Mach, F. (2000). Statins (HMG-CoA Reductase Inhibitors) as a novel type of immunosuppressor. *Nature Med.* **6,** 1399–1402.

Labarge, X. L., and McCaffrey, R. J. (2000). Multiple chemical sensitivity: A review of the theoretical and research literature. *Neuropsychol. Rev.* **10,** 183–211.

Lundholm, C. E. (1997). DDE-induced eggshell thinning in birds: Effects of p,p'-DDE on the calcium and prostaglandin metabolism of the eggshell gland. *Comp. Biol. Physiol.* **118C,** 113–128.

Maloney, E. K., and Waxman, D. J. (1999). *trans*-Activation of PPARα and PPARγ by structurally diverse environmental chemicals. *Toxicol. Appl. Pharmacol.* **161,** 209–218.

Navarro, I., Leibush, B., Moon, T. W., Plisetskaya, E. M., Baños, N., Méndez, E., Planas, J. V., and Gutiérrez, J. (1999). Insulin, insulin-like growth factor-I (IGF-I) and glucagon: The evolution of their receptors. *Comp. Biochem. Physiology B* **122,** 137–153.

Nicholson, J. K., Lindon, J. C., and Holmes, E. (1999). "Metabonomics": Understanding the metabolic responses of living systems to pathophysiological stimuli via multivariate statistical analysis of biological NMR spectroscopic data *Xenobiotica* **29,** 1181–1189.

Ottaviani, E., Franchini, A., and Hanukoglu, I. (1998). *In situ* localization of ACTH receptor-like mRNA in molluscan and human immunocytes. *Cell. Mol. Life Sci.* **54,** 139–142.

Palinski, W. (2000). Immunomodulation: A new role for statins? *Nature Med.* **6,** 1311–1312.

Panneels, V., Eroglu, C., Cronet, P., and Sinning, I. (2003). Pharmacological characterization and immunoaffinity purification of metabotropic glutamate receptor from Drosophila overexpressed in Sf9 cells. *Protein Expr. Purif.* **30,** 275–282.

Potera, C. (2000). Drugged drinking water. *Environ. Health Persp.* **108,** 446.

Robertson, D. G., Reily, M. D., Sigler, R. E., Wells, D. F., Paterson, D. A., and Braden, T. K. (2000). Metabonomics: Evaluation of nuclear magnetic resonance (NMR) and pattern recognition technology for rapid *in vivo* screening of liver and kidney toxicants. *Toxicol. Sci.* **57,** 326–337.

Rosen, E. D., and Spiegelman, B. M. (2001). PPARgamma: A nuclear regulator of metabolism, differentiation, and cell growth. *J. Biol. Chem.* **276**(41), 37731–37734.

Ross, P., De Swart, R., Addison, R., Van Loveren, H., Vos, J., and Osterhaus, A. (1996). Contaminant-induced immunotoxicity in harbour seals: wildlife at risk? *Toxicology* **112,** 157–169.

Scheffer, M., Carpenter, S., Foley, J. A., Folkes, C., and Walker, B. (2001). Catastrophic shifts in ecosystems. *Nature* **413,** 591–596.

Searls, D. B. (2003). Pharmacophylogenomics: Genes, evolution and drug targets. *Nature Drug Discovery* **2,** 613–623.

Spinney, L. (2003). Slime mould gets moody. BioMedNet Magazine, August 13–26, http://gateways.bmn.com/magazine

Stefano, G. B., and Scharrer, B. (1996). The presence of the μ_3 opiate receptor in invertebrate neural tissues. *Comp. Biochem. Physiol.* **113C,** 369–373.

Stefano, G. B., Bilfinger, T. V., Rialas, C. M., and Deutsch, D. G. (2000). 2-Arachidonyl-glycerol stimulates nitric oxide release from human immune and vascular tissues and invertebrate immunocytes by cannabinoid receptor I. *Pharmacol. Res.* **42,** 317–322.

Steger-Hartmann, T., Kümmerer, K., and Hartmann, A. (1997). Biological degradation of cyclophosphamide and its occurrence in sewage water. *Ecotoxicol. Environ. Safety* **36,** 174–179.

Straub, J. O. (2002). Environmental risk assessment for new human pharmaceuticals in the European Union according to the draft guideline/discussion paper of January 2001. *Toxicol. Lett.* **135,** 231–237.

Stumpf, M., Ternes, T. A., Haberer, K., Seel, P., and Baumann, W. (1996). Nachweis von Arzneimittelrückständen in Kläranlagen und Fliessgewässern (Determination of drugs in sewage treatment plants and river water). *Vom Wasser* **86,** 291–303.

Sumpter, J. P. (1998). Xenoendocrine disrupters-environmental impacts. *Toxicol. Lett.* **132–103,** 337–342.

Ternes, T. A. (1998). Occurrence of drugs in German sewage treatment plants and rivers. *Water Res.* **32**, 3245–3260.

Trosko, J. E. (2003). The role of stem cells and gap junctional intercellular communication in carcinogenesis. *J. Biochem. Mol. Biol.* **36**(1), 43–48.

Trosko, J. E., and Ruch, R. J. (2002). Gap junctions as targets for cancer chemoprevention and chemotherapy. *Curr. Drug Targets* **3**, 465–482.

Trosko, J. E., Chang, C. C., and Upham, B. (2002). Modulation of gap junctional communication by epigenetic toxicants: A shared mechanism in teratogenesis, carcinogenesis, atherogenesis, immunomodulation, reproductive- and neuro-toxicities. *In* "Biomarkers of Environmentally-Associated Diseases" (S. H. Wilson and W. A. Suk, Eds.), pp. 445–454. Lewis Publishers, Boca Raton, FL.

Upham, B. L., Weis, L. M., and Trosko, J. E. (1998). Modulated gap junctional intercellular communication as a biomarker of PAH epigenetic toxicity: Structure-function relationship. *Environ. Health Perspect.* **106**(Suppl. 4), 975–981.

Widmann, C., Gibson, S., Jarpe, M. B., and Johnson, G. L. (1999). Mitogen-activated protein kinase: Conservation of a three-kinase module from yeast to human. *Physiol. Rev.* **79**(1), 143–180.

Wolozin, B., Kellman, W., Rousseau, P., Celesia, G. G., and Siegel, G. (2000). Decreased prevalence of Alzheimer disease associated with 3-hydroxy-3-methylglutaryl coenzyme A reductase inhibitors. *Arch. Neurol.* **57**, 1439–1443.

Zuccato, E., Calamari, D., Natangelo, M., and Fanelli, R. (2000). Presence of therapeutic drugs in the environment. *Lancet* **355**, 1789–1790.

Mark Crane

Chancel Cottage
Faringdon
Oxfordshire, SN7 7AG, United Kingdom

Proposed Development of Sediment Quality Guidelines Under the European Water Framework Directive: A Critique

I. Introduction

Aquatic sediment contamination leading to toxic effects is a worldwide problem, especially in those nations with a long industrial history. However, the response to this problem has differed among jurisdictions (Apitz and Power, 2002). North American researchers and environmental regulators were the first to develop sediment toxicity testing methods and regulatory instruments for monitoring and controlling sediment contamination (Crane *et al.*, 1996). In Europe, research and regulation of contaminated sediments has been less coherent, with individual member states of the European Union developing sediment quality guidelines and monitoring strategies in a largely independent manner (Ahlf *et al.*, 2002; Environment Agency, 2002). The advent of the Water Framework Directive (WFD), the most important piece of European water legislation for many years, will change the way in which the aquatic environment is monitored and regulated, with implications for the monitoring and regulation of sediment contamination.

The following will describe the way in which the WFD is likely to require the development of what are often referred to as Sediment Quality Guidelines (SQGs), which would become mandatory Sediment Environmental Quality Standards (EQSs) in member states. Proposals to the European Commission for an approach to the development of sediment EQSs are briefly described; these proposals contain some controversial elements, such as a focus on suspended sediments. Several questions remain over both the general approach (are mandatory sediment EQSs defensible?) and the specific proposals for sediment EQSs under the WFD. The classes of chemicals currently under consideration for the development of Sediment EQSs include endocrine disrupters, such as the alkylphenols, pesticides, metals, and certain industrial chemicals. However, it is likely that other chemicals with high partition coefficients, such as some human and veterinary pharmaceuticals, could also require development of sediment EQSs in the future. Jones *et al.* (2002) identified several human pharmaceuticals among the top 25 prescribed in the UK with high partition coefficients (log $K_{ow} > 3$) that may enter the environment at high volumes, particularly through sewage treatment works. These include ibuprofen, sulphasalazine, naproxen, erythromycin, quinine sulphate, meneverine hydrochloride, and mefenamic acid. Similarly, Boxall *et al.* (2002) have identified several veterinary medicines that may also reach the aquatic environment and bind to sediments, including fluoroquinolones/quinalones, macrolides, macrolide endectins, tetracyclines, and synthetic pyrethroids.

II. Water Framework Directive

The WFD provides the basis for future EU water legislation for many years to come (European Commission, 2000a). The Directive aims to ensure the quality of EU waters and takes a holistic approach to water management. It updates existing water legislation through the introduction of a statutory system of analysis and planning based upon the river basin, the use of ecological as well as chemical standards and objectives, the integrated consideration of groundwater and surface water quality and quantity, the introduction of some new regulatory factors, and the phased repeal of several older European Directives. The Directive contains both environmental quality standards and emission limit values from point sources.

The need to understand the impacts of contaminated sediments on aquatic environmental quality is implicit in much of the WFD (Chave, 2001). However, sediments are mentioned explicitly only three times in the text of the Directive, and always in connection with the derivation of an

EQS. An EQS is a mandatory chemical threshold requiring immediate regulatory action, rather than a trigger value for further investigation. Hence, sediment EQSs developed within this framework would need to be enforced by Member States on a pass/fail basis.

Item 7 of Article 16 (Strategies against pollution of water) of the WFD states that, "the Commission shall submit proposals for quality standards applicable to the concentrations of the priority substances in surface water, sediments or biota." Item 35 of Article 2 defines an environmental quality standard as, "the concentration of a particular pollutant or group of pollutants in water, sediment or biota that should not be exceeded in order to protect human health and the environment." Finally, in Annex 5 of the Directive, the procedure is described for setting these environmental quality standards by Member States.

Annex 5 states that standards *may* be set for water, sediment, or biota. In a report to the European Commission (Fraunhofer Institute, 2002), the interpretation of this item in Annex 5 is that for some substances, such as those known *not* to accumulate in sediments, derivation of an EQS is unnecessary. Hence, it seems likely that although sediment EQSs may not need to be derived for all substances, the Commission will insist that they are derived for those that are known to accumulate in sediments, as triggered by an agreed water-to-sediment partition coefficient.

Annex 5 of the WFD states that the risk assessment approach described in the Technical Guidance Document (TGD) should be used to set EQSs for different environmental compartments, including sediments. EU Directives and regulations generally state some of the basic principles of risk assessment, but lack detail. Because of this, the European Commission, the Member States, and the European Chemical Industries produced the TGD (European Commission, 2003) to describe the risk assessment process for new and existing industrial chemicals. This rather lengthy guidance document consists of a series of deterministic equations for estimating chemical hazard and exposure in various environmental compartments, and the comparison of these by using a quotient approach. A predicted no effect concentration (PNEC, equivalent to an EQS) is usually estimated by adding an assessment factor to ecotoxicity data, usually a factor of 10, 100, or 1000 depending on the level of uncertainty and availability of data.

Under the WFD, EQSs need to be developed first for a list of 33 priority substances (European Commission, 2000b), plus those already regulated in "Daughter Directives" of the Dangerous Substances Directive (European Commission, 1976). These priority substances are listed in Table I.

A proposal has been made for setting EQSs for different environmental compartments, which is based upon accepted practice in the EU (Fraunhofer Institute, 2002). However, the implications of the proposals for setting sediment EQSs are considerable. Briefly, these proposals are that:

TABLE I Priority Substances for Which EQSs Are Required Under the WFD

Name	
Alachlor	(Benzo(g,h,i)perylene)
Anthracene	(Benzo(k)fluoroanthene)
Atrazine	(Indeno(1,2,3-cd)pyrene)
Benzene	Pentachlorobenzene
Brominated diphenylether	Simazine
(Bis(pentabromophenyl)ether)	Pentachlorophenol
(Diphenyl ether, octabromo deviate)	Tributyltin compounds
(Diphenyl ether, pentabromo derivative)	(Tributyltin-cation)
Cadmium and its compounds	Trichlorobenzenes
C_{10-13}-chloroalkanes	(1,2,3-Trichlorobenzene)
Chlorfenvinphos	(1,2,4-Trichlorobenzene)
Chlorpyrifos	(1,3,5-Trichlorobenzene)
Dichloromethane	Trichloromethane
1,2-Dichloroethane	Trifluralin
Di(2-ethylhexyl)phthalate (DEHP)	Tetrachloroethene
Diuron	Trichloroethene
Endosulfan	Tetrachloromethane
(alpha-endosulfan)	Drins
Fluoroanthene	Aldrin
Hexachlorobenzene	Endrin
Hexachlorobutadiene	Isodrin
Hexachlorocyclohexane	Dieldrin
(gamma-isomer, Lindane)	DDT
Isoproturon	DDT, 4,4′-isomer
Lead and its compounds	DDT, 2,4′-isomer
Mercury and its compounds	Dioxins (PCDD)
Naphthalene	Furans (PCDF)
Nickel and its compounds	PCB
Nonylphenols	(PCB 28)
(4-(para)-nonylphenol)	(PCB 52)
(4-nonylphenol, branched)	(PCB 101)
Octylphenols	(PCB 118)
(para-tert-octylphenol)	(PCB 138)
Polyaromatic hydrocarbons	(PCB 153)
(Benzo(a)pyrene)	(PCB 180)
(Benzo(b)fluoroanthene)	Copper and its compounds

1. The trigger for deriving a sediment EQS for a priority substance is a sediment-water partition coefficient (log K_p) ≥ 3 for organic substances or metals or, in the absence of a K_p value, a log $P_{ow} \geq 3$ for organic substances.

2. Contaminant concentrations in suspended sediments rather than in settled sediments should be measured and compared with EQS values. The justification for this is:

- Contaminants in suspended sediments represent "current" rather than historical pollution, as they will ultimately lead to new deposits of contamination;
- Newly settled material is the main food source for detritivorous benthic organisms.

3. The assessment factor approach described in the TGD should be used to determine a PNEC for sediment-dwelling organisms, but a species sensitivity distribution (SSD) approach (Aldenberg and Jaworska, 2000) may be used for data-rich substances.

4. Empirical data from sediment toxicity tests are preferred, but equilibrium partitioning (EP) models may be used in the absence of such data. If empirical data are available, the following assessment factors should be applied:

- 100 if one long-term test result is available;
- 50 if two long-term tests are available for species with different life history and feeding characteristics;
- 10 if three long-term tests are available for species with different life history and feeding characteristics.

If an EP approach is used, the sediment PNEC should be calculated from the following equation:

$$\mathrm{PNEC_{sed}}[\mathrm{mgkg}] = \left(\frac{\mathrm{Kp_{sed-water}}[\mathrm{LL}]}{\mathrm{bulk\ density_{sed}}[\mathrm{kgL}]}\right) \times \mathrm{PNEC_{water}}[\mathrm{mgL}]$$

For substances with a log $K_{ow} > 5$, an additional safety factor of 10 is added if the EP approach is used, to account for ingestion of sediment.

5. For substances such as metals, which occur naturally in the environment, an "added risk" approach should be taken to determine the maximum permissible addition of the substance to the environment. However, approaches to examine the bioavailability of metals, such as measurement of acid volatile sulphide (AVS) and simultaneously extracted metals (SEM), should not be used because of uncertainties in the assumptions underpinning this approach, and its lack of relevance to aerobic, suspended sediments.

III. Outstanding Questions on European Union Sediment EQSs

Several theoretical and practical questions emerge from current proposals for sediment EQSs that have implications for the regulation of priority substances and others, such as certain pharmaceuticals, that may be added to this list. These are examined in the following sections.

A. What Should Be the Trigger for Requiring a Sediment EQS?

A log K_p of 3 is not an unreasonable trigger for requiring consideration of sediment contamination. However, K_p values are rather difficult to generate empirically (Burgess, USEPA, pers. comm.), so log K_{ow} is likely to be a standard default for organic contaminants. There is no real analog of K_{ow} for metals, so the simplest approach is to list the rather limited number of metals that will be of concern for toxicological reasons (cadmium, chromium, copper, nickel, lead, silver, and zinc).

A partitioning trigger of log $K_{ow} > 3$ is exceeded by several human pharmaceuticals, but only mefenamic acid and oxytetracycline are likely to be sufficiently toxic at environmentally realistic concentrations (Jones *et al.*, 2002) to require monitoring. More veterinary medicines may require monitoring in sediments if plausible environmental pathways from sources to receptors can be demonstrated (Boxall *et al.*, 2002).

B. Should a Sediment EQS Be a Mandatory and Legally Enforceable Pass/Fail Limit or an Early, Conservative Screening Tool in a Tiered Risk Assessment Framework?

SQGs are widely used in other regions in the first tier of a risk assessment to screen out those sediments that are almost certainly of no toxicological concern. This is the view of USEPA (Horinko, 2002) and some European researchers (Sijm *et al.*, 2001), and contrasts with the proposals presented to the EC (Fraunhofer Institute, 2002). Exceeding a SQG should lead to further, site-specific investigation before a regulatory decision is made to remediate or manage sediments.

The reason for caution in the use of mandatory sediment EQSs is that:

- The site-specific predictive accuracy for toxicity of generic SQGs is uncertain;
- SQGs may not adequately take account of bioaccumulation;
- Cause and effect relationships may be difficult to demonstrate; and
- SQGs developed on the basis of one endpoint (e.g., amphipod mortality) may not be relevant to other effects endpoints.

C. Should Suspended or Settled Sediments, or Both, Be Analyzed?

Both suspended and settled (bedded) sediments are of concern in the U.S. and other countries, with the former of particular importance when assessing the environmental effects of dredging (Engler, 1980). In the Netherlands, concentrations of contaminants in suspended solids are considered to be related to sediment concentrations through empirically derived ratios of 1.5 (suspended matter:settled sediment) for metals and 2.0 for organic compounds. This is then used to predict concentrations in settled sediments, according to the following equation:

$$C_{sed} = \frac{C_{susp}}{r} = \frac{K_{sw} \leftrightarrow C_w}{r},$$

where C_{sed} = concentration in settled sediment (mg/kg); C_{susp} = concentration in suspended sediment (mg/kg); and r = empirical ratio suspended matter:settled sediment.

In the Dutch approach, K_{sw} (solids to water partition coefficient L/kg) and the total concentration (C_{tot}) of the contaminant are estimated by the following equations:

$$K_{sw} = \frac{C_s}{C_w} = \frac{C_{tot} - C_w}{SM \leftrightarrow C_w}$$

and

$$C_{tot} = C_w \leftrightarrow (1 + K_{sw} \leftrightarrow SM),$$

where C_s = concentration in the solid phase (mg/kg); C_w = concentration in water (μg/L); and SM = concentration of suspended sediment (g/L).

Members of the Rhine Commission have also taken the view that suspended sediments should be a focus for regulation because quality objectives must include (CIPR/IKSR, 1995):

- The protection of receiving soils (when deposition occurs during flood events);
- The protection of the seas beyond the Rhine's mouth; and
- The protection of sediment-dwelling organisms.

The national experts involved in the Rhine Commission could not agree on the methodology to be applied to settled and suspended sediment, so they restricted quality objective setting to the first category of protected values (soils). As a result of this, the Commission adopted quality objectives for metals and arsenic in suspended particles derived from regulations used for the disposal of sewage sludge on agricultural soils.

The Fraunhofer Institute (2002) recommendations to the EC therefore have foundations in both Dutch and Rhine Commission regulatory practice. However, there are some serious technical difficulties inherent in sampling suspended rather than settled sediments (Ingersoll *et al.*, 1997b). Suspension of sediments relies upon the energy present in an aquatic system, so it is usually difficult to determine where and when to sample for suspended sediments and how to predict where these will ultimately be deposited. When sediments are deposited, the exposure of organisms to contaminants adsorbed to them will be influenced by many factors that cannot be predicted from measurement of those sediments when in suspension. These factors include:

- Bioturbation and bioirrigation (Levinton, 1982; Rhoads and Boyer, 1982; Valiela, 1984);
- Organic carbon and other ligand concentrations, grain size distributions, and sulphide concentrations (Ankley *et al.*, 1996; Di Toro *et al.*, 1991);
- Salinity/hardness, ionic strength, pH, temperature, and dissolved oxygen concentrations (Hamelink *et al.*, 1994); and
- Variations in the oxidation-reduction depth (Lee and Swartz, 1980).

As a result of such considerations, Ingersoll *et al.* (1997b) concluded that assessment of suspended sediments is associated with higher uncertainty than assessment of settled sediments.

D. How Valid or Relevant Is Suspended Sediment Contamination for Estimating Risks to Benthic Organisms?

Contaminated suspended sediments are clearly of importance when filter-feeding organisms are considered to be susceptible. However, the relevance of suspended sediment contamination for all benthic organisms is debatable. The Fraunhofer Institute (2002) states that benthic organisms are largely detritivores, so freshly deposited sediment is of most importance. Although the fresh deposition of sediment may be an important source of food material for some benthic organisms, it does not encompass all possible routes of contaminant transfer. In addition, it is a rather indirect measure of what organisms might be exposed to at a particular site when compared with sampling of settled sediments.

E. How Should Samples of Suspended or Settled Sediments Be Taken to Minimize Variability?

Variability in results due to sampling and sample manipulation artifacts are well known when assessing sediment and pore water contamination (Adams *et al.*, 2001; Burton, 1991; Carr *et al.*, 2001; Diamond *et al.*, 2002; Environment Canada, 1994; USEPA, 2002). Oxidation of anaerobic sediments can release bound chemicals (such as metals), and homogenization of sediment samples can dilute concentrations of contaminants found in biologically important parts of the habitat (such as the sediment surface).

Sampling should be relevant to the receptor organisms of interest; this would typically be a sample of up to 5 cm depth in settled sediments.

However, Lee (1990) suggests that while analytical chemists often select the top 2 cm of sediment for analysis, organisms may feed on either the very superficial material (fresh falling detritus, e.g., *Macoma*), or feed at depth (oligochaete or polychaete worms): The Fraunhofer Institute (2002) recommendations to the EC may encompass the former scenario, but not the latter.

If suspended sediments were to be sampled, as recommended by the Fraunhofer Institute (2002), some additional logistical problems are likely to occur. Current scientific knowledge is inadequate for predicting the erosion, transport, and deposition of sediments. This is particularly so for the fine-grained coherent sediments, which are most likely to sequester contaminants. Even intuitively obvious assumptions about sediment movement may be false. Direct measurements of suspended sediment concentrations in the New York/New Jersey Harbor show that, contrary to previous assumptions, most sediment moves into the estuary on the flood tide, resulting in a net influx of sediment. Although storms lead to large amounts of sediment moving out of the estuary, again contrary to expectation, the twice-daily flux into the system can quickly replace this. Combined with turbidity measurements associated with the salt wedge, this leads to much higher estimates of dynamic sediment transport than previously estimated (Chant *et al.*, 2001; Herrington *et al.*, 2002; Rankin *et al.*, 2001; Styles *et al.*, 2001).

F. What Analytical Measurements Should Be Made to Determine the Bioavailability of Contaminants in Sediments?

In risk-based frameworks for assessing contaminated sediments, a tiered approach is used in which considerations of bioavailability are examined only when simpler lower tiers suggest that there may be a risk. For example, in Australia for metal contamination, an acid-soluble metals concentration is compared with the SQG (Batley, CSIRO, pers. comm.). If this is exceeded, then concentrations are compared with background concentrations. Only after this would measurements that help to determine bioavailability, such as AVS/SEM for metals, be employed.

This approach is precautionary and cost-effective, but depends upon acceptance of a tiered, risk assessment framework for contaminated sediments, rather than a mandatory pass/fail EQS framework. The Fraunhofer Institute (2002) proposal to ignore AVS/SEM for metals is therefore a sensible and cost-effective approach at a first, screening tier of a risk assessment, but could lead to expensive overprotection if not considered at subsequent tiers. Similarly, pharmaceuticals bound tightly to sediments may not be biologically available to benthic organisms, a possibility that could be explored within a tiered framework.

G. To What Extent Should Background Concentrations Be Taken into Account When Developing a Sediment EQS for Metals, and at What Spatial Scale?

Rowlatt *et al.* (2002), in a report to the Environment Agency of England and Wales, suggest that background concentrations must be taken into account in SQGs and sediment EQSs. This is the view of most jurisdictions, including the U.S., Canada, and Australia, and avoids problems encountered by the Dutch authorities when initial sediment quality criteria for metals were set lower than background concentrations. As a result of this, the Dutch developed an approach to include background concentrations, the "added risk approach," which forms the basis of the Fraunhofer Institute's (2002) recommendations to the EC. The Aldenberg and Slob (1993) species sensitivity distribution approach is used to define a concentration that is theoretically safe for 95% of species and incorporates background concentrations. This is then used to derive a "maximum permissible addition," and the sediment EQS is set equal to the background concentration plus the maximum permissible addition (Crommentuijn *et al.*, 2000).

However, there is an important difference in the way that background concentrations are used in North America and Australia, and the way in which the Fraunhofer Institute (2002) recommends their use in the EU. In most other jurisdictions, background concentrations are taken into account on a site-specific basis within a tiered risk assessment framework. Since this is not what is proposed by the Fraunhofer Institute, it is hard to envisage how the many different background concentrations that prevail across the EU can be taken into account. Either one, or a limited number, of background concentrations would have to be used as surrogates for all others, with consequent under- or over-protection of many sites, or an impossibly large number of site-specific sediment EQSs that would need to be set.

H. Are Sediment EQSs Based upon Equilibrium Partitioning Methods a Sufficiently Robust Underpinning for a Mandatory Standard?

The main approaches used to develop SQGs in different regions are:

- Equilibrium partitioning approach (Di Toro *et al.*, 1991; NYSDEC, 1999; USEPA, 1997);
- Screening level concentration approach (Persaud *et al.*, 1993);
- Effects range approach (Long *et al.*, 1995; USEPA, 1996);
- Effects level approach (MacDonald *et al.*, 1996; Smith *et al.*, 1996; USEPA, 1996);
- Apparent effects threshold approach (Barrick *et al.*, 1988; Cubbage *et al.*, 1997; Ginn and Pastorok, 1992);

- Consensus-based approach (MacDonald *et al.*, 2000a,b; Swartz, 1999).

All of these approaches have strengths and weaknesses; while none is intrinsically scientifically flawed, equally, none should be solely relied upon for a regulatory decision.

Some EU Member States, such as the Netherlands, rely extensively upon equilibrium partitioning (EqP) approaches when setting sediment standards, as shown in Fig. 1 (Sijm *et al.*, 2001). Dutch regulatory scientists believe that EQSs should be harmonized among environmental compartments, and equilibrium partitioning theory is one of the few ways of doing this. Therefore, most of the published Dutch sediment standards are obtained through use of EqP, and other methods have not been used extensively.

The identification of water quality criteria is critical to the Dutch approach, and can be achieved either by using assessment factors, or species sensitivity distributions (Aldenberg and Jaworska, 2000; Aldenberg and Slob, 1993). It is therefore clear that many of the recommendations made to the EC by the Fraunhofer Institute (2002) are based upon the Dutch model, which relies upon the predictive accuracy of the EqP approach. However, there are several areas of uncertainty when applying EqP to set SQGs and sediment EQSs, including the following:

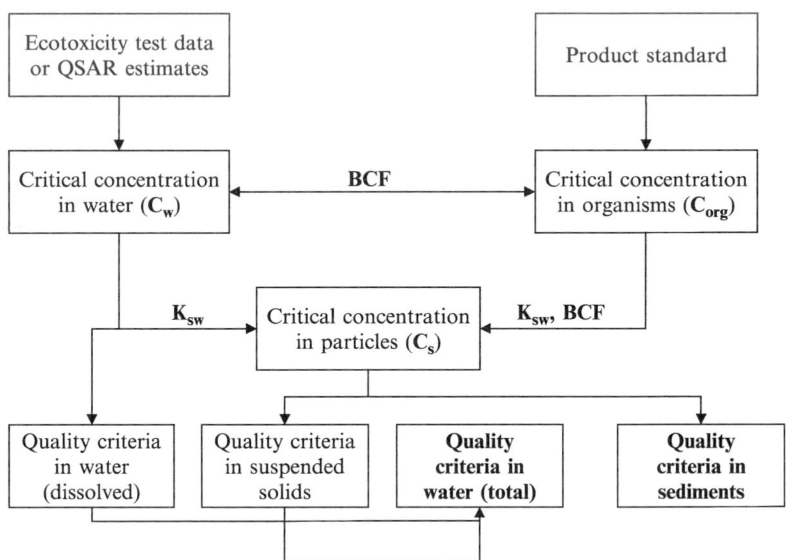

FIGURE 1 Derivation of sediment quality criteria in the Netherlands (from Van der Kooij *et al.*, 1991).

- The approach assumes that contaminants in the aqueous phase of a sediment are in equilibrium with solid sedimentary phases (Shea, 1988), which may not be the case.
- The methodology assumes that a single partitioning coefficient can characterize heterogeneous sediments.
- There may be differences in the response of water column and benthic organisms, as well as limitations in understanding the relationship of individual and population effects to assemblage-level effects (Mancini and Plummer, 1994).
- Synergistic, antagonistic, or additive effects of contaminants are not considered (except in the case of Polycyclic Aromatic Hydrocarbons [PAHs] and metals).

Because of these limitations, site-specific modifications to SQGs derived using the EqP model have been recommended to help address chemical bioavailability and species sensitivities (USEPA, 1993).

The above is not intended as a criticism of the EqP approach relative to other methods for deriving SQGs. All approaches have strengths and weaknesses, so it is important that none is used as a single, mandatory standard.

I. How Effective Is the SSD Approach in Determining No Effect Levels for Sediment Biota, and What Are the Minimum Data Requirements?

There has been extensive discussion of this subject, with much debate over the minimum sample size necessary for reliable SSD model fitting. Estimates range from 5 (Aldenberg and Jaworska, 2000), through ~10 (Crane *et al.*, 2002) and on to many more (Newman *et al.*, 2000). However, what is clear is that for most contaminants, the existence of even five reliable *sediment* toxicity values is unlikely, particularly for substances such as pharmaceuticals that tend not to have received as much environmental attention as other substances, such as pesticides. One is therefore left with a possible two-fold extrapolation method based upon the Dutch approach, involving both EqP and SSDs. The combination of these two model-based approaches, a general lack of data, and a lack of overall predictive validation means that basing sediment EQSs in the EU on this approach would be highly questionable.

In addition to this, there are general criticisms of the SSD approach for all environmental media by respected commentators (Forbes and Calow, 2002; Forbes and Forbes, 1993). These need to be addressed before SSDs are routinely used by environmental regulators to establish mandatory standards.

J. Is it Cost-Effective in Time, Money, and Materials, for Different Jurisdictions to Develop Separate SQGs? Can SQGs Developed in Different Jurisdictions, such as North America, Be Transferred to Other Jurisdictions, such as Europe, and What Criteria Should Be Satisfied to Allow Confidence in this Transfer?

SQGs derived by other jurisdictions could be used initially as SQGs in Europe, but validation with local biota would be advisable. Indeed, Australia and New Zealand have already used SQGs derived in North America as trigger values in a decision framework for assessing contaminated sediments (ANZECC/ARMCANZ, 2000). Because of a lack of local sediment effects data, the effects range low (ERL) values derived from the North American database were used as the basis of interim Australian and New Zealand guidelines, but efforts to validate these locally are under way. This could be a pragmatic way forward for the EU.

IV. Conclusions

The proposals on sediment EQSs from the Fraunhofer Institute (2002) to the EC are problematic for the regulation of industrial chemicals, including pharmaceuticals, in two main areas. The first is the mandatory, pass/fail nature of an EQS. The second is the focus on measuring contaminants in suspended sediments. Both of these proposals are out of line with approaches used elsewhere and are of doubtful use for cost-effective protection of sediment ecosystems. Instead, it is preferable to have a tiered risk assessment framework in which SQGs for settled sediments are used to screen out nonhazardous sediments, or trigger further site-specific studies on potentially hazardous ones.

If these two proposals to the EC are replaced with more operational approaches, then other debatable points, such as a reliance on EqP and SSD models, the relative advantages and disadvantages of other approaches for setting SQGs, suitable trigger values, measures of bioavailability, and incorporation of background concentrations, assume subsidiary importance. All of these latter points can be addressed through evolving experience and research, so long as a more appropriate medium (settled sediment) is assessed within a flexible, tiered risk assessment framework.

Acknowledgments

The author thanks the experts who attended a recent SETAC Pellston workshop on sediment quality guidelines (Montana, August 2002) for attending and commenting on a short presentation, and for discussing the implications of recent proposals to the EC. However, all of the opinions expressed in this paper are the author's and do not necessarily represent the opinions of the SETAC Pellston participants or SETAC. Project funding was from the Environment Agency of England and Wales.

References

Adams, W., Burgess, R. M., Gold-Bouchot, G., LeBlanc, L., Liber, K., and Williamson, B. (2001). Porewater chemistry: Effects of sampling, storage, handling and toxicity testing. *In* "Summary of a SETAC Technical Workshop: Porewater Toxicity Testing: Biological, Chemical and Ecological Considerations with a Review of Methods and Applications, and Recommendations for Future Areas of Research" (R. S. Carr and M. Nipper, Eds.). 18–22 March 2000. Society of Environmental Toxicology and Chemistry (SETAC), Pensacola, Florida.

Ahlf, W., Hollert, H., Neumann-Hensel, H., and Ricking, M. (2002). A guidance for the assessment and evaluation of sediment quality: A German approach based on ecotoxicological and chemical measurements. *J. Soils Sed.* **2**, 37–42.

Aldenberg, T., and Jaworska, J. (2000). Uncertainty of the hazardous concentration and fraction affected for normal species sensitivity distributions. *Ecotoxicol. Environ. Saf.* **46**, 1–18.

Aldenberg, T., and Slob, W. (1993). Confidence limits for hazardous concentrations based on logistically distributed NOEC toxicity data. *Ecotoxicol. Environ. Saf.* **25**, 48–63.

Ankley, G. T., DiToro, D. M., Hansen, D., and Berry, W. (1996). Technical basis and proposal for deriving sediment quality criteria for metals. *Environ. Toxicol. Chem.* **15**, 2056–2066.

ANZECC/ARMCANZ (2000). www.ea.gov.au/water/quality/nwqms

Apitz, S. E., and Power, E. A. (2002). From risk to sediment management: An international perspective. *J. Soils Sed.* **2**, 61–68.

Barrick, R., Becker, S., Pastorok, R., Brown, L., and Beller, H. (1988). Sediment Quality Values Refinement: 1988 Update and Evaluation of Puget Sound AET. Prepared by PTI Environmental Services for Environmental Protection Agency, Bellevue, Washington.

Boxall, A. B. A., Fogg, I., Blackwell, P. A., Kay, P., and Pemberton, E. J. (2002). Review of Veterinary Medicines in the Environment. R&D Technical Report P6-012/8/TR. Environment Agency, Bristol, United Kingdom.

Burton, G. A., Jr. (1991). Assessing the toxicity of freshwater sediments. *Environ. Toxicol. Chem.* **10**, 1585–1627.

Carr, R. S., Nipper, M., Berry, W. J., Burton, G. A., Ho, K. I., MacDonald, D., Scroggins, R., and Winger, P. V. (2001). Summary of a SETAC Technical Workshop: Porewater Toxicity Testing: Biological, Chemical and Ecological Considerations with a Review of Methods and Applications, and Recommendations for Future Areas of Research. 18–22 March 2000. Society of Environmental Toxicology and Chemistry (SETAC), Pensacola, Florida.

Chant, R., Glenn, S., Hunter, E., Rankin, K., Bruno, M. S., Styles, R., and Hires, R. (2001). Circulation and mixing in a complex estuarine environment: Effects on the transport and fate of suspended matter. Presented at the SETAC 22nd Annual Meeting Nov. 11–15, 2001, Baltimore, Maryland.

Chave, P. (2001). The EU Water Framework Directive: An Introduction. IWA Publishing, London, United Kingdom.

CIPR/IKSR (1995). Rhine Action Programme—Data sheets for the quality objectives. International Commission for the Protection of the Rhine, Koblenz.

Crane, M., Everts, J., Van de Guchte, K., Heimbach, F., Hill, I., Matthiessen, P., and Stronkhorst, J. (1996). Research needs in sediment bioassay and toxicity testing. *In*

"Development and Progress in Sediment Quality Assessment: Rationale, Challenges, Techniques and Strategies," (M. Munawar and G. Dave, Eds.), pp. 49–56. SPB Academic Publishing, Amsterdam, The Netherlands.

Crane, M., Whitehouse, P., Comber, S., Watts, C., Giddings, J., Moore, D., and Grist, E. (2002). Aquatic ecotoxicological risks from application of chlorpyrifos to top fruit in the UK. *Pesticide Manag. Sci.* **59**, 512–526.

Crommentuijn, T., Sijm, D., De Bruijn, J., Van den Hoop, M., van Leeuwen, K., and vand de Plassche, E. (2000). Maximum permissible concentrations for metals and metalloïds in the Netherlands, taking into account background concentrations. *J. Env. Manage.* **60**, 121–143.

Cubbage, J., Batts, D., and Briedenbach, S. (1997). Creation and analysis of freshwater sediment quality values in Washington State. Environmental Investigations and Laboratory Services Program, Washington Department of Ecology, Olympia, Washington.

Diamond, J., Burton, G. A., Jr., and Scott, J. (2002). Sediment collection, manipulation and characterization for toxicity testing. *In* "Sediment Contamination" (R. Whittemore, Ed.). Water Environment Federation Publ., Fairfax, Virginia. In Press.

Di Toro, D. M., Zarba, C. S., Hansen, D. J., Berry, W. J., Swartz, R. C., Cowan, C. E., Pavlou, S. P., Allen, H. E., Thomas, N. A., and Paquin, P. R. (1991). Technical basis for establishing sediment quality criteria for non-ionic organic chemicals using equilibrium partitioning. *Env. Toxicol. Chem.* **10**, 1541–1583.

Engler, R. M. (1980). Prediction of pollution potential through geochemical and biological procedures: Development of regulation guidelines and criteria for the discharge of dredged and fill material. *In* "Contaminants and Sediments" (R. H. Baker, Ed.). Ann Arbor Press, Ann Arbor, Michigan.

Environment Agency (2002). Sediments in England and Wales: Nature and extent of the issues. Environment Agency, Bristol, United Kingdom.

Environment Canada (1994). Guidance Document for Preparation of Sediments for Physicochemical Characterization and Biological Testing. Environment Protection Series, Report EPS 1/RM/29. Ottawa, Canada.

European Commission (1976). Council Directive 76/464/EEC of 4 May 1976 on Pollution Caused by Certain Dangerous Substances Discharged into the Aquatic Environment of the Community. Official Journal of the European Communities L129, 23–29, May 18, 1976.

European Commission (1996). Technical Guidance Document on Risk Assessment for New and Existing Substances, Part 2, Environmental Risk Assessment. Office for Official Publications of the European Community, Luxembourg.

European Commission (2000a). Directive 2000/60/EC of the European Parliament and of the Council of 23 October 2000 establishing a framework for community action in the field of water policy. *Official Journal of the European Union.* L327, 1–72, December 22, 2000.

European Commission (2000b). Proposal for a European Parliament and Council Decision Establishing the List of Priority Substances In the Field of Water Policy. COM(2000) 47 Final. European Commission, Brussels, Belgium.

Forbes, V. E., and Calow, P. (2002). Species sensitivity distributions revisited: A critical appraisal. *Hum. Ecol. Risk Assess.* **8**, 473–492.

Forbes, T. L., and Forbes, V. E. (1993). A critique of the use of distribution-based extrapolation models in ecotoxicology. *Functional Ecol.* **7**, 249–254.

Fraunhofer Institute (2002). Towards the Derivation of Quality Standards for Priority Substances in the Context of the Water Framework Directive. 2. Final Report of the Study: Identification of Quality Standards for Priority Substances in the Field of Water Policy. EAF(3)-06/06/FHI. Fraunhofer-Institute Environmental Chemistry and Ecotoxicology, Germany.

Ginn, T. C., and Pastorok, R. A. (1992). Assessment and management of contaminated sediments in Puget Sound. *In* "Sediment Toxicity Assessment" (G. A. Burton, Jr., Ed.), pp. 371–397. Lewis Publishers, Boca Raton, Florida.

Hamelink, J. L., Landrum, P. F., Bergman, H. L., and Benson, W. H. (1994). Bioavailability: Physical, Chemical, and Biological Interactions. CRC Press, Boca Raton, FL.

Herrington, T. O., Rankin, K. L., and Bruno, M. S. (2002). Frequency of sediment suspension events in Newark Bay. Presented at Protection and Restoration of the Environment IV, 1–5 July 2002. Skiathos Island, Greece.

Horinko, M. L. (2002). Principles for Managing Contaminated Sediment Risk at Hazardous Waste Sites. OSWER Directive 9285.6-08. U.S. Environmental Protection Agency Washington, D.C.

Ingersoll, C. G., Ankley, G. T., Baudo, R., Burton, G. A., Lick, W., Luoma, S. N., MacDonald, D. D., Reynoldson, T. B., Solomon, K. R., Swartz, R. C., and Warren-Hicks, W. J. (1997b). Workgroup summary report on uncertainty evaluation of measurement endpoints used in sediment ecological risk assessment. *In* "Ecological Risk Assessments of Contaminated Sediment" (C. G. Ingersoll, T. Dillon, and R. G. Biddinger, Eds.), pp. 297–352. SETAC Press, Pensacola, Florida.

Jones, O. A. H., Voulvoulis, N., and Lester, J. N. (2002). Aquatic environmental assessment of the top 25 English prescription pharmaceuticals. *Water Res.* **26**, 5013–5022.

Lee, II, H., (1990). A clam's eye view of the bioavailability of sediment-associated pollutants. *In* "Organic Substances and Sediments in Water, Volume 3, Biological" (R. A. Baker, Ed.), pp. 73–93. Lewis Publishers, Boca Raton, Florida.

Lee, H., and Swartz, R. C. (1980). Biological processes affecting the distribution of pollutants in marine sediments. Part II. Biodeposition and bioturbation. *In* "Contaminants and Sediments: Volume 2 Analysis, Chemistry, Biology" (R. A. Baker, Ed.), pp. 533–553. Ann Arbor Science, Ann Arbor, Michigan.

Levinton, J. S. (1982). Marine Ecology. Prentice-Hall, Inc., Englewood Cliffs, New Jersey.

Long, E. R., MacDonald, D. D., Smith, S. L., and Calder, F. D. (1995). Incidence of adverse effects within ranges of chemical concentrations in marine and estuarine sediments. *Env. Manage.* **19**, 81–97.

MacDonald, D. D., Carr, R. S., Calder, F. D., Long, E. R., and Ingersoll, C. G. (1996). Development and evaluation of sediment quality guidelines for Florida coastal waters. *Ecotoxicology* **5**, 253–278.

MacDonald, D. D., Ingersoll, C. G., and Berger, T. (2000a). Development and evaluation of consensus-based sediment quality guidelines for freshwater ecosystems. *Arch. Env. Contam. Toxicol.* **39**, 20–31.

MacDonald, D. D., DiPinto, L. M., Field, J., Ingersoll, C. G., Long, E. R., and Swartz, R. C. (2000b). Development and evaluation of consensus-based sediment effect concentrations for polychlorinated biphenyls (PCBs). *Env. Toxicol. Chem.* **19**, 1403–1413.

Mancini, J. L., and Plummer, A. H., Jr. (1994). A review of EPA sediment criteria. Proceedings of the Water Environment Federation 67th Annual Conference and Exposition, October 15–19, Chicago, Illinois.

Newman, M. C., Ownby, D. R., Mézin, L. C. A., Powell, D. C., Christensen, T. L., Lerberg, S. B., and Anderson, B. A. (2000). Applying species-sensitivity distributions in ecological risk assessment: Assumptions of distribution type and sufficient numbers of species. *Env. Toxicol. Chem.* **19**, 508–515.

NYSDEC, New York State Department of Environmental Conservation (1999). Technical Guidance for Screening Contaminated Sediments. Division of Fish and Wildlife, Division of Marine Resources, Albany, New York.

Persaud, D., Jaagumagi, R., and Hayton, A. (1993). Guidelines for the Protection and Management of Aquatic Sediment Quality in Ontario. Water Resources Branch, Ontario Ministry of the Environment, Toronto, Ontario, Canada.

Rankin, K. L., Chant, R., Glenn, S., Bruno, M. S., Styles, R., Fullerton, B., Harrington, T. O., and Hires, R. I. (2001). Integrated collection of hydrodynamics and surface/water quality data in NY Harbor. Presented at the SETAC 22nd Annual meeting, Nov 11–15, 2001. Baltimore, Maryland.

Rhoads, D. C., and Boyer, L. F. (1982). The effects of marine benthos on physical properties of sediments: A successional perspective. *In* "Animal-Sediment Relations" (P. McCall and M. Tevesz, Eds.), pp. 3–52. Plenum Press, New York, New York.

Rowlatt, S., Matthiessen, P., Reed, J., Law, R., and Mason, C. (2002). Recommendations on the Development of Sediment Quality Guidelines. Environment Agency, Bristol, United Kingdom.

Shea, D. (1988). Developing national sediment quality criteria. *Env. Sci. Tech.* **22,** 1256–1261.

Sijm, D., De Bruijn, J., Crommentuijn, T., and van Leeuwen, K. (2001). Environmental quality standards: Endpoints or triggers for a tiered ecological effect assessment approach? *Env. Toxicol. Chem.* **20,** 2644–2648.

Smith, S. L., MacDonald, D. D., Keenleyside, K. A., Ingersoll, C. G., and Field, J. (1996). A preliminary evaluation of sediment quality assessment values for freshwater ecosystems. *J. Great Lakes Res.* **22,** 624–638.

Styles, R., Chant, R., Glenn, S., Rankin, K., Pence, A., Burke, P., Herrington, T., Bruno, M. S., and Haldeman, C. (2001). Particle concentration and size distributions: Implications for contaminant transport in NY-NJ Harbor New York Bight. Presented at the SETAC 22nd Annual meeting, Nov 11–15, 2001. Baltimore, Maryland.

Swartz, R. C. (1999). Consensus sediment quality guidelines for polycyclic aromatic hydrocarbon mixtures. *Env. Toxicol. Chem.* **18,** 780–787.

USEPA (1993). Guidelines for Deriving Site-Specific Sediment Quality Criteria for the Protection of Benthic Organisms. EPA-822-R-93-017. U.S. Environmental Protection Agency, Office of Science and Technology, Health and Ecological Criteria Division, Washington, D.C.

USEPA (1996). Calculation and Evaluation of Sediment Effect Concentrations for the Amphipod Hyalella azteca and the midge Chironomus riparius. EPA 905/R-96/008, U.S. Environmental Protection Agency, Chicago, Illinois.

USEPA (1997). The Incidence and Severity of Sediment Contamination in Surface Waters of the United States, Volume 1: National Sediment Quality Survey. EPA 823/R-97/006, U.S. Environmental Protection Agency, Washington, D.C.

USEPA (2002). Methods for Collection, Storage and Manipulation of Sediments for Chemical and Toxicological Analyses. U.S. Environmental Protection Agency, Office of Water, Washington, D.C.

Valiela, I. (1984). Marine Ecological Processes. Springer-Verlag, New York, New York.

Van Der Kooij, L. A., Van De Meent, D., Van Leeuwen, C. J., and Bruggeman, W. A. (1991). Deriving quality criteria for water and sediment from the results of aquatic toxicity tests and product standards: Application of the equilibrium partitioning method. *Wat. Res.* **25,** 697–705.

PART V

Risk Assessment

Alistair B. A. Boxall[*]
Lindsay A. Fogg[*]
Paul Kay[*]
Paul A. Blackwell[*]
Emma J. Pemberton[†]
Andy Croxford[†]

*Cranfield Centre for EcoChemistry, Cranfield University
Shardlow, Derby, DE72 2GN, United Kingdon

[†]Environment Agency, National Centre for Ecotoxicology and Hazardous Substances
Wallingford, Oxon, OX10 8BD, United Kingdom

Prioritization of Veterinary Medicines in the UK Environment

I. Introduction

Veterinary medicines are widely used across Europe to treat disease and protect the health of animals. Dietary enhancing feed additives (growth promoters) are also incorporated into the feed of animals reared for food in order to improve their growth rates. Under Directive 81/852/EEC as amended by 92/18/EEC, veterinary medicinal products must be assessed for their quality, efficacy, and safety (to both humans and the environment). Only products approved for use by the regulatory authority may be used.

Release of veterinary medicines to the environment occurs both directly, (by the use of medicines in fish farms, for example) and indirectly (via the application of animal manure containing excreted products to land). A number of groups of veterinary medicines, primarily sheep dip chemicals

(Environment Agency, 1998, 2000, 2001; SEPA, 2000), fish farm medicines (Davies *et al.*, 1998; Jacobsen and Berglind, 1988), and anthelmintics (Madsen *et al.*, 1990; McCracken, 1993; McKellar, 1997; Ridsill-Smith, 1988; Strong, 1993; Wall and Strong, 1987) have been extensively studied, and a large body of data is available for these substances. Information is also available on selected antibiotics (Halling-Sørensen *et al.*, 1999; Holten Lützhøft *et al.*, 1999). However, there are scant data available in the public domain on the environmental fate, behavior, and effects of other generic groups of veterinary medicines, so their potential environmental impacts are less understood (Jørgensen and Halling-Sørensen, 2000).

The large number and wide variety of veterinary medicines available means that it is difficult to identify those substances that should be included in national monitoring programs or investigated in further detail in terms of environmental fate and effects. There is therefore an urgent need to identify substances likely to have the greatest potential to impact the environment. If this could be achieved, then future monitoring programs and experimental studies could be targeted at substances of concern.

The impact of a veterinary medicine on the environment will be determined by a range of factors including the quantity used, the degree of metabolism in the animal, degradation during storage of manure prior to land spreading, and the toxicity of the substance to terrestrial and aquatic organisms. The following describes the application of a straightforward prioritization scheme that incorporates these factors, to identify veterinary medicines that have the potential to impact aquatic and terrestrial systems. The scheme has been applied by the Environment Agency of England and Wales (EA) to veterinary medicines in use in the UK. The results will be used by the EA to guide policy direction, ensure that their monitoring program is effectively targeted, and identify the need for pollution prevention measures.

II. Prioritization Approach

A. Collation of Data

Data on amounts and/or sales of veterinary medicines in the UK were obtained from a number of sources, including: data obtained from Intercontinental Medical Statistics (IMS) Health and summarized in Boxall *et al.* (2004); the Veterinary Medicines Directorate (VMD) data on the sales of antimicrobial substances and sheep dip chemicals in the UK (VMD, 2001); and data in the published literature on the use of sheep dip chemicals (Liddel, 2001). Information on the metabolism, usage pattern, and ecotoxicity of veterinary medicines in use in the UK was also obtained; detailed information is reported in Boxall *et al.* (2004).

B. Outline of Prioritization Approach

The prioritization exercise considered data on usage, exposure routes, and environmental effects of all generic groups of veterinary medicines. As the focus of the study was on potential environmental impacts, the issue of microbial resistance (which may threaten the future effectiveness of antibiotic treatments for livestock and humans) was not considered. An overview of the prioritization process is illustrated in Fig. 1. The prioritization process was performed in two stages; further details are provided below.

1. Potential to Reach the Environment in Significant Amounts

Data on usage, pathways of entry to the environment, and metabolism for those veterinary medicines considered to reach the environment in potentially significant amounts were identified. Groups of substances were initially ranked using the compiled data as high ($\geq$10 tons per year [tpy]), medium ($\geq$1–$<$10 tpy), low ($<$1 tpy), or unknown usage. The potential for the substance to enter the environment was then assessed using information on the target treatment group, route of administration, metabolism in the animal, and the potential for the substance to be degraded in slurry or manure during storage. Substances were classified as having high, medium, low, or unknown potential to enter the environment using the criteria detailed in Table I.

Using the classifications determined for usage and potential to enter the environment, those substances considered to have the greatest potential to enter the environment, and therefore requiring hazard assessment, were identified using the matrix detailed in Table II. Compounds identified as high usage with a high potential to enter the environment were considered to potentially represent the greatest risk to the environment, and hence were deemed to be the first priority for further assessment.

For those compounds regarded as having low potential to enter the environment, it was considered unnecessary to assess their intrinsic hazard in the prioritization exercise, since they are likely to represent a low risk to the environment relative to other veterinary medicines. This group included those compounds administered either orally or by injection (nontopical applications) as herd treatments that are significantly metabolized, as well as compounds used to treat companion or individual food production animals by nontopical routes. In addition, compounds with a medium potential to enter the environment (for example, those used as herd treatments that are moderately metabolized, as well as those used to treat companion or individual animals by application to the skin) were excluded from hazard assessment when usage was less than one ton per year.

For compounds that are used on more than one target treatment group, the potential to reach the environment was assessed separately for each target group (companion/individuals, herds, and aquaculture) as this may

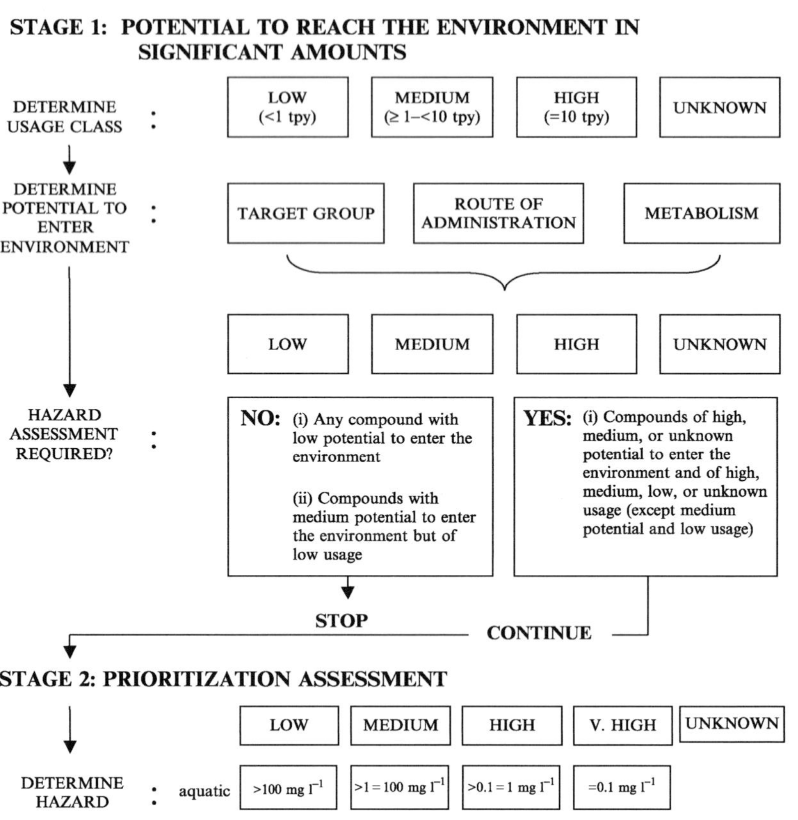

FIGURE 1 Schematic presentation of the prioritization process.

affect the potential for environmental impact. For example, for compounds that are used to treat all three target groups, the potential to reach the environment in significant amounts is considered high when used in aquaculture, but low when used to treat individuals. Likewise, compounds are

TABLE I Criteria Used to Assess the Potential for the Environment to Be Exposed to an Individual Veterinary Medicine

Classification	*Target group*	*Route of administration*	*Metabolism**	*Rationale*
High	Aquaculture	Topical/other	n/a	Substances typically applied directly into the aquatic environment.
	Herd	Topical	n/a	As the substances are applied topically, there is the potential for wash-off from the animal. Topical treatments used in herds are likely to enter the environment in higher amounts than topical treatments used to treat individual or companion animals because of the quantities used.
	Herd	Other	L	Potential impact from substances used as herd treatments that are not significantly metabolized.
Medium	Herd	Other	M	Potential impact from substances used as herd treatments that are moderately metabolized.
	Companion/ individual	Topical	n/a	Potential for direct entry to the environment in excreta. However, since only individuals are treated, the environmental impact is considered to be lower than for herd treatments. Topical treatments have a higher potential to reach the environment than other routes of administration.
Low	Herd	Other	H	Low potential for substances used as herd treatments to enter the environment because of significant metabolism.
	Companion/ individual	Other	n/a	Negligible environmental impact on the basis that it is individuals that are treated rather than herds; therefore, metabolism is not considered.
Unknown	Herd	Other	U	Unknown potential to enter environment because of insufficient data on metabolism.

* Metabolism: H, >80%; M, 20–80%; L, <20%; U, unknown.
n/a, not applicable; other, orally or by injection; individual, individual food production animals.

TABLE II Matrix Used to Identify Substances Requiring Hazard Assessment

Usage	*Potential to enter environment*	*Hazard assessment required?*
H	H	
H	M	
H	L	✓
H	U	
M	H	
M	M	
M	L	✓
M	U	
L	H	
L	M	✓
L	L	✓
L	U	
U	H	
U	M	
U	L	✓
U	U	

H, high; M, medium; L, low; U, unknown.

classified as having a higher potential to enter the environment when used as topical herd treatments than when used topically to treat companion animals or individuals.

2. *Hazard Assessment*

For those compounds identified as having the potential to enter the environment in significant quantities, a simple assessment of hazard was conducted using the toxicity data provided in Boxall *et al.* (2004). This enabled identification of those compounds with a high potential to enter the environment and which were the most toxic (and thus represented potentially the highest risk to the environment). These compounds were considered to be the highest priority for further consideration of their impact on the environment and the possible need for control measures, such as pollution reduction programs.

Substances were classified as having very high, high, medium, or low aquatic and/or terrestrial ecotoxicity using the criteria detailed in Table III. The hazard classification "unknown" was assigned to those compounds where no data for aquatic toxicity or terrestrial toxicity was available.

As an indication of the relative completeness of the available data on which the hazard classification was determined, a score was assigned (given in the footnote of Table V). For aquatic hazard classifications, the score took into account the number of trophic levels tested, as well as the type of tests conducted. Chronic tests for three different trophic levels were regarded as

TABLE III Classification Criteria for Ecotoxicity

Hazard classification	*Aquatic toxicity* (mg l^{-1})*	*Terrestrial toxicity† (mg kg^{-1})*
VH	≤ 0.1	≤ 10
H	$>0.1 \leq 1$	$>10 \leq 100$
M	$>1 \leq 100$	$>100 \leq 1000$
L	>100	>1000

* Based on harmonized system for the classification of chemicals that are hazardous for the aquatic environment; OECD (1998).

† Based on a proposed EU hazard assessment scheme for the terrestrial environment.

VH, very high; H, high; M, medium; L, low.

being more comprehensive than a mixture of chronic tests for one or two trophic levels and several acute toxicity tests. A simpler system was adopted for the terrestrial data than for aquatic toxicity data because there were comparably fewer toxicity data available for terrestrial species.

Considering both the potential to reach the environment (stage 1) and hazard classification (stage 2), substances were then assigned to one of five groups using the matrix detailed in Table IV. Compounds assigned to group one were considered to have the greatest potential for environmental impact and thus are the highest priority for further work. These were compounds that had a combination of high or medium usage, together with high or medium potential to enter the environment and very high or high toxicity to either aquatic or terrestrial organisms. Compounds that were considered to have low potential to enter the environment in significant amounts and thus did not require a hazard assessment were assigned to the lowest score of 5. Where there was uncertainty in any one of the three criteria used, such as unknown data (U) or incomplete data, the worst case classification was assumed.

III. Results

After stage 1, a number of therapeutic groups were identified that were considered to have sufficiently low potential to enter the environment and therefore not requiring a hazard assessment. These included general anaesthetics for companion animals and therapeutic groups where usage was less than 1 ton per year. Fifteen individual substances from other groups were excluded from further assessment, including some compounds that were considered to be high usage but that had a high potential for metabolism (including sulphadimidine, dimetridazole, narasin, and avilamycin).

The general anesthetics are typically gaseous and unlikely to reach water or land in significant quantities. Furthermore, the release of gaseous compounds to the atmosphere will be subject to significant dissipation in air, and

TABLE IV Matrix Used to Determine the Priority Classification of a Substance

	Priority classification											
	1			*2*					*3*			*4*
Potential to enter environment	H/M/U	H/U	H/U	M	M	H/U	H/U	H/U	M	M	H/U	All other combinations
Usage	H/U/H*/ M*/L*	H/U/H*/ M*/L*	M/U/H*/ M*/L*	M	H/U/H*/ M*/L*	L	M/L	H/U/H*/ M*/L*	M	H/U/H*/ M*/L*	M	
Hazard	VH/U	H	VH/U	VH/U	H	VH/U	H	M	H	M	M	

* Usage data incomplete.
VH, very high; H, high; M, medium; L, low; U, unknown.

as a result, aerial exposure is likely to be minimal. Therapeutic groups where usage was less than one ton per year included some antifungals, neurological preparations, and anti-inflammatory preparations. Several other therapeutic groups were also considered as low priority despite usage being unknown, because they were used to treat individual animals (companion or food production). These included the anti-inflammatory steroids, diuretics, cardiovascular and respiratory treatments, and locomotor treatments.

Compounds identified as having a high potential to enter the environment and of high usage included a number of antimicrobial compounds (tetracyclines, sulphadiazine, trimethoprim, amoxicillin, tylosin, dihydrostreptomycin, neomycin, and apramycin) and diazinon, an ectoparasiticide commonly used in sheep dip preparations (Table V). Hazard assessment of these substances resulted in a total of 56 compounds being assigned to the high priority category (group one).

There was only sufficient data available to fully characterize the potential risk for eleven of these compounds (Table VI). For two of these substances (oxytetracycline and amoxicillin), the classification was obtained for both herd and aquaculture treatment scenarios; for the other two substances (suphadiazine and sarafloxicin), the classification was obtained for the aquaculture scenario. The remaining substances (chlortetracycline, tetracycline, diazinon, tylosin, dihydrostreptomycin, apramycin, and cypermethrin) were assigned to the high priority class as a result of their use as herd treatments. For the remaining 45 compounds, some of the data required for the prioritization exercise were either unavailable or incomplete, so the prioritization exercise has incorporated one or more worst-case assumptions. Compounds identified as potentially high risk (group one), but requiring further data are also shown in Table VI.

Six compounds were assigned to group two (Table VI). These compounds are considered to potentially represent a risk to the environment, but are of less concern than the group one compounds discussed above. None of these compounds had complete data sets for the purposes of the prioritization exercise.

IV. Discussion

A pragmatic and scientific approach has been developed to enable an initial identification and prioritization of those veterinary medicines of environmental concern, using available data. The exercise has identified those compounds considered to have the greatest potential to cause environmental impacts as a consequence of their use. However, it is important to recognize that many compounds identified as high priority in this exercise may not actually cause adverse impacts on the environment. The

TABLE V Prioritization Assessment for Veterinary Medicinal Products that Have the Potential to Enter the Environment

Therapeutic group	*Chemical group*	*Major usage products (where data available)*	*Potential to reach environment*	*Relevant target group(s)*	*Usage class*	*Hazard assessment*		*Priority classification*
						Aquatic[‡]	Terrestrial[§]	
Antimicrobials	Tetracyclines	Oxytetracycline	H	H, A	H	H_3	L_3	1
		Chlortetracycline	H	H		VH_4	VH_3	1
		Tetracycline	H	H		VH_4	U	1
Antimicrobials	Potentiated sulphonamides	Sulphadiazine	H	A	H	H_4	H_3	1
		Trimethoprim	H	A		M_4	U	1
		Baquiloprim	U	H		U	U	1
Endoparasiticides—coccidiostats		Amprolium*	M	H	H[†]	U	VH_3	1
		Clopido*	U	H		U	U	1
		Lasalocid sodium*	U	H		U	U	1
		Maduramicin*	M	H		U	VH_2	1
		Nicarbazin*	U	H		U	U	1
		Robenidine hydrochloride*	U	H		U	U	1
Antimicrobials	β-lactams	Amoxicillin	H	H, A	H	VH_4	U	1
		Procaine penicillin	U	H		U	VH_3	1
		Procaine benzylpenicillin	U	H		VH_4	U	1
		Clavulanic acid	U	H		U	U	1
Ectoparasiticides—sheep dips	Organophosphates	Diazinon	H	H	H	VH_4	VH_3	1
Antimicrobials	Macrolides	Tylosin	H	H	H	VH_4	L_3	1
Growth promoters		Monensin	U	H	H[†]	U	VH_2	1
		Salinomycin sodium*	U	H		U	VH_2	1
		Flavophospolipol*	U	H		U	U	1

Antimicrobials	Aminoglycosides	Dihydrostreptomycin	H	H	H	VH_4	U	1
		Neomycin	H	C, H		L_4	U	1
		Apramycin	H	H		U	VH_1	1
		Flavomycin*	U	H		U	U	1
Endoparasiticides—wormers	Pyrimidines	Morantel	M	H	M†	U	U	1
Ectoparasiticides—sheep dips	Pyrethroids	Cypermethrin	H	H	M	VH_4	U	1
		Flumethrin	H	H		U	U	1
Endoparasiticides—wormers	Azoles	Triclabendazole	M	H	M†	U	U	1
		Fenbendazole	U	H		U	U	1
		Levamisole	U	H		U	U	1
Endoparasiticides—wormers	Macrolide endectins	Ivermectin	M	H	M†	VH_3	VH_2	1
Antimicrobial—other antibiotics		Cephalexin	U	H	M†	U	U	1
		Florfenicol	H	A		U	VH_3	1
		Tilmicosin	U	H		U	U	1
		Oxolinic acid*	H	A		VH_4	U	1
Neurological preparations—local anaesthetics		Procaine hydrochloride	U	H	M†	M_4	U	2
		Lido/lignocaine Hydrochloride	U	H		U	U	1
Antimicrobials	Pleuromutilins	Tiamulin	U	H	M†	VH_3	M_2	1
Antimicrobials	Lincosamides	Lincomycin	U	H	M	M_4	VH_1	1
		Clyndamycin	U	H		U	U	1
Antimicrobials—antifungals	Azoles	Miconazole	M	C	M	U	U	2
Endoparasiticides—wormers	Others	Nitroxynil	U	H	M†	U	U	1
Antimicrobials	Fluoroquinolones	Enrofloxacin	H	H	M	U	U	1
		Sarafloxacin		A		VH_4	VH_1	1
Sex hormones		Altrenogest	U	H	L	U	U	2
		Progesterone	U	H		U	U	2
		Medroxyprogesterone	U	H		U	U	2
Enteric preparations		Dimethicone	U	H	L†	U	U	1
		Poloxalene	U	H		U	U	1

(Continues)

TABLE V *(Continued)*

Therapeutic group	*Chemical group*	*Major usage products (where data available)*	*Potential to reach environment*	*Relevant target group(s)*	*Usage class*	*Hazard assessment*		*Priority classification*
						Aquatic[‡]	Terrestrial[§]	
Endoparasiticides—antiprotozoals		Toltrazuril	U	H	L[†]	U	U	1
		Decoquinate	U	H		U	U	1
		Diclazuril	U	H		U	U	1
Endectocides	Macrocyclic lactone injections	Moxidectin	U	H	L	U	U	2
Ectoparasiticides—others		Phosmet	H	H	U/L[†]	U	U	1
		Piperonyl butoxide	M	C		U	U	1
Ectoparasiticides—sheep dips	Amidines	Amitraz	H	H	U	M_2	U	1
Ectoparasiticides – spray and pour—ons for sheep		Deltamethrin	H	H	U	VH_4	H_3	1
		Cypromazine	H	H		VH_4	U	1
Ectoparasiticides—aquaculture treatments		Emamectin benzoate	H	A	U	VH_4	na	1
Antiseptics		?	H	C/I	U	U	U	1
Immunological products		?	U	C, H	U	U	U	1

* Specific usage data unavailable; however, compound considered to be potentially major usage.

† Usage data incomplete.

‡ *Aquatic scores*: subscript 1, 3 trophic levels, chronic test; subscript 2, 3 trophic levels, acute or chronic test; subscript 3, 3 trophic levels, acute test; subscript 4, less than 3 trophic levels, acute or chronic test or both.

§ *Terrestrial scores*: Subscript 1, 3 trophic levels; microbes, invertebrate and plants; Subscript 2, any 2 of 3 trophic levels; Subscript 3, any 1 of 3 trophic levels.

TABLE VI Substances Assigned to Groups 1 and 2 During the Prioritization Exercise*

Group 1 substances		
Oxytetracycline (H, A)	Amoxicillin (H, A)	Cypermethrin (H)
Chlortetracycline (H)	Diazinon (H)	Sarafloxicin (A)
Tetracycline (H)	Tylosin (H)	
Sulphadiazine (A)	Dihydrostreptomycin (H)	
Group 1 possible substances		
Trimethoprim	Morantel	Dimethicone
Baquiloprim	Flumethrin	Poloxalene
Amprolium	Triclabendazole	Toltrazuril
Clopidol	Fenbendazole	Decoquinate
Lasalocid sodium	Levamisole	Diclazuril
Maduramicin	Ivermectin	Phosmet
Nicarbazin	Cephalexin	Piperonyl butoxide
Robenidine hydrochloride	Florfenicol	Amitraz
Procaine penicillin	Tilmicosin	Deltamethrin
Procaine benzylpenicillin	Oxolinic acid	Cypromazine
Clavulanic acid	Lido/ligocaine HCL	Emamectin benzoate
Monensin	Tiamulin	Antiseptics
Salinomycin sodium	Lincomycin	Immunological products
Flavophospolipol	Clindamycin	
Neomycin	Nitroxynil	
Flavomycin	Enrofloxacin	
Group 2 substances		
Procaine HCL	Altrenogest	Medroyprogesterone
Miconazole	Progesterone	Moxidectin

* The treatment scenario giving rise to a potential risk to the environment is indicated in parentheses (H, herd treatment; A, aquaculture treatment).

prioritization exercise is simply a way of assessing the relative potential for veterinary medicines to cause harm, thus enabling those compounds likely to be of greatest concern to be identified and monitored.

For those compounds where sufficient data was available, the list provides a system of relative ranking on the basis of potential environmental impact. Eleven substances were assigned to group one, on the basis of a "complete" data set, and thus considered to be the highest priority. These substances include a number of antimicrobials widely used as herd treatments and/or in aquaculture (oxytetracycline, chlortetracycline, tetracycline, sulphadiazine, amoxicillin, tylosin, dihyrostreptomycin, and apramycin). A further antimicrobial compound, sarafloxacin, used exclusively in aquaculture treatments, was also identified as a high priority, as were diazinon and cypermethrin, two compounds used extensively in sheep dips.

Both cypermethrin and diazinon are known to cause environmental pollution; a significant body of data on their environmental fate, behavior, and ecotoxicity is available. Pollution incidents caused by poor sheep-dipping practices can result in ecological damage over several kilometers of watercourse (SEPA, 2000). Sheep dip chemicals are routinely monitored, and in the UK each year, there are a relatively high number of sites failing the Environmental Quality Standards (EQS), which are derived using appropriate uncertainty factors from available ecotoxicity data, for both cypermethrin and diazinon (Environment Agency, 2000, 2001).

However, with the exception of a few studies (Boxall *et al.*, 2002; Hamscher *et al.*, 2001; Kolpin *et al.*, 2002) the chemicals (other than cypermethrin and diazinon) identified as a high priority have not been looked for in the environment; only a few published studies have investigated environmental effects (Halling-Sørensen, 1999; Holten Lützhøft, 1999; Wollenberger *et al.*, 2000). Further assessment and limited targeted monitoring is therefore recommended to ascertain whether these chemicals are present in the environment at ecologically significant levels. Ideally, this would involve an integrated chemical and biological monitoring program.

The prioritization exercise highlighted the fact that there are many veterinary medicines for which little or no data are available in the public domain. Classifications of many of the compounds were based on limited data and worst-case assumptions. Forty-five substances were provisionally ranked as a high priority, including many other antimicrobial, coccidiostat, endoparasiticide and ectoparasiticide, antifungal, antiprotozoal, and growth, promoting substances. However, for many of these compounds, either accurate usage information was unavailable or their potential to enter the environment or intrinsic hazard was unknown. It is considered a priority for any future work that data should be obtained for these compounds in order to refine and extend the current work. This is required in order to ascertain whether such chemicals are correctly classified in terms of their potential risk to the environment in the current exercise. Those that have been correctly classified can then be added to the list of 11 substances described above for further consideration of their environmental impact.

It should be recognized that the work has focused exclusively on the parent compound. However, following injection or oral administration to an animal, compounds may be metabolized and subsequently excreted, in part or completely, as transformation products. In addition, if excreted as the unaltered parent compound, they may degrade on reaching the environment. The potential environmental impact of any metabolites or degradation products should be assessed, especially for those compounds considered to be low priority on the basis of this prioritization exercise, because they are extensively metabolized following administration. Data on metabolism and environmental degradation were very limited, and consequently detailed consideration in the prioritization exercise was not possible.

Several veterinary medicines for which there were no usage data were included in the prioritization exercise, as they may be distributed via routes other than those covered by the information available to this study and are therefore potentially major usage compounds. Likewise, two therapeutic groups (antiseptics and immunological compounds) for which individual compounds have not been identified are also included, on the basis that there could also be major usage compounds.

While the prioritization exercise has focused on the UK situation, other studies (Jørgensen and Halling-Sørensen, 2000; Pellican *et al.*, 2001) indicate that many of the substances identified during this study are also high usage in other countries. It is therefore likely that the results of this study could be used to design monitoring programs and set priorities in countries other than the UK.

V. Conclusions

There are a large number of veterinary medicines used in the UK. While the concentrations, behavior, and effects of selected groups of veterinary medicines has been well characterized, limited information is available on the potential impacts of the other substances. A scientifically sound and pragmatic approach has therefore been developed for identifying substances that may pose a risk to terrestrial and aquatic systems in the UK. The approach has been applied using information on tonnage sold, typical usage regimes, metabolism, and toxicity to aquatic and terrestrial organisms. Eleven substances, including antibiotics and ectoparasiticides, have been identified as high priority, and a further 45 substances have been identified as potentially high priority but requiring further data. It is recommended that in the future, the data gaps are addressed and that the high-priority substances are further assessed. Targeted monitoring and fate and effects studies should then be performed to determine the impacts, if any, that these substances may have on the environment.

Acknowledgments

The authors would like to thank the UK Environment Agency for funding this work. Part of the review phase of the study was performed during the EU Framework V Project ERAVMIS (project number EVVK1-CT-1999–00003), and the authors would like to thank the European Commission for their financial support.

References

Boxall, A. B. A., Fogg, L., Blackwell, P. A., Kay, P., Pemberton, E. J., and Croxford, A. (2004). Veterinary medicines in the environment. *Rev. Environ. Contam. Toxicol.* **180**, 1–91.

Boxall, A. B. A., Blackwell, P., Cavallo, R., Kay, P., and Tolls, J. (2002). The sorption and transport of a sulphonamide antibiotic in soil systems. *Toxicol. Lett.* **131**, 19–28.

Davies, I. M., Gillibrand, P. A., McHenery, J. G., and Rae, G. H. (1998). Environmental risk of ivermectin to sediment-dwelling organisms. *Aquaculture* **163**, 29–46.

Environment Agency (1998). Pesticides 1998: A summary of monitoring of the aquatic environment in England and Wales. National Centre for Ecotoxicology and Hazardous Substances. Environment Agency, Wallingford, United Kingdom.

Environment Agency (2000). Welsh sheep dip monitoring programme. Environment Agency, Cardiff, Wales.

Environment Agency (2001). Pesticides 1999/2000: A summary of monitoring of the aquatic environment in England and Wales. National Centre for Ecotoxicology and Hazardous Substances, Environment Agency, Wallingford, United Kingdom.

Halling-Sørensen, B. (1999). Algal toxicity of antibacterial agents used in intensive farming. *Chemosphere* **40**, 731–739.

Hamscher, G., Sczesny, S., Hoper, H., and Nau, H. (2001). Determination of persistent tetracycline residues in soil fertilized with liquid manure by high-performance liquid chromatography with electrospray ionization tandem mass spectrometry. *Anal. Chem.* **74**, 1509–1518.

Holten Lützhøft, H. C., Halling-Sørensen, B., and Jørgensen, S. E. (1999). Algal toxicity of antibacterial agents applied in Danish fish farming. *Arch. Env. Contam. Toxicol.* **36**, 1–6.

Jacobsen, P., and Berglind, L. (1988). Persistence of oxytetracycline in sediment from fish farms. *Aquaculture* **70**, 365–370.

Jørgensen, S. E., and Halling-Sørensen, B. (2000). Drugs in the environment. *Chemosphere* **40**, 691–699.

Kolpin, D. W., Furlong, E. T., Meyer, M. T., Thurman, E. M., Zaugg, S. D., Barber, L. D., and Buxton, H. T. (2002). Pharmaceuticals, hormones and other organic wastewater contaminants in US streams, 1999–2000: A national reconnaissance. *Env. Sci. Tech.* **36**, 1202–1211.

Liddel, J. S. (2001). Sheep ectoparasiticide use in the UK: 1993, 1997 and 1999. Paper presented to the 5th International Sheep Veterinary Congress, Stellenbosch, South Africa. January 21–25, 2001.

Madsen, M., Overgaard-Nielsen, B., Holter, P., Pedersen, O. C., Brocchener-Jespersen, J., Vagn Jensen, K.-M., Nansen, P., and Grovold, J. (1990). Treating cattle with ivermectins: Effects on fauna and decomposition of dung pats. *J. Appl. Ecol.* **27**, 1–15.

McCracken, D. I. (1993). The potential for avermectins to affect wildlife. *Vet. Parasitol.* **48**, 273–280.

McKellar, Q. A. (1997). Ecotoxicology and residues of anthelmintic compounds. *Vet. Parasitol.* **72**, 413–435.

OECD (1998). Harmonised integrated hazard classification for human health and environmental effects of chemicals. As endorsed by the 28th joint meeting of the Chemicals Committee and the Working Party on Chemicals in November 1998. OECD, Paris.

Pellican, C. H. P., van Turnhout, J., and Pijpers, A. (2001). Verbruikscijfers van antibacteriele middelen bij landbouwhuis diren. Poster obtained from the University of Utrecht, The Netherlands.

Ridsill-Smith, T. J. (1988). Effects of avermectin residues in cattle dung on dung beetle (Coleoptera: Scarabaeidae) reproduction and survival. *Vet. Parasitol.* **48**, 127–137.

SEPA (2000). Long term biological monitoring trends in the Tay system 1988–1999. Scottish Environmental Protection Agency, Eastern Region, Scotland.

Strong, L. (1993). Overview: The impact of avermectins on pastureland ecology. *Vet. Parasitol.* **48**, 3–17.

VMD (2001). Sales of antimicrobial products used as veterinary medicines and growth promoters in the UK in 1999. Veterinary Medicines Directorate, United Kingdom.

Wall, R., and Strong, L. (1987). Environmental consequences of treating cattle with the antiparasitic drug ivermectin. *Nature* **327**, 418–421.

Wollenberger, L., Halling-Sørensen, B., and Kusk, K. O. (2000). Acute and chronic toxicity of veterinary antibiotics to *Daphnia magna*. *Chemosphere* **40**, 723–730.

Mark H. M. M. Montforts and Joop A. de Knecht
National Institute for Public Health and the Environment (RIVM)
NL-3720BA Bilthoven
The Netherlands

European Medicines and Feed Additives Regulation Are Not in Compliance With Environmental Legislation and Policy

I. Introduction

For the product categories of medicines, veterinary medicines, and feed additives, the environmental risk assessment (ERA) procedure at registration is currently under development. The purpose of this paper is twofold: first, it investigates what limitations environmental legislation sets to the use of medicinal products and how an environmental assessment within the registration process help achieve environmental quality standards; second, it investigates whether the registration process and assessment meet these expectations. As a case study, special attention is given to veterinary medicines. For a general overview of the knowledge, problems, and research concerning sources, fate, and effects of medicines in the environment, we refer the reader to Kümmerer (2001).

II. EU Environmental Legislation and the Relation to Product Registration

The European Commission (EC) has issued several directives on the protection of the environment. The EC, national authorities, and multilateral commissions (e.g. International Commission for the Protection of the Rhine) are the competent authorities that ought to enforce a program in order to reduce existing pollution and set specific quality and emission standards in binding law, according to the 76/464/EEC, 80/68/EEC, 2000/60/EC, and 98/83/EEC directives on water, groundwater, and drinking water, respectively (Van Rijswick, 2001; Wösten *et al.*, 2001). Specific substances of concern have been identified (also known as List I and II chemicals). There is no European legislation on soil quality; however, because sediment and river banks are considered part of the water system and soil contains groundwater, quality of sediment and soil can also be considered an objective of environmental policy.

Within the Framework Directive on Water (2000/60/EEC), all acts of discharging or spreading of waste material, and polluting and deleterious substances, that might lead to contamination of surface water and groundwater are forbidden unless the relevant authority has granted authorization (a permit). The permit specifies the receiving water body, the discharged substance(s), and the measures to be taken to prevent further pollution, for the legal person (e.g. a farmer) on a case-by-case basis. The use of a product (pesticide, fertilizer) or the use of a treated product (impregnated piling, treated cattle) are such acts that have to be authorized by the appropriate authority (Ministry of Public Works, provinces, municipalities). These authorities are facing a huge administrative burden that would be relieved if all individual permits could be replaced by one single authorization. Although registration of a product cannot be regarded as a permit nor as such a legal authorization, registration can be a useful initial measure because it reduces the need for regulating emission; general conditions and restrictions have already been identified and certain substances or uses will not be allowed. The authority can now focus on site-specific circumstances.

It is possible that environmental directives on the quality of water already contain qualitative standards for substances used in medicines and feed additives, even though the products groups "medicines" and "feed additives" are not named in the environmental directives. The use of the terms *pesticide* and *biocide* in these directives do not refer to the product categories, but to the nature of the substances reaching environmental compartments after production, use, or disposal of products. If an active substance in a medicine or feed additive should be denoted pesticidal or biocidal because of its properties, the standards in these directives do apply. Are medicines to be categorized as pesticidal? The Dutch government published a document in 1989 on quality criteria for substances in soil and groundwater (TK, 1989). It was considered necessary to specialize the criteria for pesticides and biocides "that constitute a special group of environmental hazardous substances: they are developed to repel organisms, modify the growth and development of organisms, or kill organisms, and are by definition biologically active. Also by their use they discern themselves, because these substances—especially the agricultural applications—are brought into the environment directly and cannot be regained." These criteria (repel, modify, kill, biologically active, direct introduction, not regain) apply to many medicinal products as well. Also, the Netherlands Health Council advised the Ministers to treat medicines in a way comparable to pesticides and biocides because they are pharmacologically active, are spread continuously, and little is known about their effects (Health Council, 2001).

The quality of drinking water is protected under the Directive 98/83/EC. This directive aims at protecting public health by setting quality objectives to drinking water. Within the Netherlands' environmental policy, it has been the rule since 1989 that with respect to xenobiotics, groundwater should also comply with the standards for drinking water, as it often concerns soluble compounds that cannot be sufficiently removed using common purification techniques (TK, 1989). This point of view is reflected in the directives on pesticides and biocides (91/414/EEC) and (98/8/EC) where the allowable concentration in groundwater (irrespective of use as drinking water) is 0.1 μg/L.

Based on this reasoning, competent authorities have to set water quality standards to medicinal substances and feed additives that can be assigned to List I and II of the water directives. Also, they have to develop action plans to control the pollution; medicines are acknowledged as a specific group of substances in the Netherlands' 4th Water Action Program (NW4, 1998). Furthermore, to all substances that qualify as pesticidal, a qualitative standard is already available for drinking water and, at least in The Netherlands, also for groundwater (0.1 μg/L). And last but not least, the ERA should assess the risk to both surface water and groundwater for every use in order to fulfill its role as a tool in environmental protection.

The registration process should thus meet the following demands in order to be an effective tool for environmental policy:

1. An ERA is performed at every registration or renewal in order to take new data or methodologies into account.
2. Decision criteria are in compliance with the environmental directives.
3. Principles and practical procedures for the assessment at registration are operational.
4. The methodology for the setting of environmental quality criteria should be harmonized with the ERA methodology for products.

III. Product Registration and the Relation with EU Environmental Policy and Laws

The EC unfolded its vision on chemicals in Europe in the White Paper (EC, 2001c). The European Union chemicals policy must ensure a high level of protection of human health and the environment as enshrined in the Treaty both for the present generation and future generations, while also ensuring the efficient functioning of the internal market and the competitiveness of the chemical industry. Fundamental to achieving these objectives is the Precautionary Principle. Whenever reliable scientific evidence is available that a substance may have an adverse impact on human health and the environment, but there is still scientific uncertainty about the precise nature or the magnitude of the potential damage, decision-making must be based on precaution in order to prevent damage to human health and the environment. Another important objective is to encourage the substitution of dangerous substances where suitable alternatives are available. The White paper puts particular focus on substances that are carcinogenic, mutagenic, or toxic to reproduction, and on substances that are PBT (persistent, bioaccumulative, and toxic) or that otherwise give rise to high concern. The EC White Paper on existing substances is to be regarded as the underlying principle for the regulation of substances. To underline the importance, we point out that the European Council already incorporated the principles of the White paper in their reaction to the evaluation of the pesticide directive 91/414: "The Council calls on the Committee to develop a new pesticides policy in line with the relevant aspects of the forthcoming EU Chemicals Policy based on the principles endorsed by the Council Conclusion in June 2001.... In principle, these (PBT) substances should be avoided in plant protection products" (EC, 2001a).

Although the regulation of existing substances does not apply to products that are regulated in other frameworks, the same principles should (eventually) be applied in these frameworks. An ERA should be performed in order to determine the likelihood of effects in the environment. This implies that at product registration, environmental data should be available (at minimum, a risk classification should be made) and, depending on the mode and scale of use, some substances are not wanted. In order to accomplish this in an effective manner, it should be clear what the protection goals (criteria) of the assessment should be; what is acceptable (standards or levels) and what is not; and how the assessment should be performed (methodology).

A regulatory problem arises when a product registration procedure is harmonized at a European level by the Communautarian authority, while the authorities at the national level are responsible for maintaining the desired environmental quality. This may lead to a less effective implementation of the ERA as a tool for environmental policy. The product registration process also determines the availability of a product on the common market; therefore, the registration process should meet the following additional demands:

1. In order to perform an ERA at registration for the common market, common (harmonized) environmental protection goals are required. European environmental legislation provides a common basis for environmental goals for all products.
2. The ERA should be developed under the supervision of competent authorities with respect to environmental quality (for example, through national interdepartmental steering groups that prepare the national points of view).
3. Implementation of the ERA procedure is an act that will have legal consequences for stakeholders (producers, users, and third parties). Formalization of the contents and the procedure should be transparent and open to input by regulators, scientists, industry, and other interested parties; this is a view shared by the EC (EC, 2001b).

IV. Product Directives on Medicines and the Environmental Assessment

Medicines and feed additives can reach the environment at production, at use, after use (excretion), and as waste material. Only emission at production is outside the scope of the registration and is not dealt with here. Given the elaborate risk assessment schemes and methodologies formalized in a regulation on existing substances (793/93), and directives for new substances (93/67/EEC), plant protection products (91/414/EEC) and biocides (98/8/EC), where both the organization (competent authorities, technical meetings, and working groups) and the deliverables (uniform principles, dossier requirements, guidance for decision making and listing, guidance on risk assessment, guidance on models, guidance on preparing a monograph) are comparable and have been tested in practice, one would expect a similar system for medicines.

The Directive on human medicines (65/65/EEC) recognizes that an application for the marketing authorization for a medicinal product for human use must be accompanied, if applicable, by reasons for any precautionary and safety measures to be taken for the storage of the medicinal product, its administration to patients, and for the disposal of waste products, together with an indication of any potential risks represented by the medicinal product for the environment.

The directive on veterinary medicines (2001/82/EC), that replaces the 81/851 and 81/852 directives, states that with a request for registration of a veterinary medicinal product, information is to be provided to enable an assessment of the safety for the environment. Both administration and excretion of the products, and the disposal of unused material or waste, should be assessed. The assessor is free to determine what information should be delivered.

The directive on feed additives (2001/79/EC) considers that the existing regulations on feedstuffs should be supplemented by the establishment of criteria for the assessment of the risk of the additive having an adverse effect on the environment. In contrast to the medicines, the decision-making criteria for feedstuff are fastened down in law, but hardly any methodology is provided.

The directives of the three product groups all require an environmental assessment at registration. The quality of the assessment depends not only on the information in the directives. The directives may have been elaborated upon in national law and guidance documents, and the availability of operational procedures, assessment tools, and expertise of the evaluating and decision-making authorities are of importance. A further investigation into the process, the actors, and the deliverables is made. The veterinary medicinal products are explored in most detail as a case study.

As discussed above in relation to environmental policy making, the registration process (with respect to the environmental assessment) should have the following five features:

1. Formalization of the contents and the procedure should be transparent and open to input by regulators, scientists, industry, and other interested parties.
2. European environmental legislation provides a common basis for environmental goals.
3. Standards and harmonized methodology are (made) available.
4. An ERA is performed at every (re)registration.
5. Principles and practical procedures are operational.

Therefore, not only the details of the scheme and the methodology are of importance, but also the organization (who), implementation (what), and operationalization (how) of the risk assessment procedure are to be considered.

V. The Development of the ERA for Veterinary Medicinal Products in Europe

Figure 1 depicts the organization of the registration process of Veterinary Medicinal Products (VMPs), in which administrative, scientific, and regulatory responsibilities are separated. The Directorate-General (DG) Enterprise is responsible for the European legislation on the registration of VMPs. The registration process is mandated to the European Agency for the

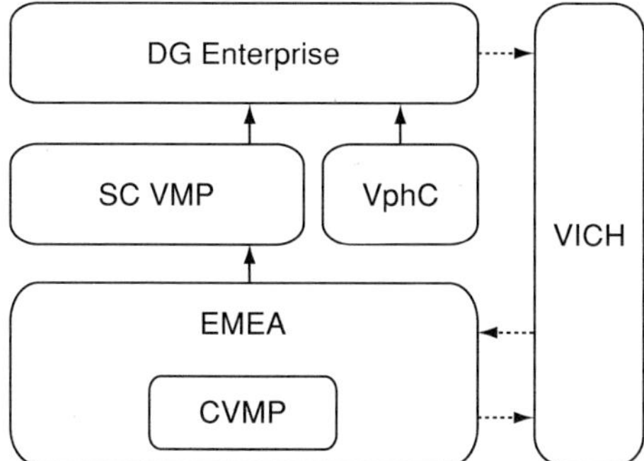

FIGURE 1 The relations between the EC, EMEA, CVMP, VICH, and SCVMP. Dotted lines, information and advice; straight lines, proposals and decisions on Marketing Authorization for products.

Evaluation of Medicinal Products (EMEA).[1] Within EMEA, the scientific committees advise on the requests for Marketing Authorization with respect to quality, efficacy, and safety of the products. These committees are the Committee for Veterinary Medicinal Products (CVMP) for veterinary medicines and the Committee for Proprietary Medicinal Products (CPMP) for human medicines. The Standing Committee on VMPs, as part of the DG Enterprise, decides on the proposal of the CVMP and turns it into binding law. DG Enterprise has also the Veterinary Pharmaceutical Committee (VPhC) at its disposal, which was installed by Directive 75/320, for advice on interpretation of the directives, compulsory consultation when changing directives, and other issues. Member states are involved in the registration process through their representations in the SC VMP and VPhC. Member states also appoint two independent experts to the CVMP.

The EMEA was mandated by DG Enterprise to elaborate on the old 81/851 and 81/852 directives (Blasius and Cranz, 1998). This has resulted in guidance documents for performing the environmental risk assessment of veterinary medicines (EMEA, 1996, 1997), but was not the end of the process. After the final draft of the EMEA (1997) guidance, an international harmonization between the EU, US, and Japan was started by the International Cooperation on Harmonization of Technical Requirements for Registration of Veterinary Medicinal Products (VICH),[2] to which both DG Enterprise and the EMEA are committed (DG Enterprise, 2000). The guidance document on Phase I was completed and finalized (15 June 2000) for implementation by July 2001 in the European Union and United States (VICH, 2000) and replaces the EMEA 1997 guidance on Phase I. This

[1]Commonly referred to as the European Medicines Evaluation Agency.
[2]Commonly referred to as the Veterinary International Conference on Harmonisation.

guidance document is at this moment leading for the registration procedure. To understand the contents of this guidance, one has to understand the organization that created it. The VICH Steering Committee (VICH SC) authorized formation of a working group to develop harmonized guidelines for conducting environmental impact assessments (EIA) for veterinary medicinal products. The mandate of this VICH Ecotoxicity/Environmental Impact Assessment Working Group (VICH Ecotox WG), as set forth by the VICH SC, is as follows:

> To elaborate tripartite guidelines on the design of studies and the evaluation of the environmental impact assessment of veterinary medicinal products. It is suggested to follow a tiered approach based on the principle of risk analysis. Categories of products to be covered by the different tiers of the guideline should be specified. Existing or draft guidelines in the US, the EU and Japan should be taken into account.

The VICH working group consists of three representatives of industry and three of the regulatory authorities, one from each continent. The VICH Ecotox WG elected to develop harmonized guidance in two phases (Phase I and Phase II). Phase I identifies VMPs that require a more extensive investigation of their potential to have environmental effects on non target organisms. The VICH SC recommended to the WG that Phase II should include a list of studies needed for VMPs that enter Phase II, and that decision-making or interpretative criteria should be included in Phase II. The working group was advised not to incorporate risk management options into Phase II.

The working group had to deliver harmonized guidance using limited resources (six experts) defending primarily industrial and governmental interests, secondarily environmental and regulatory interests. The interests of the three industrial representations are very close: maximum result (registration) for minimum costs (in terms of both dossier requirements and consumer image). The interests of the three governmental representations were however more different; apart from opening the markets and removing trade barriers, the existing registration procedures should not be compromised too much, the availability of products should not be hampered, and the environment in the three continents (with different legislative criteria) should be protected. The product of such a setting is likely to be clear on intentions, but not on details and procedures.

The draft guidance documents published by the VICH are circulated for consultation to members of industry, the CVMP, and the DG Enterprise. These interested parties have the opportunity to provide comments to their respective representatives. The EMEA working party member has to deal with (conflicting) comments from experts from all member states (represented in the CVMP), and from various experts from governmental science laboratories. Not until October 2001 (after the Phase was approved) was a working group on ecotoxicology established under the CVMP, to advise on matters related to preparation of guidelines on environmental risk assessment, in particular providing comments on the VICH phase II guidelines, which are in preparation. This group was also to provide advise on issues not covered by existing guidelines (such as developing agreed exposure calculation models) and to provide further advice at the request of the CVMP on other issues related to environmental risk assessment of VMPs (in particular for facilitating a harmonized implementation of guidelines). After adoption by the Steering Group of the VICH, the guidance is published by the EMEA and distributed to the member states and national agencies for the authorization of VMPs. Although the EC is informed on the progress of the work and contents of the guidance, it is not involved in regulatory approval of the guidance. Therefore the guidance has no legal status, even though it is an elaboration of the directive and it influences the registration process to a large extent.

The policy making process on the ERA for VMPs has the following characteristics:

- Consensual approach between industry and registration agencies;
- No participation of other interested parties;
- Little or no involvement of policy makers from DG Enterprise or DG Sanco, a concern expressed by the European Parliament (EC, 2001b);
- The guidance and interpretative criteria have no legal status;
- Until recently no scientific opinion was formed that represented the CVMP point of view;
- There was no formal exchange between the VICH and EMEA working groups and steering groups, or technical meetings for biocides or pesticides, even though these product groups share active substances and emission routes to the environment.

Thus, the first two of five features for a proper assessment are not present. The formalization of the contents and the procedure is not public or open to input by scientists and other interested parties such as competent authorities (with respect to environmental quality); the formalization has no legal status; and European legislation cannot provide common protection goals.

VI. Contents of the ERA for Veterinary Medicines

A proper assessment at this point should involve environmental goals based on European environmental legislation, and should contain standards and harmonized risk assessment methodology. The ERA for medicines and feed additives consists of two phases. In Phase I, products are assessed on their intrinsic hazard and their level of exposure (Fig. 2).

If the product fails to meet the triggers or exemptions, Phase II testing is needed. This step is not rigid, as some VMPs that might otherwise stop in Phase I may require additional environmental information to address particular concerns associated with their activity and use (VICH, 2000).

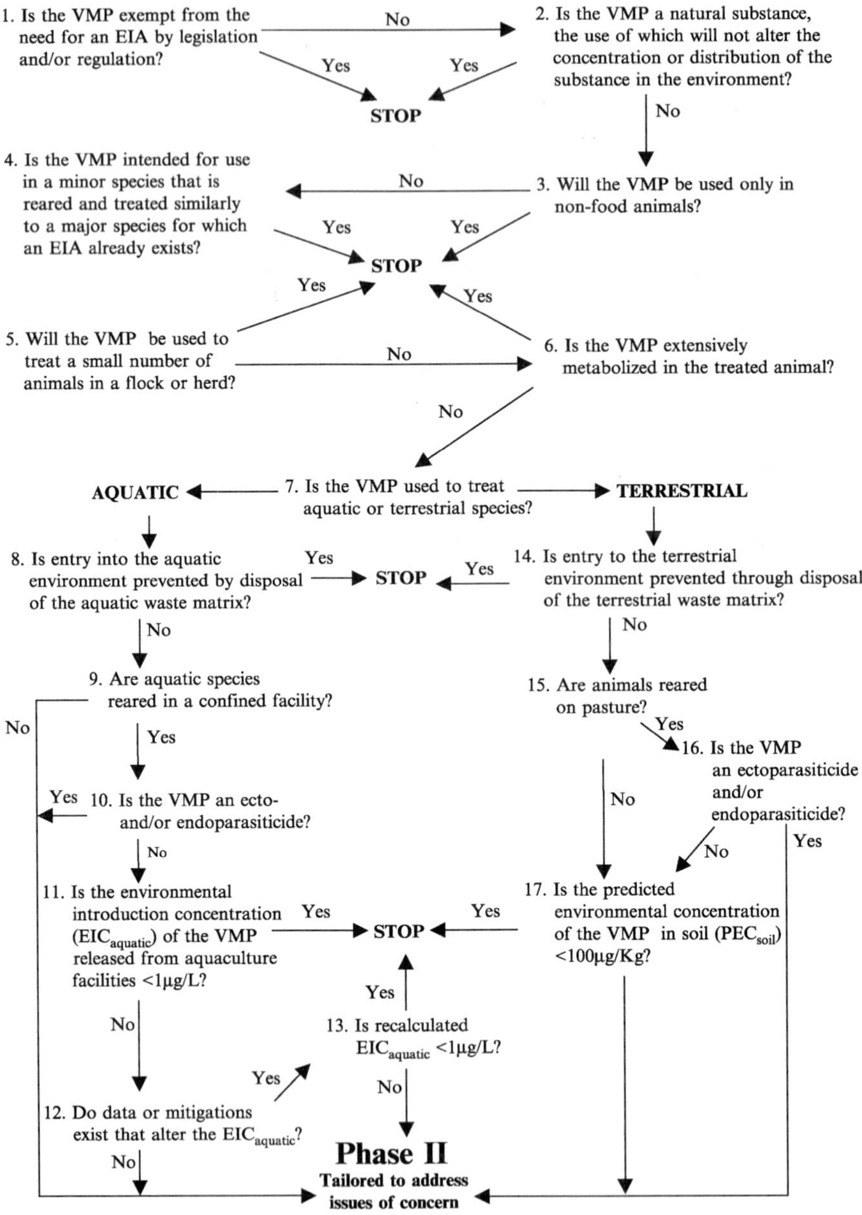

FIGURE 2 Phase I decision tree.

In Figure 2, question 8 highlights a crucial component of the decision scheme underlying the decisions also made in questions number 3, 5, 7, and 12: what are the emission routes at and after administration, and can one be certain that the emission is absent or insignificant? The burden of proof is on the applicant, and the decision is made by the assessor. In the guidance, emission is limited to four routes: no emission, emission to surface water, emission to soil, and direct emission into the environment (pasture animals). Emission to water and groundwater via soil and direct emission at application are not considered. Emission in the waste stage of the product is not included in the guidance, although is should be part of the safety assessment as required by Directive 2001/82; notably, there is no EU policy on the quality of soil (which is assessed), but there is legislation on the quality of groundwater (which is not assessed).

A total residue approach on the active substance is adopted, relieving the applicant from performing degradation route studies (animal, manure, soil, and water). Although this is a worst-case approach concerning the effect exerted by the parent compound, the fate of different metabolic fractions is not considered.

The exposure level considered irrelevant is quantified both for water and soil for antibiotics. For certain compounds (non-parasiticides), trigger values for exposure are introduced of 1 μg/L in water and 100 μg/kg in soil. These triggers are substantiated with an assessment of a dataset of toxicity values of several antibiotics, assuming:

- The data set (substances) is representative for all substances with the same pharmacological mode of action.
- The data set (endpoints) is representative for the aquatic respectively the terrestrial environment.
- Safety factors on the lowest experimental effect value are redundant due to the availability of substances in the presence of soil (due to sorption); the functional redundancy of microbes in soil; or the mitigating influence of degradation.

The approach as such (determining the level of toxicity that has a very small likelihood of being present in a product) and thus the level of exposure that can be considered an acceptable risk considering all products, is in fact equivalent to the practice in the exposure assessment where the "insignificant" emission routes are not considered in view of all other emission routes. However, the assessment to determine a safe level is found to be of poor quality from an ecotoxicological point of view (De Knecht and Montforts, 2001). The soil trigger for feed additives is, on the contrary, 10 μg/kg in the 2001/79 directive.

The VICH has not yet agreed upon a Phase II guidance. The European registration process has to rely on the available guidance on Phase II as published (EMEA, 1997) that consists of the hazard quotient approach, where predicted exposure and effect are combined. Breaching of risk ratio triggers (PEC/EC_{50}) leads to refinement of the assessment and inclusion of field studies. The guidance is however ambiguous on the decision schemes:

1. What compartments have to be assessed. Surface water and groundwater are not assessed directly in Phase I. Do they have to be assessed in Phase II?
2. What data are compulsory? What exposure models and effect models must be used, and how should field studies be designed or interpreted?
3. How should persistency and accumulation be expressed in exposure modelling and effect assessment?

Methodological problems resulting from the lack of guidance have been discussed where (Halling-Sorensen *et al.*, 2001; Montforts, 2001; Montforts *et al.*, 1999). The available guidance is ambiguous on the ERA. There are no clear data requirements, testing protocols are lacking, and decision making criteria are not clearly defined. Also, there are no clear standards expressing acceptability, the methodology has not been elaborated to a satisfactory extent, justification of the trigger values is scientifically unsound, and the ERA does not cover all Communautarian environmental legislations (i.e. groundwater). The contents of the VICH Phase I and the EMEA Phase II guidance do not bear the required characteristics of a proper assessment. It does not contain all Communautarian environmental quality criteria, clear acceptability standards, or harmonized methodology.

VII. Implementation of the ERA

A proper assessment at this point should be performed at every registration to be effective. Because science is developing as well, the ERA should be reconsidered at every renewal. The authority should apply environmental expertise at assessment and at decision-making, have methodology and criteria to its disposal, know how to weigh different interests, have expertise in the realism of risk mitigation measures, and should be able to deal with gaps in knowledge. Let us see if these requirements are met.

In Article 13 of directive 2001/82, exemptions are made to the dossier requirements and the extent of the safety assessment, as a result of which all member states (except the Netherlands and the UK) do not assess existing substances on environmental safety (De Knecht *et al.*, 2001). If this interpretation is juridical correct, the registration process cannot function as a tool in environmental policy and is not in accordance with the White Paper intentions.

In order to prepare and make sound decisions on the safety of a product, ecotoxicological expertise is required at preparation and at decision making. Neither the members of the Netherlands Committee and Working Group on the Authorization of VMPs, nor of the CVMP, reflect the fact that ecotoxicology is a safety aspect in its own right. In several member states, the ERA is not performed by qualified environmental chemists or ecotoxicologists, but by staff with a veterinary background (De Knecht *et al.*, 2001), although dossier evaluation is a sensitive step in the registration process (Boesten, 2000; Mensink *et al.*, 1995; Pontolillo and Eganhouse, 2001; Tiktak, 2000). The authorities do not have, as discussed above, clear methodology and criteria to their disposal.

The authorities have little experience with risk mitigation measures and the available guidance does not, as advised by the VICH, deal with this matter. Risk mitigation measures at registration usually target the emission of the product to the environment. Measures that target the necessity or redundancy of the product are not expressed in the risk assessment and are also outside the scope of the registration (which considers the use, and should also consider the disposal, of the product.). Product labelling intended to reduce risk can only influence the use and disposal of the (prepared) product and the treated animal, but not the use and disposal of the contaminated manure and slurry. Two examples are found in the literature (Greiner and Rönnefahrt, 2001). The first example is on the restriction on spreading of manure from animals treated with the product. A label to keep treated animals stabled is enforcable, but a label to spread the contaminated slurry some distance from the ditch is not. At the moment the veterinary practitioner or farmer uses the product, he cannot foresee the eventual spreading of the manure. The inspector can check the quality of the slurry, but not whether the medicinal product had been applied according to the label or not. The second example is the registration of a product containing alkylphenols. It is EU policy to abandon the use of these compounds (Footitt *et al.*, 1999), but this cannot be considered a risk mitigation measure at registration. Not only does the product *in casu* contain this substance, but also the use in medicines is exempted from the general risk reduction strategy and the safety of the products with these compounds should be assessed at registration.

The authorities and have not made explicit how to weigh different interests nor how to deal with gaps in knowledge. The guidance on human medicines states "since for medicinal products the benefit for humans has relative precedence over any environmental risks, the environmental risk management procedures adopted for industrial chemicals and pesticides (i.e. prohibiting or restricting their use if an unacceptable risk to the environment is evident) is neither possible nor desirable in this case. Precautionary measures through product labelling are therefore the recommended risk management procedures for medicinal products, when concerns for the environment are present." This guidance indicates that environmental risk is at most a reason to suggest risk mitigation measures and undermines the legitimacy of the ERA. Why impose a burden on the producers that will not discriminate the products? The guidances for veterinary medicines and feed additives do contain decision-making or interpretative criteria, which only lead to requests for further assessment. No information is provided on weighing of risks vs benefits, on provisional approval given an expected (low) level of risk. As there are not strict dossier requirements, it is not clear how lack of information should be included in the decision making.

The ERA at registration does not bear the last features for a proper assessment. Assessments are not made for all products, and the decision-making principles and practical procedures are not operational. It is therefore unlikely that any result of an ERA can be taken into consideration.

VIII. Discussion and Conclusions

It has been argued that environmental protection in itself is not an issue to be dealt with at EU level. There are considerable problems that cross borders (groundwater depletion) or arise on a specific location due to actions of several states, but the level of quality desired on each location and the specific member states involved in each case are not uniform. Environmental legislation should thus be a case for member states only, or multilateral negotiation (Golub, 1996). To illustrate this line of reasoning, the EC has issued directives on water and groundwater (transnational relations), but not on soil. The framework directive on water (2000/60/EC) is a fine example of the awareness of the EC of both the subsidiarity principle (regulate at the lowest appropriate level) and the complexity of the existing regulations on water quality. The framework directive formulates common objectives, leaving a great deal of decision making to the member states, and provides for a coherent approach of water management, recognizing the relationships with other policy areas, such as environment, nature, spatial planning, agriculture, and product policy. It is argued here that the subsidiarity issue on environmental regulations is not only defined by scale of the revelation of the effects, but also by the mechanics that lead to the effects, including the working of the common market for products. This is why environmental legislation serves a purpose in European registration procedures for products.

A regulatory problem arises when a product registration procedure is harmonized at a European level by the central registration authority, while the authorities at the national level are responsible for maintaining the desired environmental quality. This may lead to a less effective implementation of the ERA as a tool for environmental policy. This problem can be tackled in two ways: (1) the ERA should be based on common principles based on EU regulations and policy that steer the national authorities; or (2) the ERA should be developed under the supervision of competent authorities (for example, through national interdepartmental steering groups that prepare the national points of view).

Both options are not reflected in the forging of the ERA for medicines and feed additives. The formalization of the contents and the procedure is not transparent nor open to input by scientists and other interested parties. The formalization has no legal status, and European legislation cannot provide common protection goals in a global setting. The VICH Phase I and the EMEA Phase II guidance do not contain all Communautarian environmental quality criteria, nor clear acceptability standards, nor harmonized methodology. The scientific validity of the registration procedure is compromised (Heyvaert, 1999). Assessments are not made for all products, and the decision-making principles and practical procedures are not operational. It is therefore unlikely that any result of an ERA can be taken into consideration at registration, which undermines the legitimacy of the process.

What is the ultimate effect of these developments? Assessors at the registration agencies do not know if and how to perform or conclude an ERA. Applicants do not know what effort the authorization process will place upon them, which makes it difficult to take management decisions on the development of new products, or the renewal of old products. It is not the ERA as such, but the lack of clarity in procedure and requirements, that may ultimately compromise product availability. Products that pose a threat to environmental quality at or after use or disposal may now be registered, forcing the authorities responsible for water and land quality to regulate and enforce product use and slurry use on a casebycase basis.

The major efforts recently made by the regulators and scientists within EMEA and VICH need to be founded on clear policy decisions and embedded in a uniform and public decision-making procedure. It should take little effort to postulate Communautarian decision-making criteria together with their levels of acceptability. These will provide a solid basis for the implementation of the existing risk assessment methodologies, and subsequently help to clarify the (compulsory) data requirements and (realistic) risk mitigation measures. These five elements (criteria, standards, methodology, data requirements, and mitigation measures) will then provide a reference for deciding on the environmental acceptability, both for the producers and for the decision-makers.

Acknowledgments

This publication was funded by the EU research project ERAVMIS, EVKI-CT9-00003.

References

2000/60/EC. Directive 2000/60/EC of the European Parliament and of the Council of 23 October 2000 establishing a framework for Community action in the field of water policy.

2001/79/EC. Commission Directive 2001/79/EC of 17 September 2001 amending Council Directive 87/153/EEC fixing guidelines for the assessment of additives in animal nutrition (Text with EEA relevance.).

2001/82/EC. Directive 2001/82/EC of the European Parliament and of the Council of 6 November 2001 on the Community code relating to veterinary medicinal products.

65/65/EEC. Council Directive 65/65/EEC of 26 January 1965 on the approximation of provisions laid down by Law, Regulation or Administrative Action relating to proprietary medicinal products.

76/464/EEC. Council Directive 76/464/EEC of 4 May 1976 on pollution caused by certain dangerous substances discharged into the aquatic environment of the Community.

793/93. Council Regulation (EEC) No 793/93 of 23 March 1993 on the evaluation and control of the risks of existing substances.

80/68/EEC. Council Directive 80/68/EEC of 17 December 1979 on the protection of groundwater against pollution caused by certain dangerous substances.

91/414/EEC. Council Directive 91/414/EEC of 15 July 1991 concerning the placing of plant protection products on the market.

93/67/EEC. Commission Directive 93/67/EEC of 20 July 1993 laying down the principles for assessment of risks to man and the environment of substances notified in accordance with Council Directive 67/548/EEC.

98/8/EC. Directive 98/8/EC of the European Parliament and of the Council of 16 February 1998 concerning the placing of biocidal products on the market.

98/83/EEC. Council Directive 98/83/EC of 3 November 1998 on the quality of water intended for human consumption.

Blasius, H., and Cranz, H. (1998). Arzneimittel und Recht in Europa. WVG, Stuttgart.

Boesten, J. J. T. I. (2000). Modeller subjectivity in estimating pesticide parameters for leaching model using the same laboratory data set. *Ag. Wat. Manag.* **44,** 389–409.

De Knecht, J. A., and Montforts, M. H. M. M. (2001). Environmental risk assessment of veterinary medicine products: An evaluation of the registration procedure. *SETAC Globe* **2,** 29–31.

De Knecht, J. A., Van Vlaardingen, P. L. A., Montforts, M. H. M. M., and Linders, J. B. H. J. (2001). Report on the European Workshop on the ERA of VMPs, 25–26 January 2001. RIVM Bilthoven.

DG Enterprise (2000). Pharmaceuticals in the European Union. Office for Official Publications of the European Communities. Luxembourg ISBN 92-828-9589-0.

EC (2001a). Draft Council conclusions on the report of the Commission on the evaluation of active substances of plant protection products. Council of the European Union 15046/01.

EC (2001b). Question E-3411/00 of Erik Meijer (GUE/NGL) to the Commission. C163E/111.

EC (2001c). WHITE PAPER Strategy for a future Chemicals Policy. COM(2001) 88 final.

EMEA (1996). Note for Guidance: Environmental Risk Assessment for Immunological Veterinary Medicinal Products. EMEA/CVMP/074/95.

EMEA (1997). Note for Guidance: Environmental Risk Assessment for Veterinary Medicinal Products Other Than GMO-Containing and Immunological Products. EMEA/CVMP/055/96.

Footitt, A., Virani, S., Corden, C., and Graham, S. (1999). Nonylphenol Risk Reduction Strategy. Prepared for Department of the Environment, Transport and the Regions by Risk & Policy Analysts Limited, UK. Project: Ref/Title J227D/Nonylphenols.

Golub, J. (1996). Sovereignty and subsidiarity in EU environmental policy. *Political Studies* **XLIV,** 686–703.

Greiner, P., and Rönnefahrt, I. (2001). Environmental risk assessment of veterinary medicinal products from a regulatory perspective. Presentation 412. SETAC Europe 11th Annual meeting. May 6–10, 2001.

Halling-Sørensen, B., Jensen, J., Tjørnelund, J., and Montforts, M. H. M. M. (2001). Worst-case estimations of predicted environmental soil concentrations (PEC) of selected veterinary antibiotics and residues in Danish agriculture. *In* "Pharmaceuticals in the Environment" (K. Kümmerer, Ed.). Springer Verlag, Berlin.

Health Council (2001). Milieurisico's Van Diergeneesmiddelen. Signalement 2001/17. The Hague. The Netherlands.

Heyvaert, V. (1999). The changing role of science in Environmental regulatory decision making in the European union. *Law Euro. Affairs* **9**, 426–443.

Kümmerer, K. (2001). *Pharmaceuticals in the Environment.* Springer Verlag, Berlin.

Mensink, B. J. W. G., Montforts, M., Wijkhuizen-Maslankiewicz, L., Tibosch, H., and Linders, J. (1995). Manual for summarising and evaluating the environmental aspects of pesticides. RIVM, Report 679101022, Bilthoven, The Netherlands.

Montforts, M. H. M. M. (2001). Regulatory and methodological aspects concerning the risk assessment of medicinal products. *In* "Pharmaceuticals in the Environment" (K. Kümmerer, Ed.). Springer Verlag, Berlin.

Montforts, M. H. M. M., Kalf, D. F., Van Vlaardingen, P. L. A., and Linders, J. B. H. J. (1999). The exposure assessment for veterinary medicinal products. *Sci. Tot. Env.* **225**, 119–133.

NW4 (1998). Fourth Action Plan Watermanagement. Ministry of Traffic and Public Works. The Netherlands.

Pontolillo, J., and Eganhouse, R. P. (2001). The search for reliable aqueous solubility (S_w) and octanol-water partition coefficient (K_{ow}) data for hydrophobic organic compounds: DDT and DDE as a case study. Department of the Interior, U.S. Geological Survey, Reston, Virginia.

Tiktak, A. (2000). Application of nine pesticide leaching models to the Vredepeel dataset. Pesticide fate. *Ag. Wat. Manag.* **44**, 119–134.

TK (1989). Milieucriteria ten aanzien van stoffen ter bescherming van bodem en grondwater, TK21021. Proceedings of the Dutch parliament 21021.

Van Rijswick, H. F. M. W. (2001). De Kwaliteit Van Water (The Quality of Water). Utrecht University, The Netherlands.

VICH (2000). Environmental Impact Assessment (EIAs) for Veterinary Medicinal Products (VMPs) – Phase I. CVMP/VICH/592/98-final.

Wösten, M. A. D., Blok, J., and Van de Plassche, E. (2001). International Environmental Quality Standard Setting. RIVM/Royal Haskoning, Nijmegen. H1757.AO/R0007/EVDP/TL.

Carol Long* and Mark Crane[†]

*Veterinary Medicines Directorate
New Haw, Addlestone
Surrey, KT15 3LS, United Kingdom

[†]Crane Consultants
Faringdon
Oxfordshire, SN7 7AG, United Kingdom

Environmental Risk Assessment of Veterinary Pharmaceuticals in the EU: Reply to Montforts and de Knecht

I. Introduction

This chapter has been prepared in response to the previous chapter by Montforts and de Knecht. Montforts and de Knecht criticize the current regulatory framework in the European Union (EU) for the environmental risk assessment of veterinary medicinal products (VMPs) and feed additives (FAs). Their major criticisms are:

1. The marketing authorization procedure for VMPs and FAs is inferior to that for existing substances, new substances, pesticides, or biocides, and legislation on VMPs should be subsidiary to other European environmental legislation.
2. The VICH Working Group on Ecotoxicity, currently responsible for harmonizing environmental impact assessment procedures in the EU, North

America, Japan, Australia, and New Zealand, and (since October 2002) Canada has insufficient breadth or depth of expertise.

3. Because of its composition, the VICH Working Group on Ecotoxicity primarily defends industrial and governmental interests, with environmental interests of only secondary importance.

4. The VICH process is not open to input from scientists and other interested parties and, in any case, has no legal status under European law.

5. Phase I of the VICH guidance is technically flawed for several reasons, including a lack of consideration of emissions of VMPs to groundwater, lack of consideration of direct emissions when applied to target species, and use of unsafe trigger values.

6. Member States of the European Union, with few exceptions, do not perform adequate assessments because of a lack of available expertise.

Most of these criticisms (1–4) center upon the legal status of VMP marketing authorizations, and the way in which the VICH process is largely unaccountable and may be biased against environmental concerns. We address these two main classes of criticism below, concentrating, like Montforts and de Knecht (2002), on VMPs. We do not address their criticism 5 above, partly because in the light of experience since implementation of the Phase I guidance we have some sympathy with it, and partly because we believe that there are clear mechanisms within the VICH process for addressing such concerns. We also do not address criticism 6, which if true would clearly undermine the implementation of any legislation, but is not a criticism of the legislation itself.

II. Marketing Authorization for VMPs in the European Union

VMPs were originally regulated in the EU by means of Directive 81/851/EEC and Directive 81/852/EEC. A product had to meet the criteria of Quality, Safety, and Efficacy, according the requirements set out in the annex to Directive 81/852/EEC before it could be placed on the market. The requirement for an assessment of ecotoxicity was introduced by means of Directive 92/18/EEC and thus extended the criteria of safety from operator and consumer safety to include environmental safety. These provisions have since been consolidated in Directive 2001/82/EC; the provisions of this Directive are currently under review.

The existence of such legislation means that veterinary pharmaceutical products are outside the scope of the chemicals legislation referred to by Montforts and de Knecht, which relates to notification of new and existing substances. This is equivalent to the situation for pesticides, which are regulated under Directive 91/414/EEC, and in each case is because the legislation relating to the marketing authorization for these types of chemicals includes an effective requirement for an environmental risk assessment. Other chemical substances legislation specifically excludes these types of product.

Veterinary pharmaceutical products are not exempted from the various water provisions cited by Montforts and de Knecht (80/778/EEC; 75/440/EEC). When a veterinary medicine can be considered to come within the scope of List I or List II, then these provisions will also apply. An example might be cypermethrin, which as a synthetic pyrethroid might also be considered to be an organo-halogen. Therefore, classification of a veterinary medicine as within the scope of List I or List II would be on the basis of chemical group, while there is not a classification within List I and II of veterinary medicine. Examples of the influence of other legislation on use of veterinary medicines can be given from the UK. Veterinary medicinal products authorized for use in fish farms are also subject to the requirement to have a valid consent to discharge before the product can be used on individual farms, and disposal of used sheep dip to land is subject to provisions under the UK Groundwater Regulations.

The veterinary legislation sets out the requirements for environmental risk assessments for veterinary medicinal products in the Annex to Directive 2001/82/EC. This requires an assessment of ecotoxicity, the purpose of which is to assess potential harmful effects and to identify precautionary measures. This assessment is compulsory (except for generic products) and comprises two phases. According to the legislation, Phase I determines the potential extent of exposure to the environment of the product, its active ingredients, or relevant metabolites. This takes into account the pattern of use of the product, the likely extent to which there is direct exposure of the environment, excretion of the product and its persistence in excreta, and disposal. The second phase takes account of the extent of exposure to the environment and existing data within the dossier to determine whether further tests are necessary. If Phase II is applicable, then a more complete environmental assessment is required, which involves submission of data on environmental fate and ecotoxicological effects.

The legislation set the framework for the guidance in that it specified a Phase I exposure assessment as a first step in the ecotoxicity assessment. The legislation effectively produced a mechanism for ruling out some products from the need to produce data on fate and effects in the environment. It also made clear that whole herd treatments would be expected to require a Phase II assessment, as would those veterinary medicines whose use resulted in direct entry to the environment.

Following implementation of the legislation, the European Commission formed a working group in 1994, with representatives from several Member States, to develop Phase I and Phase II guidance on environmental risk assessment for both veterinary and human pharmaceuticals in the European

Union. The guidance drafted at that time had more in common with the requirements for pesticides than those for chemicals, which reflected the experience prevalent within the group at that time. Industry was consulted on the draft guidance document that was produced. With the founding of the European Medicines Evaluation Agency (EMEA) in 1995, the production of Phase I and Phase II guidance on ecotoxicity assessments for veterinary medicinal products was transferred to the Committee for Veterinary Medicinal Products (CVMP). There was a further consultation phase, at the end of which the CVMP formed an Ad Hoc Group of Experts, many of whom had been involved in the earlier work, to deal with the comments. The final guidelines were released in January 1997 (EMEA, 1997).

The CVMP Phase I guidance set out the triggers for assessment in Phase II, mostly based on predicted environmental concentrations (PEC) in the environment. These were PEC_{soil} equal to or greater than 10 μg/kg; $PEC_{groundwater}$ greater than 0.1 μg/l; PEC_{dung} at pasture equal to or greater than 10 μg/kg; and products with direct entry into the aquatic environment.

Where exposure of soil or aquatic environments was identified, then the guidance set out basic data requirements that were similar to those for first tier assessment of pesticides. The guidance also set requirements for areas of assessment particular to veterinary medicines, such as effects on coprophagous insects in dung and data for fish medicines used in marine environments. Also similar to other areas of ecotoxicology is the tiered approach to environmental risk assessment, where if risk cannot be characterized on the basis of the initial dataset, then further requirements are identified in order to complete the risk assessment.

III. The VICH Process

During the final stages of production of the CVMP guidance, ecotoxicity and environmental impact assessment were identified as key topics for international harmonization efforts, under the umbrella of the International Cooperation on Harmonization of Technical Requirements for Registration of Veterinary Medicinal Products (VICH). VICH activities are carried out under the auspices of the Office International de Epizooties (OIE).

The purpose of VICH is to provide a forum for a constructive dialog between regulatory authorities and the veterinary medicinal products industry on both real and perceived differences in the technical requirements for product registration in the EU, Japan, and North America. Australia and New Zealand, and more recently Canada, also participate as observers. The VICH process identifies areas where modifications in technical requirements or greater mutual acceptance of research and development procedures could lead to a more economical use of human, animal, and material resources, without compromising safety. It makes recommendations on practical ways to achieve harmonization in technical requirements affecting registration of veterinary medicinal products. Once adopted, the VICH recommendations will replace corresponding regional requirements. In the EU, the VICH procedure is administered by the EMEA through the CVMP. Although the resulting documents are not incorporated into legislation, they do become CVMP guidance and therefore have the same status as any guidance document developed by the CVMP.

The process is organized according to a nine-step procedure:

Step 1: VICH Steering Committee defines a priority topic from a position paper produced by one of its members. It establishes a working group if necessary. A Topic Leader is appointed and given a clear mandate to do the expected work.

Step 2: The appropriate working group draws up a draft recommendation.

Step 3: The draft recommendation is submitted to VICH Steering Committee to approve its release for consultation.

Step 4: Once adopted by the VICH Steering Committee, the draft recommendation is circulated to all interested parties for consultation.

Step 5: The comments received are directed to the working group for consideration. At this step the Topic Leader must be a representative of a regulatory authority. The working group then prepares a revised draft.

Step 6: The revised draft recommendation is submitted to the VICH Steering Committee for approval.

Step 7: Once approved by the VICH Steering Committee, the final recommendation and a proposed date for its implementation are circulated to the regulatory authorities represented on the committee.

Step 8: The VICH Steering Committee members report to the committee on the implementation of the recommendations in their respective regions.

Step 9: Upon the request of a VICH Steering Committee member, a recommendation can be revised in order to take account of new scientific information.

The VICH Working Group on Ecotoxicity has membership drawn from both industry and government. Table I lists the veterinary pharmaceutical representatives that make up the working group and provides a brief summary of their expertise. It is only possible to provide a very brief summary in this table.

It is important to recognize that each representative does not work in isolation, but is a focal point for each of the interests represented and beyond. Each VICH Working Group member discusses and agrees on a position with their stakeholders and presents that position in discussion with the VICH Working Group. For instance, in the EU, the CVMP AHGERA (Committee on Veterinary Medicinal Products Ad Hoc Group

TABLE I Composition of VICH Working Group in 2002

Representation	*Qualifications & experience*
EU regulator	Biologist specializing in ecology with 14 years experience of ecotoxicology/environmental risk assessment of pesticides and veterinary drugs. Membership of VICH WG ceased from October 2002.
EU industry	Veterinary surgeon with 8 years experience of biostatistical and epidemiological work and 12 years experience of preclinical studies, including toxicology, metabolism, residues, and environmental issues.
Japanese regulator	Pharmacist and analytical chemist with experience in chemical assay, immunoassay, and bioassay. Has also researched antimicrobial resistance and ecotoxicology and has been involved in reviewing veterinary medicinal products.
Japanese industry	Qualified in aquaculture, specializing in fish pathology, with 21 years experience of product development in the veterinary pharmaceutical industry.
U.S. regulator	Biologist specializing in environmental sciences. Has 8 years experience of assessing the safety and efficacy of human drugs, and for the last 17 years has produced environmental risk assessments for veterinary drugs.
U.S. industry	Microbiologist with specialism in microbial ecology. Has 18 years experience of the veterinary pharmaceutical industry, working in drug development and discovery. Currently working in the areas of ecotoxicity assessments, pharmacodynamics, and pharmacokinetics.
Aus/NZ regulator + industry	Organic chemist with 18 years experience of ecotoxicology/environmental risk assessment of pesticides, industrial chemicals, and veterinary drugs. For 8 years, National Coordinator for OECD Test Guidelines Programme.
Canadian regulator	Cell biologist with 12 years experience of environmental fate and effects assessment of pesticides, currently responsible for environmental assessment of new drugs, including veterinary medicines.
Canadian industry	Joined the VICH WG in October 2002.

on Environmental Risk Assessment) was formed during discussion of the Phase II guidance to support the regulatory representative. The AHGERA consists of regulatory representatives from EU Member State agencies involved in the regulation of veterinary medicinal products. Prior to formation of the AHGERA in November 2001, consultation with ecotoxicologists in the various Member States was achieved through individual CVMP members. This was the procedure during the development of the VICH Phase I guidance and in the early stages of discussion of Phase II. However, it became clear that for the technical detail involved in Phase II, it would be beneficial for the EU experts to be able to discuss the issues in a working group. The individual AHGERA members do not work in isolation but consult with regulatory colleagues within their Member State when forming a view. The position of the AHGERA is endorsed by the CVMP before it can be put forward to the VICH Working Group. In addition to representatives from each Member State and the EMEA (European Medicines Evaluation Agency), the CVMP has a representative from the European Commission present at all meetings.

In addition to these safeguards, there is also a six-month period of open consultation on the draft VICH document, details of which are publicized on the VICH website. The CVMP is pro-active during the consultation process and, in addition to having a link from the EMEA website to the VICH website, circulates draft documents to all interested parties within Europe. These open consultation periods are required in the development of all VICH documents and occurred during development of the Phase I guidance on ecotoxicity.

Therefore, all reasonable attempts are made to engage in a process that is both representative of all the relevant interests and open to wider appraisal and input.

IV. VICH Phase I Guidance on Ecotoxicity

The approach to the VICH guidance was based on the EU model of a Phase I and a Phase II assessment. The development of the Phase I guidance was informed by the experience of the US in assessing veterinary medicinal products since the 1980s. Summaries of ecotoxicity data on veterinary medicinal active ingredients registered in the US were made available to the VICH Working Group and used to set the PEC_{soil} trigger value. As a result the PEC_{soil} trigger was revised upwards of equal to or greater than 100 μg/kg. A further difference is that the groundwater trigger is no longer in Phase I, as the 0.1 μg/l value in various EU legislation is considered not applicable to veterinary medicines; therefore, this issue will only be dealt with for products entering Phase II. If the soil trigger is exceeded in Phase I, this will lead to an assessment of exposure of surface water and groundwater in Phase II.

Endoparasiticides and ectoparasiticides are identified in VICH Phase I as of particular concern in relation to their use in animals at pasture and in fish medicines, and these products require Phase II assessment regardless of their predicted concentration in the environment. In addition to a Phase II assessment being required for fish medicines with direct entry to the aquatic environment, there is now a specific trigger for those fish medicines where the effluent from the fish farm is treated before reaching the aquatic environment. We recognize that there are genuine concerns about some of these technical details (Boxall *et al.*, 2002), but believe that the checks and balances described in the previous section provide ample opportunity for improved scientific understanding to be fed into the VICH process.

V. VICH Phase II Guidance

The VICH Phase II Guidance is still under discussion and development. There have been a number of drafts produced within the Working Group. Initially, there were separate documents for assessment of aquaculture products, veterinary medicines used in animals intensively reared, and for products used in animals reared at pasture. The separate authorship of these documents led to problems with consistency of format and approach, with an accompanying loss of transparency. The most recent meeting of the VICH Working Group has focused on bringing transparency to the merged document. The Working Group has worked to ensure that the data requirements and approach used are consistent with documents produced by the OECD, where these are available. There are still areas of assessment specific to veterinary medicines and these are addressed in the document.

VI. Conclusions

Montforts and de Knecht are fully entitled to question some of the technical details of VMP authorization, but it is wrong to lay the blame for any technical deficiencies at the door of legislation or the VICH process. All European environmental legislation of which we are aware has required clarification through subsequent guidance documents (Crane *et al.*, 2002), and there is by no means unanimity of opinion about how other chemical substances, such as pesticides, should be regulated in Europe (Crane, manuscript submit). Harmonization of VMP authorization across most of the developed world through the VICH process is an ambitious project, which carries the hope that animal welfare will be improved through increased availability of VMPs and a reduction in redundant product testing. The VICH process is as open as possible and requires a lengthy public consultation period before any procedures are agreed. Decisions that are taken are also open to scientific challenge and subsequent change. Like Montforts and de Knecht, we agree wholeheartedly with the FEDESA view that clarity of procedure and regulatory requirements are necessary if product availability is not to be compromised. We also believe that the VICH process is an appropriate vehicle for ensuring that European legislation on the environmental risk assessment of VMPs is both transparent and operational to the benefit of the environment, farm animals, industry and the general public.

References

2001/82/EC. Directive 2001/82/EC of the European Parliament and of the Council of 6 November 2001 on the Community code relating to veterinary medicinal products.

92/18/EEC. Commission Directive 92/18/EC of 20 March 1992 modifying the Annex to Council Directive 81/852/EEC on the approximation of the laws of Member States relating to analytical, pharmatoxicological and clinical standards and protocols in respect of the testing of veterinary medicinal products (O.J. No L 97 of 10.4.92).

81/852/EEC. Council Directive 81/852/EEC of 28 September 1981 on the approximation of the laws of the Member States relating to analytical, pharmatoxicological and clinical standards and protocols in respect of the testing of veterinary medicinal products (O.J. No L 317 of 6.11.81).

81/851/EEC. Council Directive 81/851/EEC of 28 September 1981 on the approximation of the laws of the Member States relating to veterinary medicinal products (O.J. No L 317 of 6.11.81).

80/778/EEC. Directive 80/778/EEC: Quality of water intended for human consumption 80/68/EEC. Directive 80/68/EEC: Protection of Groundwater against pollution caused by certain dangerous substances.

75/440/EEC. Directive 75/440/EEC: Quality of Surface Water Intended for the Abstraction of Drinking Water in the Member States.

Boxall, A. B. A., Fogg, L., Blackwell, P. A., Kay, P., and Pemberton, E. J. (2002). Review of Veterinary Medicines in the Environment. R&D Technical Report P6-012/8/TR. Environment Agency, Bristol, United Kingdom.

Crane, M., and Gidding, J. M. (2004). 'Ecologically Acceptable Concentrations' when assessing the environmental risks of pesticides under European Directive 91/414/EEC? *Human Ecol. Risk Assess.* **10,** 1–15.

Crane, M., Sorokin, N., Wheeler, J., Grosso, A., Whitehouse, P., and Morritt, D. (2002). European approaches to coastal and estuarine risk assessment. *In* " Coastal and Estuarine Risk Assessment" (M. Newman, Ed.), pp. 15–39. Lewis Publishers, Boca Raton, Florida.

EMEA (1997). Note for guidance: Environmental risk assessment for veterinary medicinal products other than GMO-containing and immunological products. EMEA/CVMP/055/96.

Montforts, M. H. M. M., and de Knecht, J. A. (2002). European medicines and feed additives regulation are not in compliance with environmental legislation and policy. *Toxicol. Lett.* **131,** 125–136.

VICH (2000). Environmental impact assessement (EIAs) for veterinary medicinal products (VMPs)—Phase I. CVMP/VICH/592/98-final.

Jürg Oliver Straub
EurProBiol CBiol MIBiol
Corporate Safety & Environmental Protection CSE
F. Hoffmann-La Roche Ltd
CH–4070 Basel, Switzerland

Environmental Risk Assessment for New Human Pharmaceuticals in the European Union According to the 2003 Draft Guideline*

I. Introduction

In 1965, the European Economic Community started regulating proprietary medicinal products through Directive 65/65/EEC (CEEC, 1965). In the course of amendments to this directive and with the concomitant rise in importance of environmental concerns, Directive 93/39/EEC introduced the necessity to give "indications of any potential risks presented by the medicinal product for the environment" (amendment to article 4.6; CEEC, 1993a). In the same year, EEC Council Regulation 2309/93/EEC (CEEC, 1993b) detailed procedures for the registration of medicines, including an environmental risk assessment (ERA), and the establishment of a centralized European Medicines Evaluation Agency (EMEA). A technical

*This contribution is adapted from an earlier presentation on the 2001 ERA draft (Straub, 2002); it has been substantially enlarged and revised based on the recent 2003 ERA draft.

guideline for the registration of new drugs, the so-called Notice to Applicants, was first issued in 1986. As of the 1995 version following 93/39/EEC, an ERA was requested as part of the registration dossier, known today as the Common Technical Document (current version of the relevant volume of the Notice to Applicants: CEC, 2001).

As of 1994, a European Union (EU) guideline for the ERA for human pharmaceuticals containing or consisting of genetically modified organisms (GMO) was published (CEC, 1998a). Various drafts for an ERA guideline for non-GMO medicinal products, both for human and veterinary use, circulated from the early to mid-1990s. The early drafts covered both human and veterinary pharmaceuticals, but over time a progressive disentanglement of the two fields took place. In 1997, an ERA Guideline for non-GMO, non-immunological veterinary medicinal products was published (EMEA, 1997). It separately covers medication for agricultural and aqua-/piscicultural pharmaceuticals with detailed flow charts, decision criteria, and concise lists of necessary test data for the respective tiers. With a view on international harmonization with the United States and Japan, the veterinary ERA guidelines of the three political entities were streamlined in 2000 (EMEA, 2000). Meanwhile, on the human side, the corresponding U.S. Guideline (FDA, 1998) was published in 1998.

The subsequent EU draft ERA Guideline for non-GMO human pharmaceuticals (EMEA, 2001) presented a simplified, straightforward tiered procedure focusing on the aquatic compartment (Straub, 2002). Similar to earlier drafts and the U.S. guideline, it incorporated an initial crude predicted environmental concentration (PEC) threshold of 0.01 μg/l, below which no further assessment was necessary. This draft attracted a lot of criticism, particularly by the Scientific Committee on Toxicity, Ecotoxicity, and the Environment (CSTEE, 2001). The main points of criticism were that the threshold of 0.01 μg/l was not scientifically based (nor, by extension, is the U.S. entry into the environment threshold of 1.0 μg/l). The U.S. guideline (FDA, 1998) was considered more appropriate as a general design for an ERA because it has a detailed decision tree and testing requirements scheme primarily based on release into, and further behavior within, the environment. Furthermore, in the 2001 EU draft, the relatively lax guidance on treatment of metabolites and excipients found objection. In general, the adoption of proven ERA guideline schemes was recommended, specifically the Technical Guidance Document or TGD (ECB, 1999; recently revised as ECB, 2003), because the CSTEE wanted to see all ERA guidelines streamlined for better comparability. Finally, the preparation of ERAs was desired not only for new but also for existing active pharmaceutical ingredients (APIs) including over-the-counter medicines. Subsequent to these (and other) criticisms, in 2003 a heavily revised and extended draft ERA Guideline was published by the Committee for Proprietary Medicinal Products, CPMP (EMEA, 2003), which is the focus of this contribution.

II. General Principles of the 2003 Human Pharmaceuticals ERA Draft Guideline

A. Coverage

The ERA addresses mainly the API; however, it should address any substance of concern including excipients. It covers the storage, use, and disposal of pharmaceuticals, but not the synthesis of the API or the manufacture formulation of the medicinal product. The latter are regulated by other legislation and guidelines, such as through notification, classification, and labelling of new substances including isolated intermediates (CEEC, 1992). However, in contrast to all earlier drafts that only addressed new medicinal products, the 2003 draft expressly includes renewals and so-called type II variations of existing medicines involving a major predicted increase in use, such as that due to extension of indication. Major metabolites, defined as >10% of the original API, are only addressed in the uppermost level of the ERA.

B. Tiered Assessment

All draft ERA guidelines, including the recent discussion paper (e.g., CPMP, unpublished; EMEA, 2003), as well as the published veterinary (EMEA, 2000) and the U.S. human pharmaceutical ERA (FDA, 1998) guideline, have a *tiered assessment* in common. This stepwise procedure permits a decision on whether, based on available data, the pharmaceutical in question would be unlikely to represent a risk to the environment or alternatively, if further considerations involving the generation of additional experimental data or use of progressively more complex models were necessary. The ERA only proceeds to the next tier when risk cannot be excluded based on defined criteria.

The sequential steps or phases and tiers are shown in Table I; however, based on certain experimental data or in case of specific properties (endocrine active, possibly cytotoxic, mutagenic, antibiotic) that may cause "atypical ecotoxic effects," there are direct references to higher tiers. Phase I calculates a crude initial aquatic PEC that is compared with a given threshold concentration. Phase II tier A extrapolates aquatic and microorganism predicted no-effect concentrations (PNEC) and partly refines the PEC. Based on these derived data, a risk quotient of PEC/PNEC decides on the necessity of further assessment. Additionally, experimental results from Phase II tier A may identify specific compartments of concern (organisms/bioaccumulation, sediment, soil). Phase II tier B consists of detailed risk assessment according to the TGD for the identified compartments or, in the case of an API with specific properties for the identified effects.

TABLE I Flow Chart for Tiered Assessment According to the 2003 EU ERA Draft

Phase	*Data/tests needed*	*Process*	*Decision*	*Result*
Phase I	α Medicinal Product (new, renewal, or major use increase)		Low environmental risk? (vitamin, amino acid, electrolyte, etc)	Y: finalize ERA N: go on
			Specific mode of action? (potential adverse ecotoxic effects; endocrine, cytotoxic, mutagenic, etc)	Y: go to Phase II tier B in-depth ERA N: go on
	Get maximum daily dose	Calculate initial PEC	Initial aquatic PEC ≤ 0.01 μg/l?	Y: finalize ERA N: go to Phase II tier A
Phase II tier A	Do/complete stipulated environmental fate and effects test battery	Calculate initial aquatic and micro-organism PNECs	Special concerns regarding bioconcentration (logKow > 3), adsorption (Koc > 10,000) or partitioning to sediment (OECD 308)?	Y: Go to Phase II tier B for bioaccumulation/persistence assessment, terrestrial risk assessment, or sediment risk assessment; but still finish Phase II tier A N: Go on
	Get substantiated predicted use information	Calculate penetration factor, refine PEC (based on use only)	Refined PEC/PNEC ≤ 1 for both aquatic and microorganisms?	Y: Finalize ERA N: Go to Phase II tier B for aquatic and/or microorganism risk refinement
Phase II tier B	*Aquatic PNEC refinement:* do TGD chronic tests	Calculate refined aquatic PNEC		
	Microorganism PNEC refinement: do TGD recommended test(s)	Calculate refined microorganism PNEC		

	PEC refinement: Get metabolism information, Phase II tier A fate data and ready biodegradability, run Simple Treat model	Calculate refined PEC (based on metabolism, fate and use)	Refined PEC/PNEC ≤ 1 for aquatic and/or microorganisms?	Y: finalize ERA N: finalize ERA with precautionary and safety measures
	Bioaccumulation assessment: Do fish bioaccumulation study	Do TGD/OECD bioaccumulation and persistence assessment	Bioaccumulative and/or persistent?	Y: finalize ERA with precautionary and safety measures N: finalize ERA
	Terrestrial risk assessment: Do TGD recommended terrestrial fate and effect tests	Calculate TGD terrestrial PEC, calculate TGD terrestrial PNEC	Terrestrial PEC/PNEC ≤ 1?	Y: finalize ERA N: finalize ERA with precautionary and safety measures
		Calculate groundwater PEC	Groundwater PEC/microorganism PNEC ≤ 1?	Y: finalize ERA N: finalize ERA with precautionary safety measures
	Sediment risk assessment: Do sediment toxicity tests (note: no specific tests recommended in TGD)	Calculate TGD sediment PEC, calculate TGD sediment PNEC (based on tests and/or equilibrium method)	Sediment PEC/PNEC ≤ 1?	Y: finalize ERA N: finalize ERA with precautionary and safety measures
	In-depth assessment for APIs with specific mode of action: consult CPMP, if available use OECD or other guidance	Perform special assessment according to guidance received	Specific, low-concentration, environmental risks identified?	Y: finalize ERA with precautionary and safety measures N: finalize ERA
Finalize ERA	Satisfactory information about absence of risk or characterization of identified risk(s)	Prepare ERA Expert Report with assessment and evaluation of exposure and risks, precautionary and safety measures if applicable, and proposals for labelling		ω Expert Report

1. Phase I

In the first phase, assumed non-problematic APIs (vitamins, amino acids, or electrolytes) are exempted from any assessment because they are not likely to result in significant exposure to the environment. All APIs with a specific mode of action (endocrine active, cytotoxic, mutagenic, possibly antibiotic, or other) must go through an in-depth specific assessment, no matter whether they surpass the threshold or not. For all other APIs, a crude PEC for the aquatic compartment is calculated. In contrast to all other ERA guidelines or drafts, in the 2003 draft the initial crude PEC depends only on the maximum daily per capita dose of an API. Metabolism or (bio-) degradation data is not incorporated. For a worst-case estimate, the whole maximum daily API dose is multiplied by a so-called penetration factor and divided by the default TGD sewage flow (200 l/inh/d) and the default TGD dilution factor (1:10). The penetration factor represents a statistical proportion of the population being treated daily with a specific API. A default penetration factor of 1% for the Phase I PEC derivation was determined from the cumulative frequencies of about 800 APIs on the market:

$$\mathrm{PEC}_{\text{surface water}} = \frac{\mathrm{MDD} \times \mathrm{PF}}{\mathrm{V} \times \mathrm{D}} = \frac{\mathrm{MDD\ mg/pat \cdot d} \times 1}{200\ \mathrm{l/inh/d} \times 10 \times 100},$$

where MDD is the maximum daily dose in mg/patient/day; PF is the default penetration factor of 1% (1:100); V is the TGD default volume of sewage per inhabitant per day of 200 l/inh/d; and D is the TGD dilution rate of sewage works effluent by surface water of 10. Consequently, the maximum dose that does not result in a crude PEC > 0.01 μg/l calculates to 2 mg/patient/day.

If the crude aquatic PEC is below 0.01 μg/l and no other environmental concerns are apparent from specific mode of action, then according to the 2003 draft ERA guideline, it may be assumed that the pharmaceutical is unlikely to present a risk for the environment following its prescribed usage in patients. In this case the ERA stops at Phase I and the ERA Expert Report can be finalized. Otherwise, one needs to go to Phase II tier A.

2. Phase II tier A

In Phase II tier A, both environmental fate and effects data for the API concerned must be experimentally determined. The extensive required battery of tests is listed in Table II. Certain test results refer a substance directly to in-depth assessment in Phase II tier B. For example, if the $\log K_{OW}$ is > 3, a bioaccumulation (and persistence) assessment according to the TGD and the OECD (2001) must be performed. In the case of an average adsorption/desorption coefficient $K_{OC} > 10{,}000$, which suggests partitioning to sludge and transfer to soil, a terrestrial risk assessment must follow. Finally, if

TABLE II Required Data for an ERA in Phase I and II in Comparison with the Base Set (BS) Data for Chemicals Notification

Phase	*Data requirement*	*Test guideline to be used*	*Remark*
Phase I	Maximum daily dose per patient		Mandatory
Phase II tier A			
Environmental fate data	UV/visible absorption spectrum (BS)	OECD 101	Mandatory
	Melting temperature (BS)	OECD 102	Mandatory
	Water solubility (BS)	OECD 105	Mandatory
	Absorption/desorption, batch equilibrium (BS)	OECD 106	Mandatory
	n-octanol/water partition coefficient (BS)	OECD 107/117	Mandatory
	Dissociation constant	OECD 112	Mandatory
	Aerobic/anaerobic transformation in water-sediment systems	OECD 308	Mandatory
	Vapor pressure (BS)	OECD 104	Optional
	Hydrolysis as a function of the *p*H (BS)	OECD 111	Optional
	Photodegradation	OECD Monograph no. 61, seek regulatory guidance	Optional
	Ready biodegradability (BS)	OECD 301	Optional in Phase II tier A
	Estimated consumption of API per year and defined region (EU or member state)		Substantiate with health care statistics or epidemiological data
Effects data	Algal growth inhibition (BS)	OECD 201	Mandatory
	Daphnid acute immobilization and/or reproduction (BS)	OECD 202, OECD 211	Mandatory
	Fish acute toxicity (BS)	OECD 203	Mandatory

(Continues)

TABLE II (*Continued*)

Phase	*Data requirement*	*Test guideline to be used*	*Remark*
	Activated sludge respiration inhibition (BS)	OECD 209	Mandatory
Phase II tier B			
Aquatic PEC refinement	Long-term algal study	OECD 201	Various possible tests, assessment factor depends on number of chronic/long-term studies
	Chronic/reproductive daphnid toxicity	OECD 202	
	Chronic fish toxicity	OECD 210, OECD 215	
Microorganism PNEC refinement	Biodegradation inhibition control data	OECD 301, OECD 302	Various possible tests, assessment factor depends on specific tests
	Nitrification inhibition	ISO–9509	
	Activated sludge growth inhibition	ISO–15522	
	Pilot scale activated sludge simulation	OECD 303	
	Pseudomonas putida growth inhibition	NF EN ISO–10712	
	Ciliate growth inhibition	OECD in preparation	
Bioaccumulation	Bioconcentration in fish	OECD 305	Bioaccumulation and persistence assessment according to TGD and OECD (2001) criteria

Soil PNEC extrapolation	Soil microorganisms toxicity	OECD 216	Various possible tests, assessment factor depends on number of chronic/long-term studies
	Terrestrial plant growth	OECD 208	
	Acute earthworm toxicity	OECD 207	
	Collembola reproduction test	ISO–11267	
	Other trophic level long-term soil tests		
Sediment PNEC extrapolation	Sediment toxicity tests	OECD 218/219 and other non-OECD guidelines	Various possible tests, assessment factor depends on number of studies
Surface Water PEC refinement	Fraction of active ingredient excreted	ADME studies	
	Ready biodegradability	OECD 301	Mandatory for tier B PEC
	Emission fraction to surface water	Simple/Treat model	
	Adsorption factor to supended matter	TGD, part II, pp. 43–47, 75–77	
Local soil PEC from sludge application	Ready biodegradability	OECD 301	
	Aerobic/anaerobic transformation in soil	OECD 307	
	Emission fraction to directed to sludge	Simple/Treat model	
Local groundwater PEC	Calculation of groundwater PEC according to EXPOSIT/TGD model fom/with soil PEC	EXPOSIT model, TGD part II, pp. 86, 85	
Local sediment PEC	Calculation of local PEC for sediment from amount adsorbed to suspended matter in surface water	TGD, part II, p. 78	
Specific risk assessment	Investigations for APIs with specific modes of action respectively potentially atypical ecotoxic effects	Consult CPMP for guidance	

BS, EU Chemicals Notification “Base-Set” of Required Tests (CEEC, 1992; ECB, 2003).

the sediment-water transformation test (OECD 308) demonstrates partitioning of the API into the sediment compartment, a sediment risk assessment is required.

For regular Phase II tier A procedures, a PNEC for the aquatic compartment is calculated, based on a standard set of acute ecotoxicity tests with algae, daphnia, and fish according to the TGD. The lowest of the acute 50%-effect concentrations (EC_{50} or LC_{50}; otherwise, use the lowest no-observed-effect concentration, NOEC) is divided by an assessment or uncertainty factor of 1000 (if acute data for three species are available only) to derive the PNEC. The uncertainty factor is introduced to account for extrapolation from acute to chronic toxicity and intra- as well as inter-species variability in sensitivity. Where chronic data are available, lower uncertainty factors apply as listed in the TGD.

$$\mathrm{PNEC_{water}} = \frac{\text{lowest acute } \mathrm{EC_{50}},\ \mathrm{LC_{50}} \text{ or NOEC}}{1000}$$

In parallel, the calculation of a PNEC for microorganisms, particularly those in sewage works, is required, based on a standard activated sludge respiration inhibition test (OECD 209):

$$\mathrm{PNEC_{microorganisms}} = \frac{\text{NOEC or EC10}}{10}$$

In Phase II tier A, the initial crude PEC is partly refined. The default penetration factor (PF) of 1% in Phase I may be adapted based on the estimated use or consumption of the API in a given region (the EU or a member state) per year. This use prediction must be substantiated through statistics or epidemiological studies.

$$\mathrm{PF_{refined,\%}} = \frac{\text{API mg/yr} \times 100\%}{\text{MDD mg/pat/d} \times \text{N(inh)} \times \text{365d/yr}}$$

where API is the annual predicted amount of API used in the region in mg/yr; MDD is the maximum daily dose per patient; and N(inh) is the number of inhabitants of the region. In case of proven ready biodegradability, which is an optional test in Phase II tier A, this information may be included in the ERA (although it is not stated how to factor this into the PEC). It must be noted, however, that the Phase II tier A PEC refinement is not based on any of the mandatory environmental fate data elaborated for the API, hence it is only a limited and partial PEC refinement.

If both ratios of PEC/PNEC for aquatic and microorganisms are smaller than 1 and no other environmental concerns have become apparent, then it may be assumed that the pharmaceutical is unlikely to present a risk for the environment. In this case the ERA does not progress beyond Phase II tier A and the ERA Expert Report can be finalized. Otherwise, proceed to Phase II tier B.

3. Phase II tier B

Phase II tier B asks for further considerations on the basis of compartment(s) identified as potentially being at risk. It will consist of performing a detailed assessment for the respective compartment (biota/bioaccumulation; microorganisms/sewage works; aquatic organisms/surface water; terrestrial organisms/soil and groundwater; sediment-dwelling organisms/sediment; see Table I). This involves calculating a *refined PEC* based on additional data (see Table II) regarding patient metabolism, release, distribution, and fate of the substance. Using more detailed environmental models, such as the Simple/Treat sewage works model, based on the Phase II tier A and additional test data, allows for more realistic PECs. In addition, a *refined aquatic or microorganism PNEC* may be arrived at using chronic ecotoxicity data, thereby reducing the uncertainty factor. Similar procedures apply for the terrestrial and sediment assessment, while bioaccumulation is treated in the TGD and by the OECD (2001). In general, the TGD is extensively used for guidance; however, the respective procedures cannot be fully reproduced here but are only summarized (see Table II).

For possible further refinement of both PEC and PNEC, probabilistic assessment techniques as developed by pesticide environmental toxicologists may be useful. Field studies are explicitly listed as a further means, possibly in mesocosms with a miniature ecosystem, and also in surface waters after introduction of the new pharmaceutical. Such further considerations are to be described in detail in the ERA Expert Report. In addition, safety measures should be adopted where applicable.

C. Active Substance and Metabolites

In Phase I and Phase II tier A, the assessment addresses mainly the API (the active substance proper, or the active metabolite in the case of prodrugs). For the concentration threshold or the crude PEC/PNEC comparison, it is assumed that metabolites are not more toxic or recalcitrant than the API. In Phase II tier B, however, both relevant human and environmental metabolites, corresponding to >10% of the API, should also be addressed. Human metabolites are identified in the pharmacodynamic and pharmacokinetic studies during drug development, while environmental metabolites are only identified in the Phase II tier A sediment-water transformation tests (OECD 308).

D. Precautionary and Safety Measures

In the previous EU ERA draft (EMEA, 2001), the benefit for patients was expressly given precedence over possible environmental risks caused by human medicinal products. The current 2003 draft does not make reference to the priority of patient benefit any longer. However, there is no provision that an API may be banned because of residual environmental risk. Instead, precautionary or mitigating safety measures may be prescribed in cases where environmental risks cannot be excluded. Such measures may consist in product labeling, restricted use to clinics or surgeries, restricted land-spreading of sewage sludge, environmental monitoring, or further (non-detailed) measures. In order to minimize environmental exposure from pharmaceuticals in general, the following labelling is recommended even for those products that do not need special disposal measures: "Medicines no longer required should not be disposed of via wastewater or the municipal drainage system. Return them to a pharmacy or ask your pharmacist how to dispose of them in accordance with national regulations. These measures will help to protect the environment."

E. Further Points in the Draft Guideline

Applicants may request scientific advice from the CPMP on issues related to the ERA, notably regarding assessment of APIs with specific mode of action and safety measures. Formal requirements as to the contents of the ERA Expert Report are listed. These include potential environmental exposure, fate, and effects as well as risk management strategies if applicable. Specifically, the derivation of the PEC, an assessment of the potential risks and the PNEC, an evaluation of precautionary and safety measures and any labeling proposed must be presented. All the above data and interpretations (or the omission of them) shall be justified based on sound scientific reasoning. Last, a curriculum vitae of the environmental risk assessor is to be added.

III. Discussion

The new ERA draft is much more detailed than all earlier versions, particularly the 2001 draft (Table III). The stepwise procedure has more mandatory direct links to further specific assessments based on substance properties. Moreover, the adoption of the TGD approach has considerably broadened the methodology and variety of tests. This makes the new draft more comprehensive and, at the same time, more specifically tuned (or tunable) to different substance properties. On the other hand, there seem to be some inconsistencies.

High priority is given to environmental exposure with a new way of deriving the initial crude PEC in Phase I and a partial refinement thereof in Phase II tier A, both based on a default or refined penetration factor (or frequency of use) within the population. While the Phase I threshold of 0.01 μg/l has been retained from previous versions, the new way of calculating the crude surface water PEC by use of a penetration factor does make a

TABLE III Comparison of Some Properties of the New 2003 Draft EU ERA Guideline for Pharmaceuticals with Other Guidelines and Drafts

	Pharma Draft EU 2003	*Pharma Draft EU 2001*	*Pharma USA 1998*	*Veterinary EU 1997*	*Industrial Chemicals EU 1992/1999*	*Bio-/Pesticides EU1998/1991*
Tiered ERA	+	+	+	+	(+)	(+)
threshold $PEC_{surface\ water}$	0.01 μg/l[a]	0.01 μg/l	0.1 μg/l[b]	0.1 μg/l GW; 10 μg/kg soil[c]	n/a	n/a
Phase I or tier 1 max. amount PEC/PNEC	2.76 t/yr[d]	~ 4 t/yr[e]	~ 44 t/yr[f]	n/a	n/a (10 kg/yr)[g]	n/a
Technical guidance	+ (Phase II) Straightforward Phases I and II tier A; TGD for Phase II tier B; some Phase II B guidance lacking	+ Simple tiers 1 & 2	+ Detailed flowcharts according to release	+	+	+ Highly detailed guidance documents; TGD for industrial chemicals and biocides
Comprehensive ERA	Phase II tier B	Possibly tier 3	Tier 3	+	+	++
EU base set necessary	++[b]	–	–	±	+	++[b]
ERA applicable for new substances only	–[i]	+	+	–	(+)[j]	–
Registration dependent on result of ERA	(+)[k]	–	–	(+)[k]	–[l]	+[k]

GW, groundwater; +, stipulated; –, not stipulated; (+), partly stipulated; ++, more than stipulated.

[a] Initial PEC solely dependent on maximum daily dose, no metabolism or degradation may be included; the maximum daily dose resulting in a PEC of 0.01 μg/l corresponds to 2 mg.

[b] 1 μg/l effluent concentration threshold times 10-fold standard dilution factor.

[c] Agricultural API: 10 μg/kg soil and 0.1 μg/l groundwater; no threshold for aquacultural API.

[d] 2 mg/d × 365 d/yr × 3.78E8 inh/EU × 200 l/inh×d (TGD default sewage flow) × 10 (TGD default dilution factor).

[e] 0.01 μg/l × 365 d/yr × 3.78E8 inh/EU × 300 l/inh×d × 10 (TGD default dilution factor).

difference. The EU-wide penetration-factor-based Phase I PEC corresponds to 2760 kg API/yr, which is identical to the maximal annual amount of API in the 2001 draft, when no elimination is allowed for and when TGD defaults are strictly applied. However, application of the 0.01 μg/l threshold, according to the 2003 draft, corresponds to a maximum daily per capita dose of 2 mg for a given pharmaceutical, which would then not be subject of a Phase II assessment. In contrast, in all earlier drafts there was no such limitation of the daily dose. Thus, with the 2003 method of deriving a crude PEC, the limiting factor is not only the total amount of API but primarily the maximal daily dose. The PEC is no longer derived by dividing the whole API amount per geographical entity (EU or Member State) by the whole volume of water, but instead by determining the corresponding ratio for a (statistical) single user. This translates into a shift of the Phase I threshold from an EU-wide or statewide PEC, which corresponds to a continental or regional PEC in the TGD, to a local PEC. The TGD local PEC during an emission episode is not representative of larger areas, as it represents the highest possible value. Moreover, the Phase I crude PEC is even higher, because no metabolism or degradation may be included. On the other hand, the ERA is still intended for the whole of the EU (or at least for a Member State that more or less corresponds to a region in the TGD). Hence, in comparison with the 2001 draft, the new stricter way of calculating a PEC results in having to advance earlier to Phase II assessment, with subsequent time- and cost-intensive higher-tier tests.

Phase II tier A asks for a whole battery of tests relating to environmental fate and effects. Based on the acute aquatic and microorganism tests, respective PNECs are extrapolated that are used for comparison with the refined aquatic PEC. However, refinement of the PEC in Phase II tier A is only through recalculation of the penetration factor based on projected API amounts, not through incorporation of the compulsory fate data (half-lives in water or sediment, adsorption, or possibly ready biodegradability). With the exception of a fulfilled ready biodegradability in the optional test, using available metabolism or degradation information for aquatic PEC calculation is not allowed for until Phase II tier B. There is no possible rationale for

[f] 1 μg/l × 365 d/yr × 1.214E11 l/d (total effluent of publicly owned treatment works in the U.S.).

[g] Reduced or full notification procedure triggered by amount of 10 kg/yr put on market.

[h] Testing requirements beyond EU Base Set but not congruent for human API and bio-/pesticides.

[i] ERA also stipulated for renewals and major increases in the use of a product.

[j] EU White Paper/New Chemicals Policy will require data and some, limited to extensive, assessment depending on produced/marketed volume for "old" substances.

[k] Labeling and restrictions on use may be prescribed.

[l] Classification, labeling, and packaging depending on dangerous properties.

a formal requirement for detailed fate data, in particular the elaborate (and cost-intensive) OECD 308 sediment-water transformation test that, in most cases, needs radio-labelled substance, when the actual use of these data is not foreseen respectively allowed at the same stage of the ERA. Thus, the refinement of the aquatic PEC in Phase II tier A is partial, it does not make use of all available information. This procedure is neither scientific nor efficient.

In most cases, Phase II tier B ERA closely follows the methodology detailed in the TGD. Refined aquatic or microorganism assessment is requested in the case of Phase II tier A risk ratios are >1, while terrestrial or bioconcentration assessment is directly triggered by K_{OC} >10,000 and $\log K_{OW} > 3$, respectively. Refined surface water PECs and PNECs, microorganism and soil PNECs, as well as soil and groundwater PECs, require a lot of additional testing (including time and costs) but are basically straightforward. The same holds for bioconcentration assessment, where reference is made to the TGD and (in the latter) to the OECD (2001) Guidance Document No. 27. In contrast, for substances identified at the beginning as having a specific mode of action and, possibly, causing atypical ecotoxic effects, there is no clearcut guidance but only a general offer for scientific advice from the CPMP. In fact, there is no definition of "atypical ecotoxic effects" in the 2003 draft, so conceivably every effect noted at a concentration lower than that derivable using the QSARs for aquatic minimum or baseline toxicity in the TGD (ECB 2003; Part III, pp. 17–19) might qualify. While the regulators may possibly want to keep all options open through using the expression "atypical effects," this lack of definition constitutes a clear legal uncertainty in the new draft.

While the TGD is consistently applied in Phase II tier B, this is only partly the case in Phase II tier A, namely in the case of calculation of the initial PNECs for aquatic and microorganisms. On the exposure side, however, the battery of environmental fate tests stipulated for Phase II tier A does not match the initial TGD base set (see Table II), nor does the Phase II tier A partial refinement of the crude surface water PEC correspond to the much broader way the PEC is calculated in the TGD. The call by the CSTEE (2001) for adoption of the TGD methodology for broad comparability of ERAs from chemicals to biocides and APIs has only been partly met, in the uppermost tier B. It has found no echo in the middle tier A, where additional experimental data that are only standard in pesticide assessment are requested, while the very same information is then not being used for PEC refinement.

On the other hand, no reference is made to the official EU risk assessment tool, the European Union System for the Evaluation of Substances (EUSES; ECB, 1997). EUSES calculates both PEC and PNEC values based on substance use scenarios and properties data, and determines risk quotients strictly according to the TGD. Moreover, a new EUSES version that incorporates the advances and changes in the recent 2003 TGD was released early in 2004 (ECB 2004), well before the ERA draft will become effective. With the exception of the groundwater model EXPOSIT and of the widely used sewage works model Simple/Treat, no other environmental fate models are used or recommended, even though environmental modelling has become accepted throughout.

The U.S. ERA guideline for pharmaceuticals (FDA, 1998), in spite of a higher trigger for higher-tier assessment, has a very clear and detailed flowchart and decision tree, driven mainly by the entry pathway of APIs into the environment. Similarly, the EU veterinary guideline is strongly oriented on, in fact even bifurcated according to, the release of API into environmental compartments. This is not really the case with the 2003 EU draft for human APIs, as an initial crude PEC is only partly refined, even though it is recognized that most APIs enter the environment by way of human sewage, where partitioning and fate can be reasonably predicted based on the information requested. If state-of-the-art fate modelling is considered adequate for industrial chemicals (CEEC, 1992; ECB, 1999, 2003), pesticides (CEEC, 1991), biocides (CEC, 1998b; ECB, 2003), and veterinary APIs (EMEA 1997, 2000), there is no evident reason why this should not be the case for human APIs.

The EU 2003 draft ERA guideline asks for all experimental data to be elaborated under GLP quality assurance standards. Ensuring the quality of data is crucial for all further work depends on data reliability. On the other hand, this is an EU draft that refers to work being done for use within the EU. There is a perfectly acceptable former European Norm Quality Assurance Standard, SN EN 45001, that has recently been adopted as an international ISO Standard (BS EN ISO/IEC 17025:1999; ISO, 1999). Restricting tests to only one of two international standards, while disregarding the other, does not seem to be scientifically sound. Hence, changing the requirement to either GLP or ISO quality assurance would be regarded as more neutral.

While all earlier drafts, and also the adopted U.S. ERA guideline (FDA, 1998), are directed at new medicinal products, the EU 2003 ERA draft is different. An ERA is now also stipulated for renewals of drug marketing authorizations, as well as for major variations, such as foreseeable increases in API amounts due to extensions of indication. This requirement follows both the CSTEE (2001) criticism and the spirit of the new EU chemicals legislation (CEC, 2003), where substance-specific data and a risk assessment of variable depth is asked for also for existing chemicals, independent of the amount produced or marketed. On the other hand, it may lead to problems, particularly in case of old over-the-counter APIs that are not proprietary any longer, but are produced or imported by several companies in large amounts. Fair allocation of the responsibility for an ERA and fair cooperation in this task among all involved companies may prove difficult.

Major metabolites must be considered when progressing to Phase II tier B. Major is defined as >10% of the original API and covers both human and environmental metabolites. Human metabolites are characterized in drug absorption, distribution, metabolism, and excretion studies, while environmental metabolites are identified in the OECD 308 sediment-water transformation test in tier A, probably at the end of the study (not detailed in the ERA draft). Human metabolism may be limited to enhancing the aqueous solubility of the API through appending a solubilizing moiety (such as glucuronic acid) to facilitate renal excretion. On one hand, glucuronidation enlarges the molecule, meaning that the 10% limit might be reached sooner. On the other, glucuronides are likely to be cleft by activated sludge bacteria in sewage works to the original moieties, glucuronic acid and the API. Therefore, it would probably not make much sense to study glucuronides. Borderline cases, such as major metabolites of APIs that barely surpass the Phase I 0.01 μg/l threshold, are not addressed. The question thus remains whether in such cases a full ERA would be scientifically justified. In general, more scientific guidance and also proportionality for a reasonable determination and assessment of metabolites is needed for the final EU ERA.

From an organizational viewpoint, an ERA may in the future take much longer to perform. As the further detailed assessment depends on the outcome of earlier decision points, long-term tests may be triggered only after results from the Phase I tier A test battery become available. However, the OECD 308 sediment-water transformation test is itself a study of relatively long (>2 months) duration. Subsequent studies, from chronic aquatic or terrestrial toxicity to bioaccumulation or soil transformation tests, will add considerably to the total time needed for elaborating basic data. Moreover, all experimental studies are further prolonged by the time needed for preparation and setup, and for writing and finalizing the report. In all, one year or more may be needed just for the serial testing. This must be anticipated by integrating the preparation of an ERA into the drug development schedule at a relatively early stage, in order to prevent delays to drug application.

Comments on the 2003 draft were sent to the EMEA by January 2004. No detailed reactions from the EU authorities have been made public by mid-2004. It is then expected from both regulators and industry that a solid, dependable, proportional, and practicable set of ERA procedures will be passed as a European guideline, probably in the year 2005.

In conclusion, the 2003 draft EU guideline for ERA of human APIs is much stricter in its approach than all its predecessors. It does have the potential to assess both environmental fate and effects in a manner that is both more scientific, more serious, and more specific for the active substances, while basically embracing the principles of the TGD. However, the draft is still fraught with several question marks or inconsistencies that will hopefully be ironed out. The manner of calculating the Phase I initial crude PEC by using maximal daily dose and a default penetration factor is a matter of political but not scientific decision. The Phase II tier A requirement for a multitude of tests is not in line with the TGD. Moreover, these tests—if they have to be done—only make sense if the results can also be used in the same tier, or else the refinement of the PEC remains a partial patchwork and is not consistent with the TGD. The thresholds for all further decisions are again political matters, but can be defended. The Phase II tier B assessment closely follows the TGD. However, in matters of assessing in depth those APIs with specific modes of action, no formal guidance is given or referred to because none exists so far, either from the EU or the OECD. Last, while all previous draft guidelines addressed only new APIs, the 2003 draft expressly includes renewals and major variations of existing products.

Acknowledgment

The constructive criticism and suggestions of the editor, Daniel Dietrich, for improving and finalizing this manuscript in a short time are much appreciated.

References

CEC (1998a). Environmental risk assessment of human medicinal products containing or consisting of GMOs. CEC, 1998. Guidelines, Medicinal products for human use; Safety, environment and information Office for Official Publication of the European Communities Luxembourg. pp. 135–146.

CEC (1998b). Directive 98/8/EC of the European Parliament and of the Council of 16 February 1998 concerning the placing of biocidal products on the market. Off. J. L123(24/04/1998), pp. 1–63.

CEC (2001). Notice to Applicants, Medicinal Products for Human Use. Presentation and contents of the Dossier Common Technical Document (CTD). *The Rules Governing Medicinal Products in the European Union,* vol. 2, B. European Commission, Brussels. http://dg3.eudra.org/F2/eudralex/vol-2/B/ctdoct01.pdf

CEC (2003). Consultation Document concerning the Registration, Evaluation, Authorisation and Restrictions of Chemicals (REACH); Explanatory note plus volumes I–VII. Commission of the European Communities, Luxembourg. http://europa.eu.int/comm/enterprise/chemicals/chempol/whitepaper/reach.htm

CEEC (1965). Council Directive 65/65/EEC of 26 January 1965 on the approximation of provisions laid down by Law, Regulation or Administrative Action relating to proprietary medicinal products. Off. J. 22(09/02/1965), pp. 369–373.

CEEC (1991). Council Directive 91/414/EEC of 15 July 1991 concerning the placing of plant protection products on the market. Off. J. L230(19/08/1991), pp. 1–32.

CEEC (1992). Council Directive 92/32/EEC of 30 April 1992 amending for the seventh time Directive 67/548/EEC on the approximation of the laws, regulations and administrative provisions relating to the classification, packaging and labelling of dangerous substances. Off. J. L154(05/06/1992), pp. 1–29.

CEEC (1993a). Council Directive 93/39/EEC of 14 June 1993 amending Directives 65/65/EEC, 75/318/EEC and 75/319/EEC in respect of medicinal products. Off. J. L214(24/08/1993), pp. 22–30.

CEEC (1993b). Council Regulation (EEC) No 2309/93 of 22 July 1993 laying down Community procedures for the authorization and supervision of medicinal products for human and veterinary use and establishing a European Agency for the Evaluation of Medicinal Products. Off. J. L214(24/08/1993), pp. 1–21.

CPMP (unpubl.). Assessment of potential risks for the environment posed by medicinal products for human use (excluding products containing live genetically modified organisms): Phase 1 Environmental Risk Assessment. III 5504/94 Draft 7, 10.01.95, unpublished.

CSTEE (2001). Opinion on: Draft CPMP discussion paper on environmental risk assessment of medicinal products for human use [non-GMO containing]. C2/JCD/csteeop/CPMP paperRAssessHumPharm12062001/D(01), 24th CSTEE pleanry meeting, Brussels, 12 June 2001. http://europa.eu.int/comm/food/fs/sc/sct/out111_en.pdf

ECB (1997). European Union System for the Evaluation of Substances, EUSES; version 1.00. ECB, Joint Research Centre of the European Commission, Ispra.

ECB (1999). Technical Guidance Document in Support of Commission Directive 93/67/EEC on Risk Assessment for New Notified Substances and Commission Regulation (EC) No. 1488/94 on Risk Assessment for Existing Substances. Office for Official Publications of the European Communities, Luxemburg.

ECB (2003). Technical Guidance Document on Risk Assessment in Support of Commission Directive 93/67/ EEC on Risk Assessment for New Notified Substances[,] Commission Regulation (EC) No. 1488/94 on Risk Assessment for Existing Substances [and] Directive 98/8/EC of the European Parliament and of the Council concerning the placing of biocidal products on the market. ECB, Ispra.June 2003. http://ecb.jrc.it/

ECB (2004). European Union System for the Evaluation of Substances, EUSES; version 2.0. ECB, Joint Research Centre of the European Commission, Ispra. http://ecb.jrc.it/new-chemicals/

EMEA (1997). Environmental risk assessment for veterinary medicinal products other than GMO-containing and immunological products. EMEA/CVMP/055/96. EMEA, London. http://www.emea.eu.int/pdfs/vet/regaffair/005596en.pdf

EMEA (2000). Guideline on environmental impact assessment (EIAS) for veterinary medicinal products – phase I. VICH Topic GL6 (Ecotoxicity Phase I), Step 7. EMEA/CVMP/VICH/592/98.final. http://www.emea.eu.int/pdfs/vet/vich/ 059298en.pdf

EMEA (2001). Draft CPMP Discussion Paper on Environmental Risk Assessment of Non-Genetically Modified Organism (Non-GMO) Containing Medicinal Products for Human Use. CPMP/SWP/4447/00 draft corr.

EMEA (2003). Draft. Note for Guidance on Environmental Risk Assessment of Medicinal Products for Human Use. CPMP/SWP/4447/00/draft, London, 24 July 2003. http://www.emea.eu.int/pdfs/human/swp/444700en.pdf

FDA (1998). Guidance for Industry. Environmental Assessment of Human Drugs and Biologics Applications. FDA, CDER/CBER, CMC 6, rev. 1. http://www.fda.gov/cber/gdlns/environ.pdf

ISO (1999). General requirements for the competence of testing and calibration laboratories; BS EN ISO/IEC 17025:1999. International Standardisation Organisation, Geneva.

OECD (2001). Guidance document on the use of the harmonised system for the classification of chemicals which are hazardous for the aquatic environment. OECD Environment Directorate, Series on Testing and Assessment No. 27, ENV/JM/MONO(2001)8. Organisation for Economic Co-Operation and Development, Paris.

Straub, J. O. (2002). Environmental risk assessment for new human pharmaceuticals in the European Union according to the draft guideline/discussion paper of January 2001. *Toxicol Lett.* **135**, 231–237.

Index

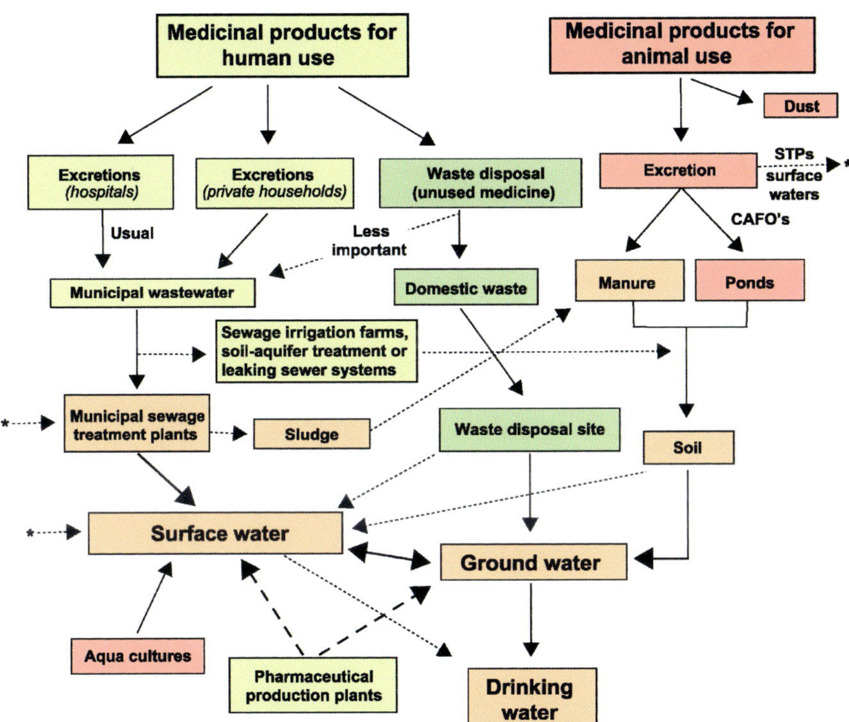

HEBERER AND ADAM, FIGURE 1 Scheme showing possible sources and pathways for the occurrence of pharmaceutical residues in the aquatic environment (modified according to Heberer, 2002a). The asterick indicates that the residues in excretions from animals may also be reaching the municipal STPs or surface waters.

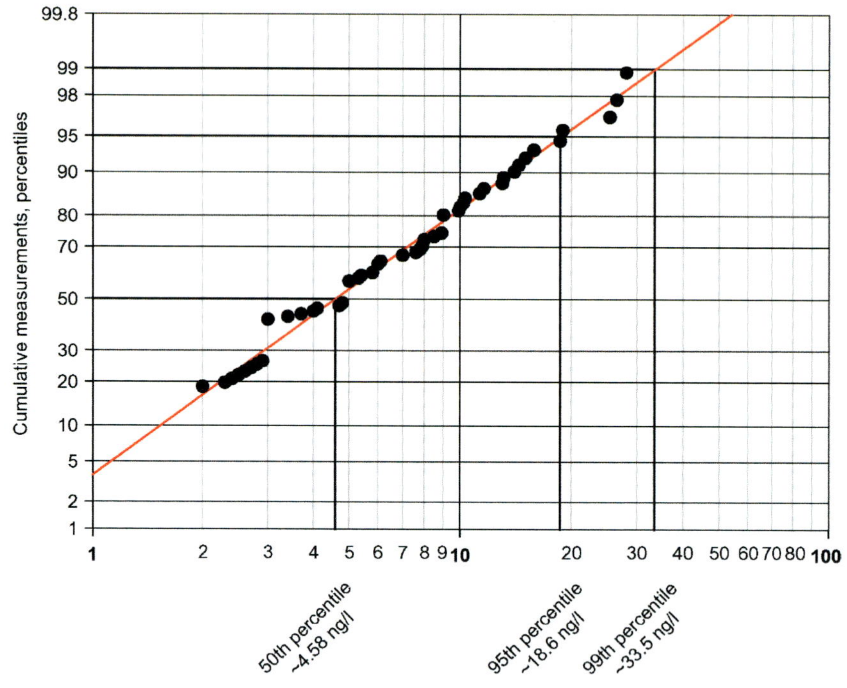

JÜRG OLIVER STRAUB, FIGURE 3 Log-probit graph of 90 EHMC monitoring data points.

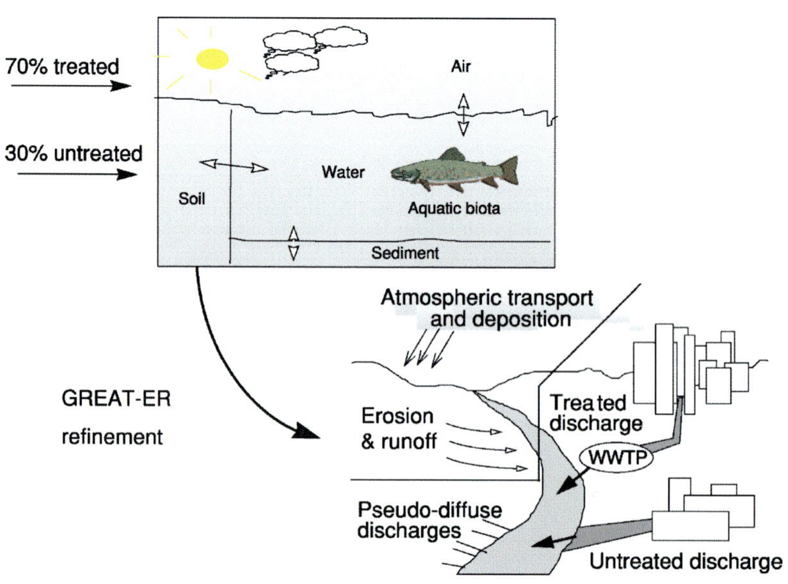

SCHOWANEK AND WEBB, FIGURE 1 Refinement of generic regional exposure models (EUSES) by taking actual discharge pathway, treatment, and river flow data into account (GREAT-ER). WWTP, wastewater treatment plant.

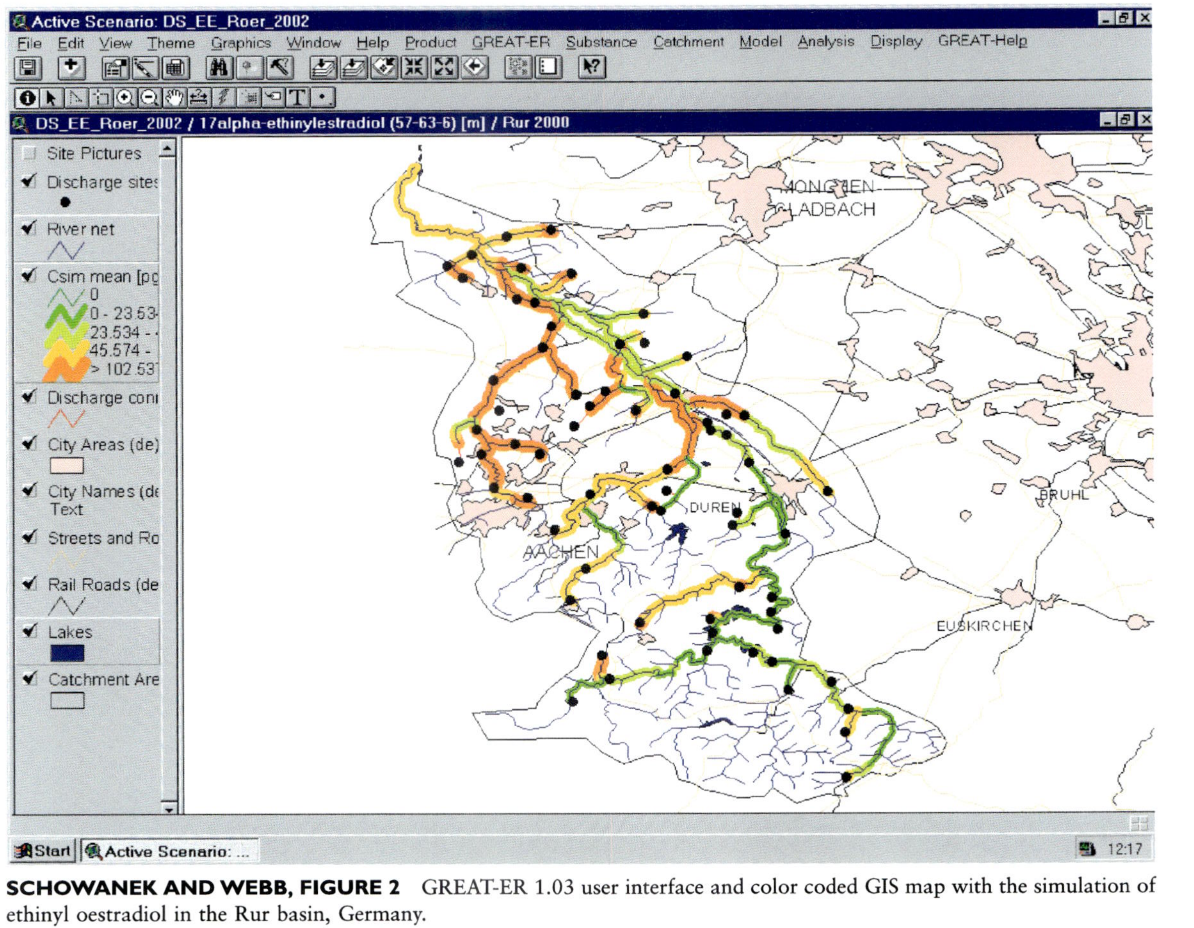

SCHOWANEK AND WEBB, FIGURE 2 GREAT-ER 1.03 user interface and color coded GIS map with the simulation of ethinyl oestradiol in the Rur basin, Germany.

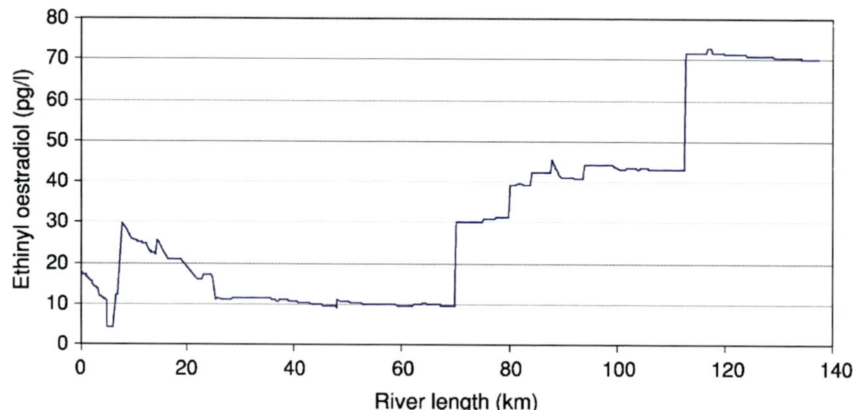

SCHOWANEK AND WEBB, FIGURE 3 Simulated PEC of ethinyl oestradiol under mean flow conditions in the Rur river, Germany. River stretch length, 300 m; km0, Kalterherberg; step changes at 70 km and 120 km represent the respective inputs from Duren and Aachen.

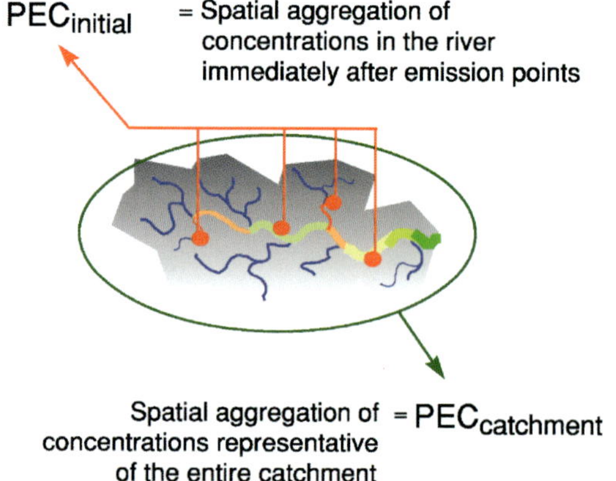

SCHOWANEK AND WEBB, FIGURE 4 Schematic illustration of the $PEC_{initial}$ and $PEC_{catchment}$ concepts, as developed within GREAT-ER for GIS-assisted regional risk assessment.